Linear Models: An Introduction

Linear Models: An Introduction

IRWIN GUTTMAN
Department of Statistics
University of Toronto

1807 175 YEARS OF PUBLISHING 1982

JOHN WILEY & SONS
New York Chichester Brisbane Toronto Singapore

Library of Congress Cataloging in Publication Data:

Guttman, Irwin.
Linear models.

(Wiley series in probability and mathematical statistics, ISSN 0271-6232)
Bibliography: p.
Includes index.
1. Linear models (Statistics) I. Title.
II. Series.

QA276.G87 519.5′352 82-2637
ISBN 0-471-09915-5 AACR2

Printed in the United States of America

10 9 8 7 6 5 4 3 2 1

To Mary, Karen and Daniel—
How Lucky Can You Get!

Preface

This book is an outgrowth of instruction notes used in two one-semester courses at the University of Toronto. The first course, covering the material in Chapters 1 through Section 3.4 of Chapter 3, is for students who have had one previous course in Statistics, know how the t, χ^2, and F distributions are defined and derived, have had some experience with p-variate distributions ($p \geq 2$), and, in addition, have had a one-semester course in Linear Algebra under their belts. In the first course, the topics covered in Chapter 2 are discussed without proof. The second course, given mainly to graduate students, deals with the proofs of the theorems given in Chapter 2, reviews the first part of Chapter 3, and discusses the remaining material in this book.

The objective of this book is to give students the essential concepts of regression, least squares, and linear models. The approach is essentially geometrical, but parallel algebraic discussion is quoted as the different topics arise. A unifying theme is the principle of orthogonalization, first introduced for the problem of nuisance parameters, but also used in discussing the analysis of covariance and orthogonal polynomials. The principle of orthogonalization rests on another unifying theme, namely, that of projection, and this is discussed first in Chapter 3 for the full-rank case, and continued, using the same essential ideas, in Chapters 4 and 5 for the non-full-rank case.

In designing first courses (and a book) for students in any field, the choice of topics and how to present them are very personal matters, involving, the author's philosophy and individual prejudices, and while this book reflects my own preferences, I would nevertheless like to acknowledge the comments and criticisms of many colleagues and students. In particular, I express my gratitude to Edmee Aragon, J. Berkowitz, Y. M. Chan, M. J. Evans, Michael S. Green, G. E. Moorehouse, K. Sykora, and R. Tibshirani of the University of Toronto; and A. D. Baccay and P. Smith of Old Dominion University. In addition, I would like to acknowledge the hospitality, encouragement, and valuable discussions afforded to me by A. P. Dawid, D. V. Lindley, and Mervyn Stone during the initial drafting of this manuscript, commenced during a sabbatical leave spent at the University College London, and to the entire Statistics group at Old Dominion University, where the final draft was achieved.

Finally, many thanks are due to Mrs. Frances Mitchell and Patricia Miller for their expert typing, done with patience, care, and much concern.

IRWIN GUTTMAN

Toronto, Ontario, Canada
April 1982

Contents

1 The Simple Linear Model

A very great percentage of Statistics involves itself with what has become known as the *linear model*, or the regression model. What constitutes a linear model, and how we analyze data taken in accordance with the linear model, is the subject of this book.

To begin our discussion, a definition is in order. We first suppose that interest lies in a certain (response) variable η, which is thought to be dependent on the functionally independent variables $z_1, z_2, \ldots, z_s$, that is, $\eta = f(z_1, \ldots, z_s)$. For example, in a certain production process, the yield of a product, η, may depend on the amount z_1 of catalyst used, the pressure z_2 maintained during the process, and the temperature z_3, at which the process is run.

We say that η obeys a *linear model* if

$$\eta = f(z_1, \ldots, z_s) = \sum_{j=1}^{k} \beta_j x_j(z_1, \ldots, z_s), \tag{1.1}$$

where the x_j are functions of the z_i only. The quantities $\beta_1, \ldots, \beta_k$ are *parameters*, usually unknown, and we note that they enter into Eq. (1.1) *linearly*, that is, to degree 1.

To gain some feel for models of type (1.1), we note some examples at this point. The first simple example is provided by the relationship

$$\eta = \alpha + \beta z, \tag{1.2}$$

which is of the form (1.1) with $s = 1$, $k = 2$, $\beta_1 = \alpha$, $\beta_2 = \beta$, and $x_1(z) = 1$,

$x_2(z) = z$ with $z_1 = z$. Sometimes the model given by (1.2) is rewritten as $\eta = \alpha + \beta x$, with the understanding that $x = x_1(z) = z$. Note that $(\beta_1, \beta_2) = (\alpha, \beta)$ enters into (1.2) linearly.

Another example that generalizes (1.2) is provided by a polynomial relationship of degree d in a simple variable z, that is,

$$\eta = \tau_0 + \tau_1 z + \tau_2 z^2 + \cdots + \tau_d z^d \tag{1.3}$$

or

$$\eta = \beta_1 x_1 + \beta_2 x_2 + \beta_3 x_3 + \cdots + \beta_{d+1} x_{d+1} \tag{1.3a}$$

with $\beta_j = \tau_{j-1}$, $x_j = x_j(z) = z^{j-1}$, and $j = 1, \ldots, d+1$ so that $s = 1$ and $k = d + 1$. Again, the important point to note is that the β_j enter the model (1.3) and (1.3a) *linearly*. We remark here that a model of nonpolynomial form may still be a linear model in the sense of the definition implied by (1.1). An example is provided by the model

$$\eta = \alpha + \beta \sin 2\pi z. \tag{1.4}$$

Since α and β enter into (1.4) linearly, this model is also a linear model because it is of the form (1.1), with $s = 1$, $k = 2$, $x_1(z) = 1$, $x_2(z) = \sin 2\pi z$, $\beta_1 = \alpha$, $\beta_2 = \beta$, and $z_1 = z$.

In contrast to the above examples, an example of a *nonlinear* model is provided by

$$\eta = (\beta_2 - \beta_1)^{-1}[\exp(-\beta_1 z) - \exp(-\beta_2 z)] \tag{1.5}$$

for $z > 0$, with $\beta_2 > \beta_1$ and where $\exp(a) = e^a$. Clearly this is not a linear model in the parameters β_1 and β_2.

We remark also that certain nonlinear models, that is, nonlinear in the parameters, may be turned into linear models by suitable *transformation*. For example, consider the response function

$$\xi = \delta \exp(\gamma z). \tag{1.5a}$$

Clearly (1.5a) is not linear in the parameter γ. However, on taking logarithms to base e, we have

$$\ln \xi = \ln \delta + \gamma z, \tag{1.5b}$$

which is of the form

$$\eta = \beta_1 + \beta_2 x \tag{1.5c}$$

with $\beta_1 = \ln\delta$, $\beta_2 = \gamma$, $x_1(z) = 1$, $x_2(z) = z$, etc., and (1.5c) is clearly of linear-model-form (1.1). We will see that care must be taken with this type of transformation. For example, if on the basis of certain data we proceed to make inference about (1.5c) using a least-squares analysis (to be discussed subsequently), then proceeding from inferences about (1.5c), inference about (1.5a) must be done in accordance with certain precise rules. We note too that many nonlinear models cannot be reduced to linear models by transformation—indeed an example of this sort is provided by the model (1.5).

1.1. A SIMPLE LINEAR MODEL AND ITS ANALYSIS BY LEAST SQUARES

To familiarize ourselves with what is entailed in a least-squares analysis, we will now discuss the analysis of a model of the form (1.1). This will be done without the use of matrices or n-dimensional geometric arguments (which will be first used in Chapter 3).

Starting with a concrete example, suppose we are interested in how changes in temperature cause changes in the volume of a gas. To investigate this, we can perform experiments in which the volumes of a gas, say y_i, in a cylinder are measured at various *preselected* settings of the (functionally) independent variable, temperature, say x_i, where $i = 1, \dots, N$. Suppose it is known that, apart from experimental error, the relationship between y and x is linear, so that we may state that the expectation of y given x, $E(y \mid x)$, the so-called *regression function* of y upon x, is such that

$$E(y \mid x) = \eta_x = \alpha + \beta x \tag{1.6}$$

and furthermore, that no matter what x is, the observed volume y incurs the *same experimental error*, which may be stated as

$$V(y \mid x) \equiv \sigma^2, \qquad \text{all } x. \tag{1.7}$$

[Assumption (1.7) is referred to in the literature as the *homoscedastic* assumption.] Two remarks are apparent if (1.6) and (1.7) hold. First, for a given value x of the independent variable, observations taken on y at the given value x will vary in accordance with a certain distribution D. (For example, D could be N, the normal distribution.) We are denoting the variance of the distribution of the y values which could be generated at x by σ^2. We summarize by stating:

$$\text{(1.7}a\text{)} \qquad \text{At } x, \quad y = D(\alpha + \beta x, \sigma^2)$$

for (1.6) says that the expectation of y at x lies on the line $\alpha + \beta x$. Second, for another value x', $x' \neq x$, of the independent variable, we summarize by noting that

$$\text{(1.7}b\text{)} \qquad \text{At } x', \quad y = D(\alpha + \beta x', \sigma^2).$$

That is, apart from location as measured by the expectations of the distributions, the distributions of y at x and y at x' differ only in location. This is pictured in Figure 1.1.1.

We may summarize (1.7a) and (1.7b) by writing

$$\text{(1.7}c\text{)} \qquad y - (\alpha + \beta x) = \varepsilon$$

for any x, where the so-called (random) error (ε) has the *same* distribution for all x, with expectation 0 and variance σ^2. We sometimes speak of σ^2 itself as the error incurred when sampling y at x, but of course we mean that the error incurred is ε, where ε has a distribution with mean 0 and variance σ^2. We further suppose that the observation y_i generated when the temperature x_i is used, is *independent* of y_j, generated when x_j is used.

The statements (1.6), (1.7), and the assumption of statistical independence constitute the *model* for our experiment. Now it could be that we have

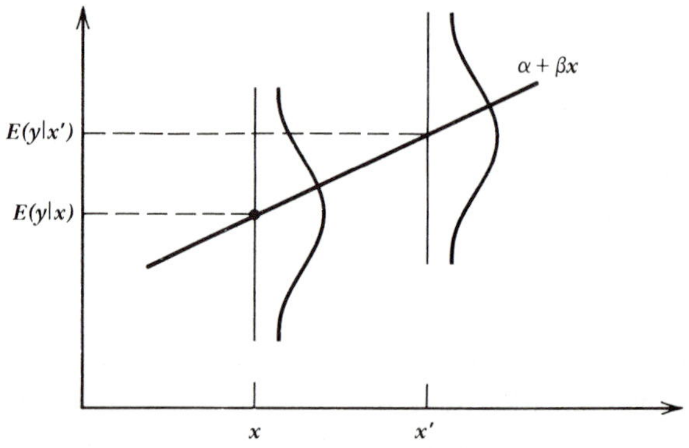

FIGURE 1.1.1. Distributions of y at x and x' have exactly the same form but differ in location.

stated (1.6) because we know that the relationship is indeed linear in x, or because in the region of usual choices of x, the relationship is to very good approximation linear in x. Or it may be that we wish to investigate the nature of the functional relationship between y and x and are using (1.6) as a natural first step, that is, as a statement to be looked at once the data are at hand.

As mentioned previously [see (1.7*c*)], we may use an alternative description of our model and write

$$y_i = \alpha + \beta x_i + \varepsilon_i, \tag{1.8}$$

where the ε_i are incurred experimental errors and are distributed such that

$$E(\varepsilon_i) = 0, \qquad V(\varepsilon_i) = \sigma^2 \tag{1.8a}$$

and the ε_i are *independent* of one another (or at least uncorrelated). A typical plot of the data (x_i, y_i), $i = 1, \ldots, N$, may appear as in Figure 1.1.2.

We note here that our notation is very general: Several x_i terms may have the same value, but we are supposing that *at least two* of the preselected x's *are different*. In fact, the question of how best to choose the x's at which to experiment is intrusive here and we discuss the topic later in Section 3.8 of Chapter 3. Suffice it to say at this point that, because we wish to estimate a slope β among other things, we surely need at least two observations taken at different values of x.

We now seek a method that will produce estimates of the unknown parameters α and β, in the functional relationship η, and of the experimental error σ^2. One way we could proceed *conceptually* is as follows: Suppose

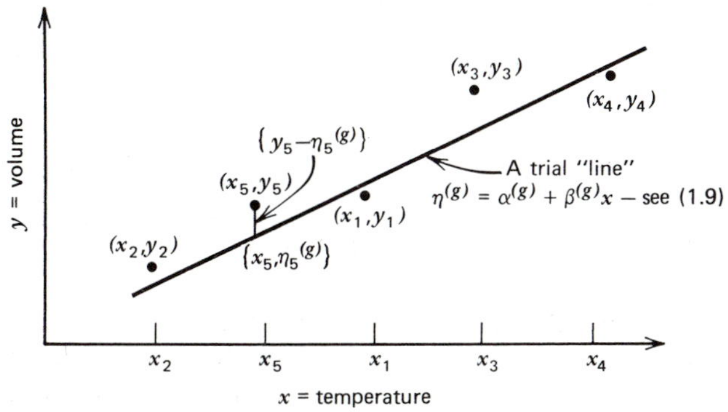

FIGURE 1.1.2. Observations y_i plotted against x_i, where x_i is the temperature used when observing the volume y_i (x_i's preselected).

that on some basis, we estimate the unknown α and β by $\alpha^{(g)}$ and $\beta^{(g)}$ respectively. (The superscript g stands for a "guess.") This implies that we are estimating the unknown regression function (assumed to be linear in x) to be

$$\eta^{(g)} = \alpha^{(g)} + \beta^{(g)}x, \tag{1.9}$$

and we note in passing that the points on the line corresponding to the preselected values of x_i have coordinates $(x_i, \eta_i^{(g)})$, where we are writing η_i for η_{x_i} and $\eta_i^{(g)}$ for $\eta_{x_i}^{(g)}$, so that

$$\eta_i^{(g)} = \alpha^{(g)} + \beta^{(g)}x_i, \qquad i = 1, \ldots, N. \tag{1.9a}$$

A way of measuring how good an estimate (1.9) is is to ask "How far is the line (1.9) from the observed data $\{(x_i, y_i)\}$." A criterion function that measures this is the sum of squares of the vertical distances of the observed (x_i, y_i) to the points $(x_i, \eta_i^{(g)})$ that lie on the estimated line, that is,

$$\begin{aligned} Q_g &= \sum_{i=1}^{N} \left(y_i - \eta_i^{(g)}\right)^2 \\ &= \sum_{i=1}^{N} \left(y_i - \alpha^{(g)} - \beta^{(g)}x_i\right)^2 \end{aligned} \tag{1.10}$$

(see Figure 1.1.2). Now if we were to compare the goodness of two estimated lines, that is, compare $(\alpha^{(g)}, \beta^{(g)})$ with another estimate of (α, β), for example, $(\alpha^{(g')}, \beta^{(g')})$, it is intuitively obvious that we would choose between $(\alpha^{(g)}, \beta^{(g)})$ and $(\alpha^{(g')}, \beta^{(g')})$ according to the following:

$$\begin{aligned} &\text{If } Q_g > Q_{g'}, \quad \text{choose } \left(\alpha^{(g')}, \beta^{(g')}\right). \\ &\text{If } Q_g = Q_{g'}, \quad \text{choose either } \left(\alpha^{(g')}, \beta^{(g')}\right) \text{ or } \left(\alpha^{(g)}, \beta^{(g)}\right). \\ &\text{If } Q_g < Q_{g'}, \quad \text{choose } \left(\alpha^{(g)}, \beta^{(g)}\right). \end{aligned} \tag{1.11}$$

In words, we would select that pair of estimates of (α, β) that gives rise to the estimated line which is closest to the data, as measured by Q.

Of course, there is an infinite number of choices that can be made on some basis or other for estimated values of (α, β), and extending the above principle, we see that we wish to find the values for α and β, for example,

$(\hat{\alpha}, \hat{\beta})$, that give rise to an estimated line

$$\hat{\eta} = \hat{\alpha} + \hat{\beta}x \tag{1.12}$$

which is *closest to the data over all possible choices of estimates of* (α, β). That is, we wish to invoke the so-called *principle of least squares* which to repeat states: Choose $(\hat{\alpha}, \hat{\beta})$ as estimates of (α, β), where

$$\min_{(\alpha,\beta)} Q = \min_{(\alpha,\beta)} \sum_{i=1}^{N} (y_i - \alpha - \beta x_i)^2 = \sum_{i=1}^{N} (y_i - \hat{\alpha} - \hat{\beta}x_i)^2. \tag{1.13}$$

There are several ways of finding the values $(\hat{\alpha}, \hat{\beta})$ which minimize

$$Q = \sum_{i=1}^{N} (y_i - \alpha - \beta x_i)^2 = Q(\alpha, \beta). \tag{1.14}$$

The first method, which we give here, employs calculus. The second method, which we detail in Appendix 1.1, is algebraic.

To find those values of α and β that minimize $Q(\alpha, \beta)$, given in (1.14), we proceed to take partial derivatives with respect to α and β, set these derivatives equal to zero, and solve for (α, β), denoting the solution by $(\hat{\alpha}, \hat{\beta})$, as stated above. We then have

$$\begin{aligned} \left.\frac{\partial Q(\alpha, \beta)}{\partial \alpha}\right|_{(\hat{\alpha}, \hat{\beta})} &= 0 \\ \left.\frac{\partial Q(\alpha, \beta)}{\partial \beta}\right|_{(\hat{\alpha}, \hat{\beta})} &= 0. \end{aligned} \tag{1.15}$$

This in turn produces the two equations

$$\begin{aligned} -2\sum_{i=1}^{N} (y_i - \hat{\alpha} - \hat{\beta}x_i) &= 0 \\ -2\sum_{i=1}^{N} (y_i - \hat{\alpha} - \hat{\beta}x_i)x_i &= 0. \end{aligned} \tag{1.15a}$$

Dividing each of these equations by -2, performing the summations, and rearranging so that the right-hand sides of the equations contain only y_i

terms, we obtain

$$(1.15b)\qquad \begin{aligned}(N)\hat{\alpha} + \left(\sum_1^N x_i\right)\hat{\beta} &= \sum_1^N y_i\\ \left(\sum_1^N x_i\right)\hat{\alpha} + \left(\sum_1^N x_i^2\right)\hat{\beta} &= \sum_1^N x_i y_i.\end{aligned}$$

Equations (1.15*b*) are usually called the *normal equations* for estimating α and β. As long as the determinant of the coefficients

$$(1.16)\qquad \begin{vmatrix} N & \sum_1^N x_i \\ \sum_1^N x_i & \sum_1^N x_i^2 \end{vmatrix} = \Delta(x_1,\ldots,x_N)$$

is not zero, the normal equations (1.15*b*) will have a unique solution with respect to $(\hat{\alpha}, \hat{\beta})$. Now this determinant has the value

$$(1.16a)\qquad \begin{aligned}\Delta(x_1,\ldots,x_N) &= N\left[\sum x_i^2 - \frac{(\sum x_i)^2}{N}\right]\\ &= N\sum_1^N (x_i - \bar{x})^2\end{aligned}$$

and this, of course, is not zero, unless all the x_i values are equal—ruled out by the assumption stated after (1.8*a*). The solution to the normal equations (1.15*b*) is called the *least-squares solution* since Eqs. (1.15*b*) are found as a result of minimizing (1.14). The solutions are given by

$$(1.17)\qquad \begin{aligned}\hat{\alpha} &= \bar{y} - \hat{\beta}\bar{x}\\ \hat{\beta} &= \frac{\sum_1^N x_i y_i - N\bar{x}\bar{y}}{\sum_1^N x_i^2 - N\bar{x}^2} = \frac{\sum_1^N (x_i - \bar{x})(y_i - \bar{y})}{\sum_1^N (x_i - \bar{x})^2} = \frac{S(\dot{x}, \dot{y})}{S(\dot{x}^2)}\end{aligned}$$

where (see Problem 1.2), denoting the set of $(x_i - \bar{x})$ by $\dot{x}$, and so on, we have

$$(1.17a)\quad \begin{aligned} S(\dot{x}, \dot{y}) &= \sum_{1}^{N} (x_i - \bar{x})(y_i - \bar{y}) = \sum_{1}^{N} x_i y_i - N\bar{x}\bar{y} \\ S(\dot{x}^2) &= \sum_{1}^{N} (x_i - \bar{x})(x_i - \bar{x}) = \sum_{1}^{N} (x_i - \bar{x})^2 = \sum_{1}^{N} x_i^2 - N\bar{x}^2. \end{aligned}$$

As indicated previously, the motivation for the terminology "least-squares solution" when referring to $(\hat{\alpha}, \hat{\beta})$ is as follows. The normal equations (1.15*b*) arise as a result of differentiating the *sum of squares* $Q(\alpha, \beta)$ given by (1.14), in the quest for the point in the α-β plane which minimizes Q. (Further discussion and justification for the rationale of minimizing Q is given in Chapter 3.) The claim is that (1.15), or equivalently (1.15*b*), gives us this minimum point, and hence $Q(\hat{\alpha}, \hat{\beta})$ would be *least* when evaluating $Q(\alpha, \beta)$ over all other choices of (α, β). We now demonstrate this (see also Appendix A1.1).

Suppose that we write Q as

$$(1.17b)\quad Q = \sum_{i=1}^{N} \left[(y_i - \hat{\alpha} - \hat{\beta}x_i) + (\hat{\alpha} - \alpha) + (\hat{\beta} - \beta)x_i \right]^2$$

where $(\hat{\alpha}, \hat{\beta})$ is given by (1.17); that is, $(\hat{\alpha}, \hat{\beta})$ satisfies (1.15*a*). Now, expanding (1.17*b*), we have

$$(1.17c)\quad \begin{aligned} Q = &\sum_{i=1}^{N} (y_i - \hat{\alpha} - \hat{\beta}x_i)^2 + \sum_{i=1}^{N} \left[(\hat{\alpha} - \alpha) + (\hat{\beta} - \beta)x_i \right]^2 \\ &+ 2\sum_{i=1}^{N} (y_i - \hat{\alpha} - \hat{\beta}x_i)\left[(\hat{\alpha} - \alpha) + (\hat{\beta} - \beta)x_i \right]. \end{aligned}$$

We note that the cross product given in the third term of Eq. (1.17*c*) is zero because it is equal to

$$(1.17d)\quad 2\left[(\hat{\alpha} - \alpha) \sum_{i=1}^{N} (y_i - \hat{\alpha} - \hat{\beta}x_i) + (\hat{\beta} - \beta) \sum_{i=1}^{N} (y_i - \hat{\alpha} - \hat{\beta}x_i)x_i \right],$$

and consulting (1.15a), we have that (1.17d) is zero. Hence

$$(1.17e)\quad Q = \sum_{i=1}^{N} (y_i - \hat{\alpha} - \hat{\beta}x_i)^2 + \sum_{i=1}^{N} [(\hat{\alpha} - \alpha) + (\hat{\beta} - \beta)x_i]^2$$

and minimizing Q over (α, β) is now seen to be accomplished by minimizing the second term of (1.17e), and this is clearly done when we choose $(\hat{\alpha}, \hat{\beta})$ for (α, β).

We also note that the above implies that the minimum value of Q is

$$(1.17f)\qquad Q(\hat{\alpha}, \hat{\beta}) = \sum_{i=1}^{N} (y_i - \hat{\alpha} - \hat{\beta}x_i)^2 = \min_{\alpha, \beta} Q(\alpha, \beta)$$

Given the estimates $(\hat{\alpha}, \hat{\beta})$ of (α, β), we have that the estimate of (1.6), that is, of $\eta = \alpha + \beta x$, is the *least-squares line* (see Figure 1.1.3)

$$(1.18)\qquad \hat{\eta} = \hat{\alpha} + \hat{\beta}x = \hat{\eta}_x,$$

with $(\hat{\alpha}, \hat{\beta})$ given in (1.17). Note that when $x = \bar{x}$, we have

$$(1.18a)\qquad \hat{\eta}_{\bar{x}} = \hat{\alpha} + \hat{\beta}\bar{x} = (\bar{y} - \hat{\beta}\bar{x}) + \hat{\beta}\bar{x} = \bar{y},$$

which is to say that the least-squares line passes through the *center of gravity* of the data, $(\bar{x}, \bar{y})$. We sometimes refer to the entire process by which we arrived at (1.18) as *fitting a straight line by least squares*, or as *regressing y on x*. We note that points on the least-squares line at the preselected values of

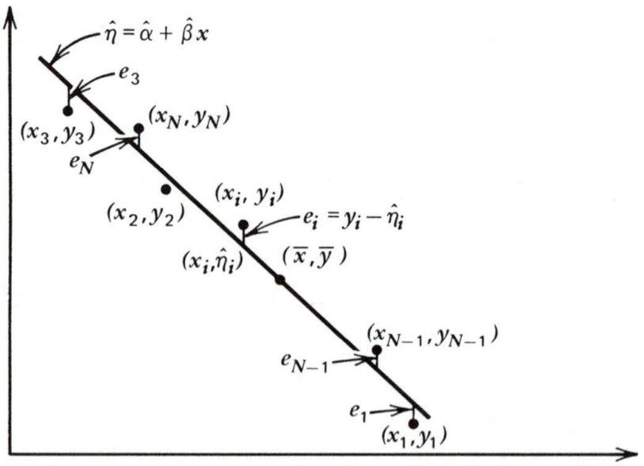

FIGURE 1.1.3. Least-squares line $\hat{\eta} = \hat{\alpha} + \hat{\beta}x$ and residuals $e_i = y_i - \hat{\eta}_i$.

x_i are $(x_i, \hat{\eta}_i)$, and indeed, recalling (1.14), (1.15), and the others, we have

$$Q(\hat{\alpha}, \hat{\beta}) = \sum_1^N (y_i - \hat{\alpha} - \hat{\beta}x_i)^2 = \sum_1^N (y_i - \hat{\eta}_{x_i})^2$$

(1.19)

$$= \min_{\alpha, \beta} Q(\alpha, \beta).$$

Often, $Q(\hat{\alpha}, \hat{\beta})$—the minimum of $Q(\alpha, \beta)$ over all possible choices of estimates of (α, β)—is denoted by SS_e, and SS_e is referred to as the sum of squares of the deviations e_i, where

$$e_i = y_i - \hat{\eta}_i = y_i - (\hat{\alpha} + \hat{\beta}x_i). \tag{1.20}$$

The quantity e_i is often referred to as the *residual* of y_i and can be thought of as that value which is "left over" on subtraction of the estimate $\hat{\eta}_i = \hat{\eta}_{x_i}$ of $\eta_{x_i} = \alpha + \beta x_i$ from the observed value y_i (see Figure 1.1.3). Indeed, SS_e is often referred to as the *sum of squares of residuals*.

We note that the sum of the residuals is zero:

$$\sum_1^N e_i = \sum_{i=1}^N (y_i - \hat{\alpha} - \hat{\beta}x_i)$$

(1.20*a*)

$$= \sum_{i=1}^N [(y_i - \bar{y}) - \hat{\beta}(x_i - \bar{x})]$$

$$= \sum_{i=1}^N (y_i - \bar{y}) - \hat{\beta} \sum_{i=1}^N (x_i - \bar{x})$$

$$= 0.$$

This property need not hold in general (see, e.g., the model of Problem 1.1) when doing a least-squares analysis, but it must hold whenever a linear model that is being fitted has in it a pure constant term, a point that will emerge in Chapter 3 [see Problem 3.20(b)]. Note that (1.20*a*) implies

$$\Sigma(y_i - \hat{\eta}_i) = 0 \quad \text{or} \quad \Sigma y_i = \Sigma \hat{\eta}_i. \tag{1.20b}$$

We note in passing that we may express the minimum sum of squares as

$$(1.20c)\qquad Q(\hat{\alpha}, \hat{\beta}) = \sum_{1}^{N} (y_i - \hat{\eta}_{x_i})^2 = \sum_{1}^{N} e_i^2 = \mathrm{SS}_e$$

In Section 1.2, we will prove that $E(\hat{\alpha}) = \alpha$ and $E(\hat{\beta}) = \beta$. Hence, we have that the expectation of a residual e_i, $E(e_i \mid x_1, \ldots, x_N)$, which we denote by $E(e_i)$, is such that

$$(1.20d)\qquad \begin{aligned} E(e_i) &= E(y_i - \hat{\alpha} - \hat{\beta}x_i \mid x_1, \ldots, x_N) \\ &= E(y_i \mid x_i) - E(\hat{\alpha}) - x_i E(\hat{\beta}) \\ &= \alpha + \beta x_i - \alpha - \beta x_i \\ &= 0, \end{aligned}$$

which is to say, the e_i terms give information about error alone. It will be seen in Section 1.3 that $\Sigma_1^N e_i^2 = \mathrm{SS}_e$ can be used as a basis to estimate σ^2, the error component of the model (1.6)–(1.7)!

Now $\hat{\alpha}$, $\hat{\beta}$, and SS_e and the quantities $\hat{\eta}_i$ and e_i have important statistical properties, and we now investigate these in Section 1.2. We will continue the description of the least-squares analysis in subsequent sections.

1.2. STATISTICAL PROPERTIES OF THE LEAST-SQUARES ANALYSIS

In this section, we remind the reader that our assumptions so far have been very minimal indeed. The first assumption is that we are observing a random variable y at various preselected values x_i of x, whose distribution D, has finite first and second moments,

$$(1.21a)\qquad E(y \mid x) = \alpha + \beta x$$

$$(1.21b)\qquad V(y \mid x) = \sigma^2$$

and that the N observations y_i, observed (or generated) at x_i, are statistically independent. We now investigate the expectation and variance of the least-squares estimates $(\hat{\alpha}, \hat{\beta})$ of (α, β).

To begin we note, from (1.17), that we may write $\hat{\beta}$ as

$$\hat{\beta} = \frac{\Sigma_1^N (x_i - \bar{x}) y_i}{S(\dot{x}^2)} \tag{1.22}$$

since it is a simple algebraic exercise to show that

$$\begin{aligned} \sum_1^N (x_i - \bar{x})(y_i - \bar{y}) &= \Sigma x_i y_i - N\bar{x}\bar{y} \\ &= \Sigma (x_i - \bar{x}) y_i . \end{aligned} \tag{1.22a}$$

Hence, from (1.22), we see that

$$\hat{\beta} = \sum_1^N c_i y_i, \qquad \text{where} \quad c_i = \frac{(x_i - \bar{x})}{S(\dot{x}^2)}. \tag{1.22b}$$

The weights c_i, $i = 1,\ldots,N$, depend only on the preselected values of the x_i's, and hence are constants, which implies that $\hat{\beta}$ is a linear combination of the y_i's. The coefficients of the linear combination are the c_i's, and these have the properties (see Problem 1.3)

$$\begin{aligned} &\text{(i)} \quad \sum_1^N c_i = 0, \\ &\text{(ii)} \quad \sum_1^N c_i^2 = \frac{1}{S(\dot{x}^2)}, \\ &\text{(iii)} \quad \sum_1^N c_i x_i = 1. \end{aligned} \tag{1.22c}$$

From (1.22b), we have upon taking expectations that $E(\hat{\beta})$, or more properly,

$$\begin{aligned} E(\hat{\beta} \mid x_1,\ldots,x_N) &= \sum_1^N c_i E(y_i \mid x_1,\ldots,x_N) \\ &= \sum_1^N c_i E(y_i \mid x_i) \\ &= \sum_1^N c_i \eta_i \end{aligned} \tag{1.22d}$$

where $\eta_i = E(y_i \mid x_i) = \alpha + \beta x_i$. Hence

$$E(\hat{\beta}) = \sum_1^N c_i(\alpha + \beta x_i)$$

(1.22e)
$$= \alpha \sum_1^N c_i + \beta \sum_1^N c_i x_i$$

$$= \beta,$$

from (1.22c). That is to say, the least-squares estimate $\hat{\beta}$ of β is *unbiased* for β. Continuing, we have upon taking variances of both sides of (1.22) the variance of $\hat{\beta}$, $V(\hat{\beta})$, or more properly,

(1.23)
$$V(\hat{\beta} \mid x_1, \ldots, x_N) = \sum_1^N c_i^2 V(y_i \mid x_1, \ldots, x_N),$$

since the y_i's are independent. Then, using the assumption of homoscedasticity,

$$V(\hat{\beta}) = \sum_1^N c_i^2 V(y_i \mid x_i)$$

(1.23a)
$$= \sigma^2 \sum_1^N c_i^2$$

$$= \frac{\sigma^2}{S(\dot{x}^2)}.$$

[We note that if the x_i's are chosen as wide apart as possible, giving rise to large $S(\dot{x}^2)$, the variance of $\hat{\beta}$ will tend to be small. This is a desired result since $\hat{\beta}$ is an estimate of β, a slope, and a good estimate of the slope is intuitively one that is based on a wide spread of x's, borne out by the result (1.23a).]

Using (1.22b) and the first line of (1.17), it is easy to see that $\hat{\alpha}$ may be written as a linear combination of the y_i's, that is,

(1.24)
$$\hat{\alpha} = \sum_{i=1}^N \left(\frac{1}{N} - \bar{x} c_i \right) y_i,$$

from which we may prove that (see Problem 1.4)

$$(1.24a) \qquad E(\hat{\alpha}) = \alpha, \qquad V(\hat{\alpha}) = \sigma^2\left[\frac{1}{N} + \frac{\bar{x}^2}{S(\dot{x}^2)}\right].$$

Also, the least-squares estimate of $\eta_x = \alpha + \beta x$—the expectation of y at x —is the linear function

$$(1.25) \qquad \hat{\eta}_x = \hat{\alpha} + \hat{\beta}x = \sum_1^N \left[\frac{1}{N} + c_i(x - \bar{x})\right] y_i$$

and the reader may verify that [see Problem 1.4(ii)]

$$(1.25a) \qquad E(\hat{\eta}_x) = \eta_x, \qquad V(\hat{\eta}_x) = \sigma^2\left[\frac{1}{N} + \frac{(x-\bar{x})^2}{S(\dot{x}^2)}\right].$$

[Note that $(1.25a)$ implies that for estimating η_x estimation near or at $\bar{x}$, the "center" of the x_i's, gives the least variance.] We note in passing that (1.24) and $(1.24a)$ give a special case of (1.25) and $(1.25a)$, because at $x = 0$, the intercept of the fitted line is $\hat{\alpha}$, and $\hat{\alpha}$ estimates the unknown intercept $\eta_{x=0} = \alpha$. That is, for $x = 0$ we have

$(1.25b)$

$$\hat{\eta}_{x=0} = \hat{\alpha}, \qquad E(\hat{\eta}_{x=0}) = \alpha, \qquad V(\hat{\eta}_{x=0}) = \sigma^2\left[\frac{1}{N} + \frac{(0-\bar{x})^2}{S(\dot{x}^2)}\right].$$

For later developments, it is convenient to have the covariance of $\hat{\alpha}$ and $\hat{\beta}$, that is,

$$(1.26) \qquad E[(\hat{\alpha} - \alpha)(\hat{\beta} - \beta)] = \text{cov}(\hat{\alpha}, \hat{\beta}).$$

To determine the value of (1.26), we note from (1.22)–$(1.22e)$ that we may write

$$(1.26a) \qquad (\hat{\beta} - \beta) = \sum_1^N c_i(y_i - \eta_i) = \sum_1^N c_i[y_i - E(y_i \mid x_i)],$$

and similarly it is easy to see that

$$(1.26b) \qquad (\hat{\alpha} - \alpha) = \sum_{j=1}^N \left(\frac{1}{N} - \bar{x}c_j\right)(y_j - \eta_j)$$

[see (1.24)]. Hence,

$$(\hat{\alpha}-\alpha)(\hat{\beta}-\beta)=\left[\sum_1^N\left(\frac{1}{N}-\bar{x}c_j\right)(y_j-\eta_j)\right]\left[\sum_1^N c_i(y_i-\eta_i)\right]$$

(1.27)
$$=\sum_{i=1}^N\left(\frac{1}{N}-\bar{x}c_i\right)c_i(y_i-\eta_i)^2$$

$$+\sum\sum_{i\neq j}\left(\frac{1}{N}-\bar{x}c_i\right)c_j(y_i-\eta_i)(y_j-\eta_j).$$

If we take expectations on both sides of (1.27), and recalling that y_i is assumed independent of y_j for $i \neq j$, we have

(1.27a)
$$\operatorname{cov}(\hat{\alpha},\hat{\beta})=\sigma^2\sum_{i=1}^N\left(\frac{1}{N}c_i-\bar{x}c_i^2\right)+0$$

$$=-\frac{\sigma^2\bar{x}}{S(\dot{x}^2)}$$

[from (1.22c)].

At this point, we may legitimately ask "Why do we perform least-squares in this situation?" The answer lies in the implications of the so-called Gauss theorem, a version of which we discuss now. The theorem in its full generality will be discussed in subsequent chapters. Roughly speaking, the Gauss theorem states that least-squares estimates $\hat{\alpha}$ or $\hat{\beta}$ of α or β have variance that is not larger than the variance of any other unbiased estimate which is linear in the y_i's, and in fact $a_1\hat{\alpha}+a_2\hat{\beta}$ is a minimum variance (unbiased) estimate for $a_1\alpha+a_2\beta$ (a_1 and a_2 are arbitrary constants). We have the following statement:

Theorem 1.2.1. *Suppose* (x_i, y_i), $i=1,\dots,N$, *are observation points obtained under the conditions stated above, namely,* $x_1,\dots,x_N$ *are preselected values of* x *at which the independent* $y_1,\dots,y_N$ *are observed, where* $E(y_i \mid x_i) = \alpha+\beta x_i$, *and* $V(y_i \mid x_i)\equiv\sigma^2$, *for all* i. *Denote the least-squares estimate of* (α,β) *by* $(\hat{\alpha},\hat{\beta})$, *given by* (1.17). *Suppose interest lies in the estimation of*

(1.28)
$$\tau=a_1\alpha+a_2\beta.$$

Then among all unbiased estimates of τ *that are linear in the* y_i, *the least-squares estimate,*

$$\hat{\tau} = a_1\hat{\alpha} + a_2\hat{\beta}$$

has minimum variance.

Proof. From (1.22e) and (1.24a), it is easy to see that $E(\hat{\tau}) = \tau$. Now suppose we have another estimate t which is unbiased for τ and linear in the y_i. Then

$$t = \sum_{i=1}^{N} d_i y_i \tag{1.29}$$

and the condition of unbiasedness states that $E(t \mid x_1, \ldots, x_N) = E(t)$ is such that

$$E(t) = \sum_{i=1}^{N} d_i E(y_i \mid x_i) = \sum_{i=1}^{N} d_i(\alpha + \beta x_i) = \tau. \tag{1.30}$$

Hence,

$$\alpha \sum_{i=1}^{N} d_i + \beta \sum_{i=1}^{N} d_i x_i = a_1\alpha + a_2\beta \tag{1.30a}$$

for all possible (α, β). We then have

$$\sum_{i=1}^{N} d_i = a_1 \quad \text{and} \quad \sum_{i=1}^{N} d_i x_i = a_2. \tag{1.30b}$$

Suppose now we write the least-squares estimate $\hat{\tau}$ as a linear combination of the y_i. We have

$$\begin{aligned} \hat{\tau} = a_1\hat{\alpha} + a_2\hat{\beta} &= \sum_1^N a_1\left(\frac{1}{N} - \bar{x}c_i\right) y_i + \sum_1^N a_2 c_i y_i \\ &= \sum_1^N \left[\frac{a_1}{N} + (a_2 - a_1\bar{x})c_i\right] y_i. \end{aligned} \tag{1.31}$$

The variance of $\hat{\tau}$ is thus (remembering that the y_i are independent)

$$V(\hat{\tau}) = \sigma^2 \sum_1^N \left[\frac{a_1}{N} + (a_2 - a_1\bar{x})c_i\right]^2, \tag{1.32}$$

and using (1.22c) we have

$$V(\hat{\tau}) = \sigma^2\left[\frac{a_1^2}{N} + \frac{(a_2 - a_1\bar{x})^2}{S(\dot{x}^2)}\right]. \tag{1.33}$$

Taking the expectations in (1.31) and subtracting, we may now write

$$\hat{\tau} - E(\hat{\tau}) = \hat{\tau} - \tau = \sum_1^N \left[\frac{a_1}{N} + (a_2 - a_1\bar{x})c_i\right](y_i - \eta_i) \tag{1.34}$$

and similarly, from (1.29) we have

$$t - E(t) = t - \tau = \sum_1^N d_j(y_j - \eta_j). \tag{1.35}$$

By using the method that led to the result (1.27a) [see Problem 1.4(iii)], the reader may verify that

$$\operatorname{cov}(\hat{\tau}, t) = \sigma^2 \sum_{i=1}^N \left[\frac{a_1}{N} + (a_2 - a_1\bar{x})c_i\right] d_i, \tag{1.35a}$$

and by using (1.30b) we thus have

$$\operatorname{cov}(\hat{\tau}, t) = \sigma^2\left[\frac{a_1^2}{N} + (a_2 - a_1\bar{x})\sum_1^N c_i d_i\right]. \tag{1.35b}$$

But

$$\begin{aligned} \sum_1^N c_i d_i &= \sum_1^N \frac{(x_i - \bar{x})\, d_i}{S(\dot{x}^2)} \\ &= \frac{[\Sigma_1^N x_i d_i - \bar{x}\Sigma_1^N d_i]}{S(\dot{x}^2)} \\ &= \frac{(a_2 - a_1\bar{x})}{S(\dot{x}^2)}. \end{aligned} \tag{1.35c}$$

Inserting (1.35c) into (1.35b) yields

$$\operatorname{cov}(\hat{\tau}, t) = \sigma^2\left[\frac{a_1^2}{N} + \frac{(a_2 - a_1\bar{x})^2}{S(\dot{x}^2)}\right], \tag{1.36}$$

and, using (1.33) and (1.36), we have the surprising result:

$$\text{cov}(\hat{\tau}, t) = V(\hat{\tau}). \tag{1.37}$$

Now consider $t - \hat{\tau}$. We have that

$$0 \leq V(t - \hat{\tau}) = V(t) + V(\hat{\tau}) - 2\,\text{cov}(\hat{\tau}, t), \tag{1.38}$$

and from (1.37) we have that

$$0 \leq V(t) - V(\hat{\tau}) \tag{1.38a}$$

or

$$V(\hat{\tau}) \leq V(t) \tag{1.38b}$$

which was to be proved. We note that equality occurs if and only if $t = \hat{\tau}$, which means that

$$t = \sum_{i=1}^{N} d_i y_i = \hat{\tau} = \sum_{1}^{N} \left[\frac{a_1}{N} + (a_2 - a_1\bar{x})c_i \right] y_i$$

or

$$d_i = \frac{a_1}{N} + (a_2 - a_1\bar{x})c_i, \qquad \text{for all } i = 1, \ldots, N, \tag{1.39}$$

and if this does not hold we have

$$V(\hat{\tau}) < V(t). \tag{1.39a}$$

We remark here that the above is very general, and, in particular, we have the following corollaries.

Corollary 1.2.1.1. ($a_1 = 0$, $a_2 = 1$). *Among all unbiased estimates of β, linear in the y_i, the least-squares estimate $\hat{\beta}$ is of minimum variance.*

Corollary 1.2.1.2. ($a_1 = 1$, $a_2 = 0$). *Among all unbiased estimates of α, linear in the y_i, the least-squares estimate $\hat{\alpha}$ is of minimum variance.*

Corollary 1.2.1.3. ($a_1 = 1$, $a_2 = x$). *Among all unbiased estimates of $\eta_x = \alpha + \beta x$, linear in the y_i, the least-squares estimate $\hat{\eta} = \hat{\alpha} + \hat{\beta}x$ is of minimum variance.*

The reader should note that all the results of this section hold under the weaker assumption that the observations, the y_j's, are uncorrelated. Indeed, very often it is the case that this assumption (uncorrelated y_j's) is made, and not the assumption of independence. Of course, if the assumptions made are that the y_j's are uncorrelated and normally distributed, then, of course the y_j's are independent. (See Chapter 2.)

1.3. CONTINUING THE ANALYSIS: ESTIMATION OF σ^2

As stated before, if the model we are fitting is correct [in our case, if $E(y \mid x) = \eta_x$ is linear in x, and we fit a line to the data], then, intuitively, residuals $e_i = y_i - \hat{\eta}_{x_i}$ should tell us "only about the error," that is, only about σ^2. In fact, it does turn out that $SS_e = \Sigma_1^N e_i^2$ carries information about σ^2 [see (1.44) below, for example, and recall (1.20c)].

To see this, we first note that we may write [see (1.20b)]

$$SS_e = \sum_1^N (y_i - \hat{\alpha} - \hat{\beta}x_i)(y_i - \hat{\alpha} - \hat{\beta}x_i)$$

$$\text{(1.40)} \qquad = \sum_1^N (y_i - \hat{\alpha} - \hat{\beta}x_i)y_i - \hat{\alpha}\sum_1^N (y_i - \hat{\alpha} - \hat{\beta}x_i)$$

$$-\hat{\beta}\sum_1^N (y_i - \hat{\alpha} - \hat{\beta}x_i)x_i.$$

But the last two terms in the above expansion of SS_e are zero, from (1.15a), so that we have

$$\text{(1.40a)} \qquad SS_e = \sum_1^N y_i^2 - \left(\hat{\alpha}\sum_1^N y_i + \hat{\beta}\sum_1^N x_i y_i\right),$$

$$= SS_t - SS_r.$$

Expression (1.40a) is important computationally—it states that the sum of squares of residuals may be computed by subtracting from the sum of squares of the observations y_i, the sum of the products of the least-squares estimates $\hat{\alpha}$ and $\hat{\beta}$ and the right-hand sides of the first and second normal equations, respectively. This latter sum is denoted by SS_r and referred to as the regression sum of squares. The normal equations are given by (1.15b).

Now using the normal equations (1.15*b*), it can be proved [see Problem 1.4(iv)] that

$$\text{(1.40}b\text{)}\quad SS_r = \hat{\alpha}\Sigma y_i + \hat{\beta}\Sigma x_i y_i = N\hat{\alpha}^2 + 2(\Sigma x_i)\hat{\alpha}\hat{\beta} + (\Sigma x_i^2)\hat{\beta}^2$$

and also that

$$\text{(1.40}c\text{)}\qquad SS_r = \Sigma \hat{\eta}_u^2.$$

The alternative form (1.40*c*) for SS_r is at this point a justification for the terminology used for it—that is, the regression sum of squares, because indeed SS_r is the sum of the squares of $\hat{\eta}_u$, the ordinates at the points x_u on the regression line $\hat{\eta} = \hat{\alpha} + \hat{\beta}x$.

It is also interesting to note the form of the right-hand side of (1.40*b*), because the coefficients of the $\hat{\alpha}^2$, $2\hat{\alpha}\hat{\beta}$, and $\hat{\beta}^2$ terms are elements obtained from the coefficients of the normal equations (1.15*b*): For the coefficient of $\hat{\alpha}^2$, we use the *first* coefficient in the *first* normal equation (i.e., the coefficient of $\hat{\alpha}$ in the first normal equation); for the coefficient of $\hat{\beta}^2$, we use the *second* coefficient in the *second* normal equation (i.e., the coefficient of $\hat{\beta}$ in the second normal equation); and for the coefficient of $2\hat{\alpha}\hat{\beta}$, we use *either* the *second* coefficient occurring in the *first* normal equation, or, the *first* coefficient occurring in the *second* normal equation. (The choice of either accounts for the factor 2 that goes with $\hat{\alpha}\hat{\beta}$.) The coefficients of the normal equations turn out to be very important for any linear (in the parameters) model, and we return to this point in subsequent chapters.

Any of the quantities mentioned in (1.40*b*) and (1.40*c*) are referred to as the *regression sum of squares*, denoted by SS_r. That is,

$$\text{(1.40}d\text{)}\quad SS_r = N\hat{\alpha}^2 + 2\Sigma x_i\hat{\alpha}\hat{\beta} + \Sigma x_i^2\hat{\beta}^2 = \hat{\alpha}\Sigma y_i + \hat{\beta}\Sigma x_i y_i = \sum_{u=1}^{N} \hat{\eta}_u^2,$$

and we inquire now into its expectation. We remind the reader that, for any random variable W with finite first and second moments, letting $\mu_W = E(W)$ and $\sigma_W^2 = E[(W - \mu_W)^2]$,

$$\text{(1.41)}\qquad E(W^2) = \sigma_W^2 + \mu_W^2$$

and similarly, for two random variables W and V, with appropriate moments existing, that

(1.41*a*)

$$E(WV) = \text{cov}(W, V) + \mu_W\mu_V, \qquad \text{cov}(W, V) = E[(W - \mu_W)(V - \mu_V)].$$

Using the extreme right-hand side of (1.40*b*), we have, from the results of

Section 1.2, that

$$E(\mathrm{SS}_r) = N\left\{\sigma^2\left[\frac{1}{N} + \frac{\bar{x}^2}{S(\dot{x}^2)}\right] + \alpha^2\right\} + 2N\bar{x}\left[-\frac{\sigma^2\bar{x}}{S(\dot{x}^2)} + \alpha\beta\right]$$

(1.42)

$$+ \left(\Sigma x_i^2\right)\left[\frac{\sigma^2}{S(\dot{x}^2)} + \beta^2\right],$$

and after some algebraic simplification, we have [essentially since $S(\dot{x}^2) = \Sigma x_i^2 - N\bar{x}^2$]

(1.42*a*) $$E(\mathrm{SS}_r) = 2\sigma^2 + \left[N\alpha^2 + 2(\Sigma x_i)\alpha\beta + \left(\Sigma x_i^2\right)\beta^2\right].$$

We make two remarks about (1.42*a*). There are two parameters α and β in the model we are concerned with, and that accounts for the factor 2 in the term $2\sigma^2$ in (1.42*a*). That term is followed by a quadratic expression in the unknowns α and β, and the coefficients of this quadratic form have been introduced and discussed in (1.40*b*) and subsequently.

It is now easy to calculate $E(\mathrm{SS}_e)$, since we have from (1.40*a*) and (1.40*b*) that

(1.43) $$\mathrm{SS}_e = \sum_1^N y_i^2 - \mathrm{SS}_r$$

so that

$$E(\mathrm{SS}_e) = \sum_1^N E\left(y_i^2\right) - E(\mathrm{SS}_r)$$

(1.43*a*)

$$= \sum_1^N \left[\sigma^2 + (\alpha + \beta x_i)^2\right] - E(\mathrm{SS}_r),$$

which implies that

(1.43*b*) $$E(\mathrm{SS}_e) = N\sigma^2 + \left(N\alpha^2 + 2\alpha\beta\Sigma x_i + \beta^2\Sigma x_i^2\right) - E(\mathrm{SS}_r).$$

Using (1.42*a*) we have

(1.44) $$E(\mathrm{SS}_e) = (N-2)\sigma^2.$$

The factor $N - 2$ appearing in (1.44) is sometimes referred to as the *degrees of freedom* (DF) for error—see Section 1.4 for a discussion. If we now let

$$(1.45) \qquad S_{y\cdot x}^2 = \frac{\text{SS}_e}{(N-2)} = \frac{\Sigma(y_i - \hat{\eta}_i)^2}{(N-2)}$$

we have that

$$(1.46) \qquad E\left(S_{y\cdot x}^2\right) = \sigma^2;$$

that is, $S_{y\cdot x}^2$ is an *unbiased* estimate of σ^2.

1.4. CONTINUING THE ANALYSIS: THE ANALYSIS OF VARIANCE, PART 1

From (1.43), we have the interesting identity

$$(1.47) \qquad \sum_{1}^{N} y_i^2 = \sum_{i=1}^{N} (y_i - \hat{\eta}_{x_i})^2 + \left[N\hat{\alpha}^2 + 2(\Sigma x_i)\hat{\alpha}\hat{\beta} + (\Sigma x_i^2)\hat{\beta}^2\right]$$

$$= \text{SS}_e + \text{SS}_r$$

The identity (1.47) states that we may *partition* the total sum of squares into two parts: one gives information on error (SS_e) and the other on the parameters of the regression function (SS_r). It is convenient to tabulate the elements of (1.47) in a summary table, called an analysis of variance table (Table 1.4.1).

TABLE 1.4.1. Analysis of Variance for the Model $E(y\,|\,x) = \alpha + \beta x$, $V(y\,|\,x) = \sigma^2$

Source	Degrees of Freedom (DF)	Sum of Squares	Mean Square[a]	Expected Mean Square
Due to regression	2	SS_r	$\text{SS}_r/2$	$\sigma^2 + \frac{1}{2}[N\alpha^2 + 2N\bar{x}\alpha\beta + (\Sigma x_i^2)\beta^2]$
Due to error	$N-2$	SS_e	$\text{SS}_e/(N-2)$	σ^2
Total	N	Σy_i^2		

[a] Mean square = sum of squares/DF

We will comment on the degrees of freedom entries in subsequent chapters. *At this point* they may be considered as follows: Before seeing the data, the data point $(y_1,\ldots,y_N)$ is free to lie anywhere in N-dimensional Euclidean space, R^N. Two constraints on the (x_i, y_i), $i = 1,\ldots,N$, are needed to calculate $\hat{\alpha}$ and $\hat{\beta}$, namely, the normal equations (1.15*b*) or equivalently the solutions given by (1.17). That is, $(\hat{\alpha}, \hat{\beta})$ are constrained to lie in a two-dimensional subspace of the N-dimensional sample space of $(y_1,\ldots,y_N)$. It will be seen, and it is most likely that the reader will intuitively guess, that the N quantities $e_i = y_i - \hat{\eta}_i$ must lie in the complement of the subspace that contains $(\hat{\alpha}, \hat{\beta})$, and its dimension is $N - 2$. After all, the e_i are N in number but must satisfy (1.15*a*), which, rewritten here, take the form

$$\Sigma e_i = \Sigma (y_i - \hat{\eta}_{x_i}) = 0$$

(1.48)

$$\Sigma e_i x_i = \Sigma (y_i - \hat{\eta}_{x_i}) x_i = 0.$$

[We will see in Chapter 3 that the above implies $(\hat{\eta}_{x_1},\ldots,\hat{\eta}_{x_N})$ are constrained to a two-dimensional subspace of R^N and $(e_1,\ldots,e_N)$ to a $(N-2)$-dimensional subspace in R^N.]

Inspection of the expected mean square column (Table 1.4.1) shows that information about the statement "$\alpha = 0$ and $\beta = 0$" may be obtained by comparing the mean square for regression $\text{SS}_r/2$ with $\text{SS}_e/(N-2)$, the so-called mean square for error, for if the assertion "$\alpha = 0$ and $\beta = 0$" is true, then both these mean squares have the same expectation—σ^2. Exactly how to compare these mean squares for the case of normality is discussed in Section 1.6 of this chapter. In particular, see (1.68).

1.5. THE ANALYSIS OF VARIANCE, PART 2

Very often when regressing y on x and concerned with the model of Section 1.1, we are interested not in the statement "$\alpha = 0$ and $\beta = 0$," but in a statement that concerns one and only one of the parameters, for example, "$\beta = 0$." Our aim then would be to find that portion of the regression sum of squares that gives information about β, or, to put it another way, that portion of the regression sum of squares that does not contain information about α. The process we use is called *orthogonalization* and it will be dealt with in full generality in subsequent chapters. For the simple model being discussed here, the general method discussed in Chapter 3, Section 3.5, requires us to rewrite the model in the following way:

$$
\begin{aligned}
E(y_i \mid x_i) &= \alpha + \beta x_i \\
&= (\alpha + \beta \bar{x}) + \beta(x_i - \bar{x}) \qquad (1.49) \\
&= \phi + \beta(x_i - \bar{x})
\end{aligned}
$$

where $\bar{x}$ is the mean of the N preselected values x_i of x. Note that $\Sigma_{i=1}^{N}(x_i - \bar{x}) = 0$. Furthermore, note that if we know the values of ϕ and β, we automatically know α since

$$\phi = \alpha + \beta\bar{x} \quad \text{or} \quad \alpha = \phi - \beta\bar{x}. \tag{1.49a}$$

Indeed (1.49) is a linear (in the parameters ϕ and β) model, and denoting $x_i - \bar{x}$ by w_i we have

$$\eta_i = E(y_i \mid x_i) = \phi + \beta w_i. \tag{1.50}$$

We may find the least-squares estimates of ϕ and β by minimizing

$$\Sigma(y_i - \eta_i)^2 = \Sigma(y_i - \phi - \beta w_i)^2. \tag{1.51}$$

[Of course (1.51) is the same quantity as Q, defined at (1.14), because $\Sigma(y_i - \eta_i)^2 = \Sigma(y_i - \phi - \beta w_i)^2 = \Sigma(y_i - \alpha - \beta x_i)^2$.] Differentiating (1.51) with respect to ϕ and β, and recalling that

$$\Sigma w_i = \Sigma(x_i - \bar{x}) = 0, \tag{1.51a}$$

we find (see Problem 1.4) the normal equations for the least-squares estimates of (ϕ, β), say $\hat{\phi}, \hat{\beta}$, are [see the form of the normal equations (1.15*b*)]

$$
\begin{aligned}
N\hat{\phi} &= \Sigma y_i \\
\hat{\beta}\Sigma w_i^2 &= \sum_{1}^{N} w_i y_i.
\end{aligned} \tag{1.52}
$$

These equations "solve themselves," and we have, since $w_i = x_i - \bar{x}$,

$$\hat{\phi} = \bar{y}, \tag{1.53}$$

and, as before,

$$(1.53a) \qquad \hat{\beta} = \frac{\Sigma(x_i - \bar{x})y_i}{S(\dot{x}^2)}.$$

Furthermore, the Gauss theorem states that among all linear (in the y_i) estimates of $\tau = (1)\phi + (-\bar{x})\beta$, the estimate

$$(1.53b) \qquad \hat{\tau} = (1)\hat{\phi} + (-\bar{x})\hat{\beta}$$

has minimum variance. We note that $\hat{\tau}$ is

$$(1.54) \qquad \hat{\tau} = (1)\bar{y} - \hat{\beta}\bar{x} = \hat{\alpha},$$

and indeed, $\hat{\tau} = \hat{\alpha}$ is the least-squares estimate of $\tau = (1)\phi + (-\bar{x})\beta = \alpha$.

In addition to the above estimates, we note that the fitted line is

$$(1.54a) \qquad \hat{\eta}_i = \hat{\phi} + \hat{\beta}w_i = \hat{\phi} + \hat{\beta}(x_i - \bar{x}) = \hat{\alpha} + \hat{\beta}x_i,$$

as before, and that the regression sum of squares due to ϕ and β is, by the rule of formation discussed after (1.40b), a quadratic in ϕ and β whose coefficients are obtained from the normal equations (1.52), so that we have

$$(1.55) \qquad \text{SS}_r = N\hat{\phi}^2 + 2(0)\hat{\phi}\hat{\beta} + \left(\Sigma w_i^2\right)\hat{\beta}^2.$$

This may be rewritten

$$(1.55a) \qquad \begin{aligned} \text{SS}_r &= N\hat{\phi}^2 + \Sigma(x_i - \bar{x})^2\hat{\beta}^2 \\ &= N\bar{y}^2 + \hat{\beta}^2 S(\dot{x}^2) \end{aligned}$$

or, as is easily verified,

$$(1.55b) \qquad = \hat{\phi}\Sigma y_i + \hat{\beta}S(\dot{x}, \dot{y}).$$

We see that SS_r has now been partitioned into a part depending on $\hat{\phi}$ alone and a part depending on $\hat{\beta}$ alone. [Equation (1.55a) states that SS_r is the sum of the products of the estimates times the corresponding right-hand sides of the normal equations (1.52)—the prescription we have met before.]

Indeed, consulting (1.43) we have [see Problem 1.5(a)]

$$\text{(1.56)} \qquad \begin{aligned} \mathrm{SS}_e &= \sum_1^N (y_i - \hat{\eta}_i)^2 = \sum_1^N y_i^2 - [N\bar{y}^2 + \hat{\beta}^2 S(\dot{x}^2)] \\ &= \sum_1^N (y_i - \bar{y})^2 - \hat{\beta}^2 S(\dot{x}^2), \end{aligned}$$

a fact that can be proved directly from (1.40a) [see Problem 1.5 and Eq. (A1.1.10) of Appendix A1.1]. We often tabulate the elements of the first line of (1.56) in an analysis of variance table (Table 1.5.1).

Problem 1.5(b) asks the reader to derive the entries of the expected mean square column in Table 1.5.1. We note that the expected sum of squares due to ϕ and due to β totals to

$$\text{(1.57)} \qquad 2\sigma^2 + N(\alpha + \beta\bar{x})^2 + \beta^2 S(\dot{x}^2)$$

which is the same, after some algebra, as (1.42a). [Problem 1.5(b) asks for the algebraic details.]

Now examination of the expected mean square column shows that comparison of the regression sum of squares due to β with the error mean square sheds some light on the assertion that "$\beta = 0$," *because if this assertion is correct*, then

$$\text{(1.58)} \qquad E(\hat{\beta}^2 S(\dot{x}^2)) = E(\mathrm{SS}_e/(N-2)) = \sigma^2.$$

Very often, in this situation where interest lies primarily in β alone, the analysis of variance table quoted is a modified version of Table 1.5.1, and constructed as in Table 1.5.1(a).

TABLE 1.5.1. Analysis of Variance for the Linear Model with Interest in β

Source	Degrees of Freedom	Sum of Squares	Expected Mean Square
Due to ϕ	1	$N\hat{\phi}^2 = N\bar{y}^2 = \hat{\phi}\sum y_i$	$\sigma^2 + N\phi^2$
Due to β	1	$\hat{\beta}^2 S(\dot{x}^2) = \hat{\beta} S(\dot{x}, \dot{y})$	$\sigma^2 + \beta^2 S(\dot{x}^2)$
Due to error	$N-2$	SS_e	σ^2
Total	N	$\sum y_i^2$	

TABLE 1.5.1(a)

Source	Degrees of Freedom	Sum of Squares	Mean Square	Expected Mean Square
Due to β	1	$\hat{\beta}^2 S(\dot{x}^2)$	$\hat{\beta}^2 S(\dot{x}^2)/1$	$\sigma^2 + \beta^2 S(\dot{x}^2)$
Due to error	$N-2$	SS_e	$\mathrm{SS}_e/(N-2)$	σ^2
Total	$N-1$	$\Sigma y_i^2 - N\bar{y}^2 = \Sigma (y_i - \bar{y})^2$		

As the reader will note, Table 1.5.1(a) is essentially Table 1.5.1 with the first line of the latter missing, and we compensate for this by subtracting from the total line in Table 1.5.1 the entries in the first line. A clue that this was to be expected lies in an alternative expression for SS_e, the residual sum of squares. From (1.40a) we may write

$$\begin{aligned} \mathrm{SS}_e &= \Sigma y_i^2 - \left(\hat{\alpha} N\bar{y} + \hat{\beta}\Sigma x_i y_i\right) \\ &= \Sigma y_i^2 - \left[(\bar{y} - \hat{\beta}\bar{x})N\bar{y} + \hat{\beta}\Sigma x_i y_i\right] \qquad (1.59) \\ &= \Sigma y_i^2 - \left[N\bar{y}^2 + \hat{\beta}\,(\Sigma x_i y_i - N\bar{x}\bar{y})\right] \end{aligned}$$

and using (1.17a),

$$\begin{aligned} \mathrm{SS}_e &= \Sigma y_i^2 - \left[N\bar{y}^2 + \hat{\beta} S(\dot{x}, \dot{y})\right] \\ &= \Sigma (y_i - \bar{y})^2 - \hat{\beta} S(\dot{x}, \dot{y}) \\ &= \Sigma (y_i - \bar{y})^2 - \hat{\beta}^2 S(\dot{x}^2) \end{aligned}$$

or, put another way,

$$\Sigma (y_i - \bar{y})^2 = \mathrm{SS}_e + \mathrm{SS}_r(\beta) \qquad (1.59a)$$

where the regression sum of squares $\mathrm{SS}_r(\beta)$ due to β is determined as

$$\mathrm{SS}_r(\beta) = \hat{\beta} S(\dot{x}, \dot{y}) = \hat{\beta}^2 S(\dot{x}^2). \qquad (1.59b)$$

We have seen [Tables 1.5.1 and 1.5.1a and Problem 1.5(b)] that $\mathrm{SS}_r(\beta)$ gives information, apart from σ^2, about β because its expected value is $\sigma^2 + \beta^2 S(\dot{x}^2)$. Furthermore, from (1.44) and (1.45), we have that $S_{y \cdot x}^2$ is unbiased for σ^2. Note too that the elements of (1.59a) are the basis for the (reduced) analysis of variance Table 1.5.1(a), appropriate for the situation discussed here, namely, interest focuses on the statement "$\beta = 0$."

We often refer to $SS_r(\beta)$ as the extra regression sum of squares due to β. Indeed, when fitting $\eta = \alpha + \beta x$, we sometimes write SS_e as $SS_e(\alpha, \beta)$, so that from (1.59*a*) we have

$$SS_r(\beta) = \Sigma (y_i - \bar{y})^2 - SS_e(\alpha, \beta). \tag{1.59c}$$

It is interesting to think of (1.59*c*) as follows. Suppose we are in the situation where independent y_i's are such that $\eta_i = E(y_i) = \alpha$ and $V(y_i) = \sigma^2$. Then as is well known (see the beginning of Appendix A1.1 for a review), if we do least-squares—fitting the line $\eta = \alpha$ by minimizing the criterion $\Sigma (y_i - \alpha)^2$—we obtain the solution

$$\hat{\alpha} = \bar{y}, \tag{1.59d}$$

and the sum of squares of residuals $SS_e(\alpha)$ for this model that gives information on σ^2 is

$$SS_e(\alpha) = \Sigma (y_i - \bar{y})^2 = \min_{\alpha} \Sigma (y_i - \alpha)^2. \tag{1.59e}$$

Indeed, if $E(y_i) = \alpha$ and $V(y_i) = \sigma^2$ for all i, then $E(\hat{\alpha}) = E(\bar{y}) = \alpha$ and $E[\Sigma (y_i - \bar{y})^2] = (N - 1)\sigma^2$. Hence, we may rewrite (1.59*c*) as

$$SS_r(\beta) = SS_e(\alpha) - SS_e(\alpha, \beta) \tag{1.59f}$$

and the phrase "extra regression sum of squares due to β" is now to be read as that part of the regression sum of squares resulting from inclusion of β (i.e., βx) in the model. We will return to this point in Chapter 3, since it is in reality a valuable general clue as to how to determine (extra) regression sum of squares resulting from particular effects being included in linear models.

Consulting Eq. (1.59*a*), we have that, since all quantities mentioned are positive,

$$\Sigma (y_i - \bar{y})^2 > SS_e, \tag{1.59g}$$

and in fact

$$\frac{SS_e}{\Sigma (y_i - \bar{y})^2} = 1 - \frac{\hat{\beta}^2 S(\dot{x}^2)}{\Sigma (y_i - \bar{y})^2}. \tag{1.59h}$$

The last term on the right-hand side is often denoted by R^2 and, by consequence of (1.59g) and (1.59h), we have

$$(1.59i)\qquad SS_e = \Sigma(y_i - \bar{y})^2(1 - R^2), \qquad 0 \le R^2 \le 1.$$

R^2 can, in view of (1.59i), be used as a measure of fit—if $R^2 = 0$, then $SS_e = \Sigma(y_i - \bar{y})^2$—and the data tell us that we should have just fitted the model $E(y_i) = \alpha$ and not have bothered to regress the y_i on x_i, since we end up with the error sum of squares that would have been obtained if we had "fit by least squares" $E(y_i) = \alpha$ when $V(y_i) = \sigma^2$ for all i, and (y_i, y_j) are independent. However, as R^2 approaches the value 1, SS_e is reduced and tends to zero, so that the data tell us that it is explained well by the model $E(y_i | x_i) = \alpha + \beta x_i$, and that we should be regressing y_i on x_i, and so on.

An alternative form for R^2 can be found as follows: we have

$$(1.59j)\qquad R^2 = \frac{\hat{\beta}^2 S(\dot{x}^2)}{\Sigma(y_i - \bar{y})^2} = \frac{[S(\dot{x}, \dot{y})]^2}{S(\dot{x})S(\dot{y})}$$

or

$$(1.59k)\qquad R = \frac{S(\dot{x}, \dot{y})}{[S(\dot{x}^2)S(\dot{y}^2)]^{1/2}} = \hat{\beta}\,[S(\dot{x}^2)/S(\dot{y}^2)]^{1/2}.$$

From (1.59i) we have

$$(1.59l)\qquad -1 \le R \le 1.$$

R is often referred to as the *correlation coefficient* between x and y. Note that it is a measure of linear fit in the sense described above.

Unfortunately, many people regard values of R near ± 1 as furnishing evidence that "x causes y." This misuse often results because high correlation between x and y is often caused by a further variable or variables. A famous example, quoted in Box, Hunter, and Hunter (1978), is that of the relationship of the number of storks (x) and the population size (y) of a certain town in Norway. The data collected, a plot of which is given in Figure 1.5.1, invite the fitting of a "straight line" and a high value of R results. But, of course, storks don't cause births, and the point here is that the populations of storks and humans are both increasing functions of time. Other examples that are similar but more subtle can be quoted. "Let the buyer beware."

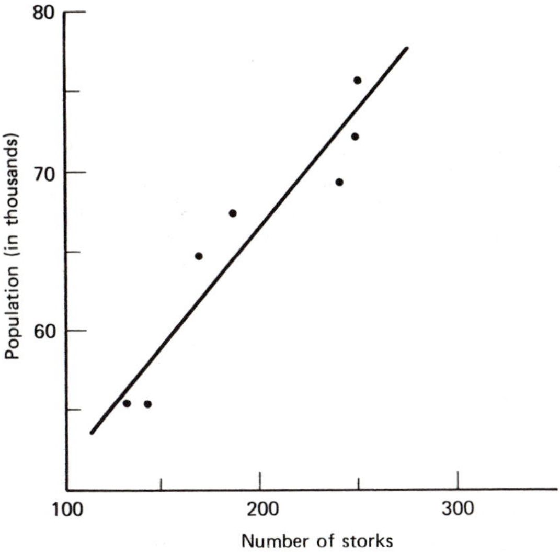

FIGURE 1.5.1. Plot of the population of Oldenburg (Norway) at the end of each year against the number of storks observed in that year, 1930–1936. [From Box, Hunter, and Hunter (1978), with permission of the authors.]

It is interesting to note that, from (1.59*j*) and using (1.59*f*),

$$R^2 = \frac{SS_r(\beta)}{\Sigma(y_i - \bar{y})^2} \tag{1.59m}$$

$$= \frac{SS_e(\alpha) - SS_e(\alpha, \beta)}{SS_e(\alpha)}.$$

This relation generalizes when we have more than one independent variable involved in a linear response function, and as will be discussed in Chapter 3, a definition of a multiple correlation coefficient can be similarly defined as in (1.59*m*) and can be used as a measure of "how good the fit is" when we regress y's on the independent variables concerned.

1.6. SOME NEEDED DISTRIBUTION THEORY: TESTS, CONFIDENCE INTERVALS

In previous sections, we have not been precise about how to use various statistics mentioned in order to examine various assertions about the parameters (and/or functions of the parameters) of our linear model

$$(1.60) \qquad E(y \mid x) = \alpha + \beta x, \qquad V(y \mid x) = \sigma^2,$$

with independence, or no correlation (at least) of the y_i observed at preselected x_i and so on. The basis for doing this is to first assume *normality* for the distribution of y given x, then to look at some procedures based on this assumption, and finally to examine how good these procedures are for non-normal distributions. Now, as is well known, if two random variables y_i and y_j are jointly distributed by the bivariate normal distribution and uncorrelated, then y_i and y_j are independent. So, adding the assumption of normality may be summarized by stating the model to be

$$(1.61) \qquad y_i = \text{ID}N\left(\alpha + \beta x_i, \sigma^2\right), \qquad i = 1, \ldots, N,$$

where the notation ID stands for *independently distributed*. [We will have occasion to use in the sequel the notation IID, standing for *independent and identically distributed*, which of course is not the case here, since for $x_i \neq x_j$, $E(y_i \mid x_i) \neq E(y_j \mid x_j)$; i.e., the means are different.] When the normality assumption holds (so that in previous notation $D = N$) we may state the following theorem.

Theorem 1.6.1. *Under the assumption* (1.61), *the least-squares estimates* $(\hat{\alpha}, \hat{\beta})$, *given by* (1.17), *have a joint distribution which is that of the normal bivariate distribution, whose density is*

$$(1.62) \qquad \frac{\left[NS(\dot{x}^2)\right]^{1/2}}{2\pi\sigma^2} \exp\left\{-\frac{1}{2\sigma^2}\left[N(\hat{\alpha} - \alpha)^2 + 2(\Sigma x_i)(\hat{\alpha} - \alpha)(\hat{\beta} - \beta) + \Sigma x_i^2(\hat{\beta} - \beta)^2\right]\right\}$$

which implies

$$E(\hat{\alpha}) = \alpha, \qquad E(\hat{\beta}) = \beta, \qquad \operatorname{cov}(\hat{\alpha}, \hat{\beta}) = -\frac{\bar{x}\sigma^2}{S(\dot{x}^2)}$$

$$(1.62a) \qquad V(\hat{\alpha}) = \frac{\sigma^2 \Sigma x_i^2}{NS(\dot{x}^2)} = \sigma^2\left[\frac{1}{N} + \frac{\bar{x}^2}{S(\dot{x}^2)}\right]$$

$$V(\hat{\beta}) = \frac{\sigma^2}{S(\dot{x}^2)},$$

and $(\hat{\alpha}, \hat{\beta})$ *are distributed independently of* SS_e, *which itself is such that*

$$SS_e = \sigma^2 \chi^2_{N-2}. \tag{1.62b}$$

This theorem will be proved in subsequent chapters. It has many important implications, which we now state, again without proof at this point. The implications will be proven in relevant sections of Chapters 2 and 3.

Corollary 1.6.1.1. *The distribution of the random variable*

(1.63)

$$N(\hat{\alpha} - \alpha)^2 + 2(\Sigma x_i)(\hat{\alpha} - \alpha)(\hat{\beta} - \beta) + (\Sigma x_i^2)(\hat{\beta} - \beta)^2 = T(\alpha, \beta)$$

is such that

$$T = \sigma^2 \chi^2_2 \tag{1.63a}$$

and T *is independent of* $SS_e = \sigma^2 \chi^2_{N-2}$.

We may use Corollary 1.6.1.1 in many interesting ways. The first way is to note that, having arrived at the estimate $(\hat{\alpha}, \hat{\beta})$ of (α, β), we may wish to inquire into the statement labeled H_0, referred to as a null hypothesis and succinctly denoted as

$$H_0: \alpha = \alpha_0 \quad \text{and} \quad \beta = \beta_0. \tag{1.64}$$

Both parameters are involved, and such a hypothesis is usually referred to as a *simultaneous hypothesis*. It implies that the distribution being sampled at $x = x_i$ has a mean $\alpha_0 + \beta_0 x_i$. Since no mention is made of the value of σ^2, the distribution is left unspecified by H_0, so that H_0 is a *composite* hypothesis.

Now we note that if H_0 *is true*,

$$\begin{aligned} T &= T(\alpha_0, \beta_0) \\ &= N(\hat{\alpha} - \alpha_0)^2 + 2(\Sigma x_i)(\hat{\alpha} - \alpha_0)(\hat{\beta} - \beta_0) + (\Sigma x_i^2)(\hat{\beta} - \beta_0)^2 \end{aligned} \tag{1.65}$$

is distributed as $\sigma^2 \chi^2_2$, independent of $SS_e = \sigma^2 \chi^2_{N-2}$, no matter what the true values of (α, β) may happen to be. Hence, in order to examine H_0 we may use as a test statistic,

$$F = \frac{T(\alpha_0, \beta_0)/2}{SS_e/(N-2)} \tag{1.65a}$$
$$= \frac{T(\alpha_0, \beta_0)/2}{S^2_{y \cdot x}},$$

because if H_0 *is true*, it is very easy to see from Corollary 1.6.1.1 that

$$F = F_{2, N-2} \tag{1.66}$$

where, in general, F_{m_1, m_2} denotes the Snedecor–Fisher random variable, with (m_1, m_2) degrees of freedom. The rejection region of significance level γ is specified by (see Problem 1.6)

$$\text{Reject } H_0\colon \alpha = \alpha_0, \beta = \beta_0 \quad \text{at level } \gamma \text{ if } F \geq F_{2, N-2; \gamma} \tag{1.67}$$

[$F_{m_1, m_2; \delta}$ denotes the point exceeded with probability δ when using the Snedecor F distribution with (m_1, m_2) degrees of freedom.]

We note that the analysis of variance Table 1.4.1 prepares for the above when $\alpha_0 = 0$, $\beta_0 = 0$, because by comparing $T(0,0)/2 = SS_r/2$ and $SS_e/(N - 2)$ by using their ratio, we have a special case of the above:

To test $H_0\colon \alpha = 0, \beta = 0$, reject at level γ if

$$F = \frac{T(0,0)/2}{S^2_{y \cdot x}} > F_{2, N-2; \gamma} \tag{1.68}$$

(see the expected mean square column of Table 1.4.1).

The procedure given in (1.68) states that to test the assertion "$\alpha = 0$ and $\beta = 0$", we compare $SS_r/2$ with $SS_e/(N - 2)$, where SS_r is given by (1.40c). It is interesting to note that the procedure used to test $\alpha = \alpha_0$, $\beta = \beta_0$, and $(\alpha_0, \beta_0) \neq (0,0)$, where F is defined at (1.65a), is of the same form; that is, it turns out that $T(\alpha_0, \beta_0)$ is a regression sum of squares found as follows. We have that

$$E(y_i \mid x_i) = \alpha + \beta x_i. \tag{1.68a}$$

Denoting $y_i - \alpha_0 - \beta_0 x_i$ by y_i^*, we then have

$$E(y_i^* \mid x_i) = \alpha^* + \beta^* x_i \tag{1.68b}$$

where

$$(1.68c)\quad y_i^* = y_i - \alpha_0 - \beta_0 x_i, \qquad \alpha^* = \alpha - \alpha_0, \qquad \beta^* = \beta - \beta_0.$$

If we regress y_i^* on x_i we find that the regression sum of squares in this situation is [Problem 1.6(b)]

$$(1.68d)\qquad \mathrm{SS}_r^* = n\hat{\alpha}^{*2} + 2\Sigma x_i \hat{\alpha}^* \hat{\beta}^* + (\Sigma x_i^2)\hat{\beta}^{*2}$$

and that the error sum of squares is

$$(1.68e)\qquad \mathrm{SS}_e^* = \mathrm{SS}_e.$$

We note too that the statement "$\alpha = \alpha_0$ and $\beta = \beta_0$" is equivalent to "$\alpha^* = 0$ and $\beta^* = 0$." Finally, it can be shown [Problem 1.6(c)] that

$$(1.68f)\qquad \hat{\alpha}^* = \hat{\alpha} - \alpha_0, \qquad \hat{\beta}^* = \hat{\beta} - \beta_0$$

so that

$$(1.68g)\qquad \begin{aligned} \mathrm{SS}_r^* &= T(\alpha_0, \beta_0) \\ &= N(\hat{\alpha} - \alpha_0)^2 + 2\Sigma x_i(\hat{\alpha} - \alpha_0)(\hat{\beta} - \beta_0) + \Sigma x_i^2(\hat{\beta} - \beta_0)^2. \end{aligned}$$

Putting all these results together, we see that we may turn any hypothesis-testing problem of the form "an intercept has value α_0 and a slope has value β_0" into a problem of the form "a certain intercept has value 0 and a certain slope has value 0," by suitable transformation on the y_i to y_i^* [see (1.68c)], regressing y_i^* on x_i, and finally by comparing a regression mean square with an error mean square.

Another interesting use of Corollary 1.6.1.1 arises when we wish to complement point estimates of α and β with a (simultaneous) confidence region for (α, β). Using Corollary 1.6.1.1, we have

$$(1.69)\qquad \frac{T(\alpha, \beta)/2}{S_{y\cdot x}^2} = F_{2,\,N-2}.$$

Hence, we have that (Pr stands for "the probability that")

$$(1.70)\qquad \Pr\left[T(\alpha, \beta) \le 2S_{y\cdot x}^2 F_{2,\,N-2;\,\gamma}\right] = 1 - \gamma,$$

because the left-hand side of (1.70) is, by using (1.69),

$$(1.70a)\quad \Pr\left[F_{2,\,N-2} \le F_{2,\,N-2;\,\gamma}\right] = 1 - \Pr\left[F_{2,\,N-2} > F_{2,\,N-2;\,\gamma}\right]$$

so that the $100(1-\gamma)\%$ *joint* or *simultaneous confidence region* for (α, β) is the set of points (α, β) in the α-β plane which satisfies the inequality in brackets on the left-hand side of (1.70). We denote such a set by

$$\{(\alpha, \beta) \mid T(\alpha, \beta) \le 2S_{y\cdot x}^2 F_{2, N-2; \gamma}\}$$

$$= \{(\alpha, \beta) \mid N(\alpha - \hat{\alpha})^2 + 2(\Sigma x_i)(\alpha - \hat{\alpha})(\beta - \hat{\beta}) + (\Sigma x_i^2)(\beta - \hat{\beta})^2 \le 2S_{y\cdot x}^2 F_{2, N-2; \gamma}\}. \tag{1.70b}$$

This set is the interior and boundary of an ellipse centered at $(\hat{\alpha}, \hat{\beta})$.

Interest very often focuses on *one* parameter; either α *or* β, for example, or on a linear combination of (α, β), such as $\tau = a_1\alpha + a_2\beta$ (e.g., if $a_1 = 0$, $a_2 = 1$, $\tau = \beta$). To deal with confidence regions and/or test statistics for such situations, we have the following:

Corollary 1.6.1.2. *Under the assumptions* (1.61), *the random variable*

$$\hat{\tau} = a_1\hat{\alpha} + a_2\hat{\beta} \tag{1.71}$$

is normally distributed, in particular,

$$\hat{\tau} = N\left\{\tau, \sigma^2\left[\frac{a_1^2}{N} + \frac{(a_2 - a_1\bar{x})^2}{S(\dot{x}^2)}\right]\right\}, \tag{1.71a}$$

and $\hat{\tau}$ *is independent of* $\text{SS}_e = \sigma^2\chi_{N-2}^2$, *and hence* $\hat{\tau}$ *is independent of* $S_{y\cdot x}^2 = \text{SS}_e/(N-2) = \sigma^2\chi_{N-2}^2/(N-2)$.

From this corollary, it is easy to verify (see Problem 1.7) that

$$\frac{\hat{\tau} - \tau}{S_{y\cdot x}\left[a_1^2/N + (a_2 - a_1\bar{x})^2/S(\dot{x}^2)\right]^{1/2}} = t_{N-2} \tag{1.72}$$

or

$$\frac{(\hat{\tau} - \tau)^2}{S_{y\cdot x}^2\left[a_1^2/N + (a_2 - a_1\bar{x})^2/S(\dot{x}^2)\right]} = F_{1, N-2}, \tag{1.72a}$$

and we may use (1.72) and (1.72*a*) for a variety of situations. To begin with, suppose interest lies in β, the slope (rate of change) of the true regression line $\eta = \alpha + \beta x$. From the results above, we find that (putting $a_1 = 0$, $a_2 = 1$)

$$\tau = (0)\alpha + (1)\beta = \beta, \qquad \hat{\tau} = \hat{\beta}, \tag{1.73}$$

and

$$\frac{\hat{\beta} - \beta}{\left[S_{y \cdot x}^2 / S(\dot{x}^2)\right]^{1/2}} = t_{N-2}, \tag{1.73a}$$

and (1.73a) may be used to construct a test statistic for the null hypotheses concerning β only, or to find (marginal) confidence intervals for β.

Specifically, suppose having arrived at $\hat{\beta}$—the least-squares estimate of β—we may wish to *test* the statement succinctly described by

$$H_0^{(1)}: \beta = \beta_0. \tag{1.74}$$

[We should point out that for any stated null hypothesis H_0, we are assuming and implying that if the stated null hypothesis is not true, then it must be that the complementary statement is true. For example, if (1.74) does not hold, then we imply that β can have any possible real value *not* equal to β_0, which we sometimes summarize by saying that the alternative hypothesis $H_1^{(1)}$: $\beta \neq \beta_0$ is true.] Now using (1.73a), it is easy to see that a test statistic for examining $H_0^{(1)}$ is

$$t = \frac{\hat{\beta} - \beta_0}{\left[S_{y \cdot x}^2 / S(\dot{x}^2)\right]^{1/2}} \tag{1.75}$$

and a test procedure for examining $H_0^{(1)}$ is

(1.75a) Reject $H_0^{(1)}$ at level γ if the observed value of t is such that

$$|t| > t_{N-2;\, \gamma/2},$$

where, in general, $t_{m;\,\delta}$ denotes the point exceeded with probability δ when using the Student's-t distribution with m degrees of freedom.

It is interesting to note that [see (1.72a) and the above] the above procedure is equivalent to

Reject $H_0^{(1)}$ at level γ if the observed value of

$$F = t^2 = \frac{(\hat{\beta} - \beta_0)^2 S(\dot{x}^2)}{S_{y \cdot x}^2} > F_{1, N-2; \gamma}. \tag{1.75b}$$

(We remind the reader that, $t_m^2 = F_{1, m}$ and $F_{1, m; \delta} = t_{m; \delta/2}^2$.) In the special case that $\beta_0 = 0$, we are concerned *with the elements of Table 1.5.1(a)*, and consulting the expected mean square column, we see that under $H_0^{(1)}: \beta = 0$, it is natural to compare $SS_r(\beta)$ and SS_e using the ratio of their mean squares, which is the statistic F defined by (1.75b) with $\beta_0 = 0$. (See also Problem 1.7.)

From (1.73a), it is easy to see that the interval (the set of values of β)

$$\left\{ \beta \,\middle|\, \hat{\beta} - \left[\frac{S_{y \cdot x}^2}{S(\dot{x}^2)} \right]^{1/2} t_{N-2; \gamma/2} \leq \beta \leq \hat{\beta} + \left[\frac{S_{y \cdot x}^2}{S(\dot{x}^2)} \right]^{1/2} t_{N-2; \gamma/2} \right\} \tag{1.76}$$

is a $100(1 - \gamma)\%$ confidence interval for β (Problem 1.8).

Now by putting $a_1 = 1$ and $a_2 = x$, we find using (1.72) that

$$\frac{\hat{\eta}_x - \eta_x}{S_{y \cdot x}\left[1/N + (x - \bar{x})^2 / S(\dot{x}^2)\right]^{1/2}} = t_{N-2} \tag{1.77}$$

where of course, $\eta_x = \alpha + \beta x$ and $\hat{\eta}_x = \hat{\alpha} + \hat{\beta} x$. We may use (1.77) as a test statistic and/or to help derive a confidence interval for η_x (Problem 1.9). The latter, for confidence level $1 - \gamma$, takes the form

$$\begin{aligned} \left\{ \eta_x \,\middle|\, \hat{\eta}_x - S_{y \cdot x}\left[\frac{1}{N} + \frac{(x - \bar{x})^2}{S(\dot{x}^2)} \right]^{1/2} t_{N-2; \gamma/2} \leq \eta_x \leq \hat{\eta}_x \right. \\ \left. + S_{y \cdot x}\left[\frac{1}{N} + \frac{(x - \bar{x})^2}{S(\dot{x}^2)} \right]^{1/2} t_{N-2; \gamma/2} \right\}. \end{aligned} \tag{1.78}$$

If we allow x to vary, we generate a confidence belt, with the upper and lower straps of the belt, when plotted against x, having the equations

$$\eta = \hat{\eta}_x \pm S_{y \cdot x}\left[\frac{1}{N} + \frac{(x - \bar{x})^2}{S(\dot{x}^2)}\right]^{1/2} t_{N-2;\,\gamma/2} \tag{1.79}$$

and, as implied when we were discussing the variance of $\hat{\eta}_x$, the belt has its shortest width at $x = \bar{x}$.

A special case of (1.77)–(1.79) is that when $x = 0$, we then have

$$\eta_x = \eta_0 = \alpha, \qquad \hat{\eta}_0 = \hat{\alpha} \tag{1.80}$$

and from (1.77)

$$\frac{\hat{\alpha} - \alpha}{S_{y \cdot x}\left[1/N + \bar{x}^2/S(\dot{x}^2)\right]^{1/2}} = t_{N-2} \tag{1.81}$$

which gives rise to the $100(1 - \gamma)\%$ confidence interval for α,

$$\left\{\alpha \,\middle|\, \hat{\alpha} - S_{y \cdot x}\left[\frac{1}{N} + \frac{\bar{x}^2}{S(\dot{x}^2)}\right]^{1/2} t_{N-2;\,\gamma/2} \le \alpha \le \hat{\alpha} \right.$$

$$\left. + S_{y \cdot x}\left[\frac{1}{N} + \frac{\bar{x}^2}{S(\dot{x}^2)}\right]^{1/2} t_{N-2;\,\gamma/2}\right\}. \tag{1.82}$$

It should be noted that, for a given set of data, postulated values of (α, β)—for example, (α_0, β_0)—could very well lie in the intervals (1.82) and (1.76), respectively, while the point (α_0, β_0) lies outside the joint confidence region for (α, β) given by (1.70b). That is to say, the postulated values α_0 and β_0 could be *marginally admissible* (in the sense that, separately, these values belong to the respective confidence intervals), and yet be *inadmissible* when viewed *simultaneously* if the point (α_0, β_0) lies outside the joint confidence region for (α, β). See Problem 1.10 for an example of the occurrence of this phenomenon. Needless to say, great care must be taken, and the purpose of the experiment must be borne in mind. For example, if the experiment is to be done because of interest in β—the rate of change of the response with respect to x—then (1.76) is of interest. If, however, the experiment is to be performed to gain knowledge of (α, β), then (1.70b) is relevant.

1.7. GOODNESS OF FIT: SOME ELEMENTARY MODEL BUILDING

So far, in this chapter we have been using assumptions that call for fitting a linear function in x to the data. Indeed, there are many situations in which we either know exactly that the response variable η is expected to behave as a linear function in x, *or*, we know on empirical grounds that to good approximation, for values of x in the experimental region commonly used, the *response function* is linear in x. However, there are many instances in which it is important that the assumption of linearity be checked, and we now describe one such method for detecting so-called *lack of fit* for this situation.

For example, it may well be that we fit a straight line, but that a straight line should not be fitted because the response function is quadratic in x, or some other nonlinear function of x, and we wish to detect that this "lack of fit" has occurred. It turns out that we may easily provide ourselves with a method for handling the above problem, if we run an experiment as follows. Suppose we preselect m places of x, $x_1,\ldots,x_m$, and at each of the x_j so selected we observe the response n_j times, independent (statistically) of one another. Let the total sample size be $N = \sum_{j=1}^{m} n_j$, and suppose we denote the n_j observations generated at x_j by $(y_{1j},\ldots,y_{ij},\ldots,y_{n_j j})$. We may think of the data when plotted as made up of the points $(x_{ij}, y_{ij}) = (x_j, y_{ij})$, $i = 1,\ldots,n_j$, $j = 1,\ldots,m$. We are assuming that for each j, $n_j \geq 1$, and there is at least one j—j'—for which $n_{j'} > 1$. We will assume here that $m \geq 3$ [see (1.88a) below]. Without loss of generality, we are also assuming that $x_1 < x_2 < \cdots < x_m$.

When a straight line is fitted by least squares, we of course obtain the least-squares line

$$\hat{\eta} = \hat{\alpha} + \hat{\beta}x \tag{1.83}$$

where here $\hat{\alpha}$ and $\hat{\beta}$ may be written as [see (1.17) and Problem 1.11]

$$\hat{\alpha} = \bar{y} - \hat{\beta}\bar{x}, \qquad \bar{y} = \frac{1}{N}\sum_{j=1}^{m}\sum_{i=1}^{n_j} y_{ij} = \frac{1}{N}\sum_{j=1}^{m} n_j \bar{y}_{\cdot j} \tag{1.84a}$$

with

$$\bar{y}_{\cdot j} = n_j^{-1}\sum_{i=1}^{n_j} y_{ij} \tag{1.84b}$$

and

$$(1.84c)\quad \hat{\beta} = \frac{\sum_{j=1}^{m}\sum_{i=1}^{n_j}(x_j - \bar{x})(y_{ij} - \bar{y})}{S(\dot{x}^2)} = \frac{\sum_{j=1}^{m} n_j(x_j - \bar{x})(\bar{y}_{\cdot j} - \bar{y})}{S(\dot{x}^2)}$$

with

$$(1.84d)\qquad S(\dot{x}^2) = \sum_{j=1}^{m} n_j(x_j - \bar{x})^2, \qquad \bar{x} = \frac{\sum_{j=1}^{m} n_j x_j}{N}.$$

Of course, the minimum sum of squares of deviations is the sum of squares of residuals

$$(1.85)\qquad \mathrm{SS}_e = \sum_{j=1}^{m}\sum_{i=1}^{n_j} e_{ij}^2 = \Sigma\Sigma(y_{ij} - \hat{\eta}_{ij})^2$$

where

$$(1.85a)\qquad \hat{\eta}_{ij} = \hat{\alpha} + \hat{\beta}x_j, \qquad i = 1,\ldots,n_j.$$

Intuitively, if some other function should have been fitted, rather than a linear function the residuals should contain information other than that of error, because the y_{ij} tell us about the "true" function, while $\hat{\eta}_{ij}$ are designed to tell us only about linear components. Hence, it is expected that SS_e will contain, in addition to information about $\sigma^2 = V(y_{ij} \mid x_j)$, $i = 1,\ldots,n_j$, $j = 1,\ldots,m$, information about departures from the assumed linear function from the correct function, and hence SS_e will be expected to be larger than information about σ^2 alone. This can be checked if we can get an estimate of error, σ^2, which does not depend on any fitting; the way we have generated the data allows us to do just that.

For we may, as a preliminary step, view the above data as having been generated so that a "one-way analysis of variance" can be used to analyze the data, where there are m groups, with n_j observations in group j, $j = 1,\ldots,m$, and where the means of the groups are hypothesized to be

$$(1.86)\qquad E(y_{ij} \mid x_j) = \alpha + \beta x_j, \qquad j = 1,\ldots,m,$$

but it is feared that (1.86) is not true and that the means of the m populations at x_j, $j = 1,\ldots,m$ lie not on a linear function but perhaps on a function of higher degree in x, for example, quadratic, cubic, or on any other nonlinear function of x, such as an exponential or sinusoidal function

TABLE 1.7.1. Preliminary Analysis of Variance for *m* Groups for Lack of Fit Investigation

Source	Degrees of Freedom	Sum of Squares
Between	$m - 1$	$SS_B = \sum_{j=1}^{m} n_j(\bar{y}_{\cdot j} - \bar{y})^2$
Within	$N - m = \sum_{j=1}^{m} (n_j - 1)$	$SS_W = \sum_{j=1}^{m} \sum_{i=1}^{n_j} (y_{ij} - \bar{y}_{\cdot j})^2$
	$N - 1 = \Sigma n_j - 1$	$SS_T = \sum_{j=1}^{m} \sum_{i=1}^{n_j} (y_{ij} - \bar{y})^2$

of x. Be that as it may, for such a one-way analysis of variance, the "error term" is the "within sum of squares," and in fact, the one-way analysis of variance table for m groups may be constructed in the usual fashion as shown in Table 1.7.1.

It is easy to see that SS_W, the so-called "within sample sum of squares," gives information on σ^2, the error, alone. From *standard elementary theory*, we have that since $y_{1j}, \ldots, y_{ij}, \ldots, y_{n_j j}$ are n_j independent and identically distributed observations, with distribution that has mean $E(y_{ij} \mid x_j) = \eta_{x_j}$ and variance $V(y_{ij} \mid x_j) = \sigma^2$ for all $i = 1, \ldots, n_j$, then

$$E\left[\sum_{i=1}^{n_j} (y_{ij} - \bar{y}_{\cdot j})^2\right] = (n_j - 1)\sigma^2 \tag{1.87}$$

or

$$E\left[\sum_{j=1}^{m} \sum_{i=1}^{n_j} (y_{ij} - \bar{y}_{\cdot j})^2\right] = \sigma^2 \sum_{j=1}^{m} (n_j - 1) = \sigma^2(N - m). \tag{1.87a}$$

Now since we are fitting a straight line, we may break up the between sum of squares with its $m - 1$ degrees of freedom into $SS_r(\beta) = \hat{\beta}^2 S(\dot{x}^2)$, the regression sum of squares due to the slope of the fitted line, with its one degree of freedom, and a quantity, which we denote by SS_L (standing for sum of squares of the lack of fit component), with $m - 2$ degrees of freedom, where

$$\text{SS}_L = \text{SS}_B - \text{SS}_r(\beta). \tag{1.88}$$

It can be shown that if the assumption of linearity is *correct*, then

$$E(\text{SS}_L) = (m-2)\sigma^2 \tag{1.88a}$$

while, if it is *not correct*, we write

$$E(\text{SS}_L) = (m-2)\sigma^2 + \Lambda^2 \tag{1.88b}$$

where Λ^2 depends on how the true function differs from linearity and is such that $\Lambda^2 = 0$ if $\eta = \alpha + \beta x$. [See Problem 3.33(b) where it is assumed that if $\eta \neq \alpha + \beta x$, then $\eta = \gamma_0 + \gamma_1 x + \gamma_2 x^2$.] Accordingly, we may draw up a second-stage analysis of variance table (Table 1.7.2).

It is apparent that to test "lack of fit," we need only compare $\text{SS}_L/(m-2)$ with $\text{SS}_W/(N-m)$. Indeed, if we postulate normality for distribution of the y_{ij}, it can be shown that

$$F_L = \frac{\text{SS}_L/(m-2)}{\text{SS}_W/(N-m)} \tag{1.89}$$

has the $F_{m-2,\,N-m}$ distribution if indeed the hypothesis

$$\eta = \alpha + \beta x \tag{1.90}$$

is true. This means that in fitting a straight line, no "lack of fit" is

TABLE 1.7.2. Lack of Fit Analysis of Variance when Fitting a Straight Line

Source	Degrees of Freedom	Sums of Squares	Expected Mean Square
Between $m-1$	╱Regression 1	$\text{SS}_r(\beta) = \hat{\beta}^2 S(\dot{x}^2)$	$\sigma^2 + \beta^2 S(\dot{x}^2) + g(\Lambda^2)^a$
	╲Lack of Fit $m-2$	$\text{SS}_L = \text{SS}_B - \hat{\beta}^2 S(\dot{x}^2)$	$\sigma^2 + \Lambda^2/(m-2)$
Within	$N-m$	$\text{SS}_W = \Sigma\Sigma(y_{ij} - \bar{y}_{\cdot j})^2$	σ^2
Total	$N-1$	$\text{SS}_T = \Sigma\Sigma(y_{ij} - \bar{y})^2$	

[a] $g(\Lambda^2)$ is such that $g(0) = 0$.

committed, and the procedure (consulting the expected mean square column of Table 1.7.2) of significance level γ is

(1.91) Reject linearity if $F_L > F_{m-2,\,N-m;\,\gamma}$; accept otherwise.

[The results quoted above, starting one line after (1.87*a*), are justified by the general results of Section 3.6, Chapter 3.]

It is interesting to note that we may recast Table 1.7.2 so as to reflect a breakup of SS_e, the sum of squares of residuals. By consulting (1.59) we have that

$$SS_e = SS_T - \hat{\beta}^2 S(\dot{x}^2), \qquad SS_T = \Sigma\Sigma(y_{ij} - \bar{y})^2. \tag{1.92}$$

But from Table 1.7.2, we see that

$$SS_T - \hat{\beta}^2 S(\dot{x}^2) = SS_L + SS_W \tag{1.92a}$$

which is to say that, putting (1.92) and (1.92*a*) together,

$$SS_e = SS_L + SS_W. \tag{1.93}$$

The new version of Table 1.7.2 now takes the form of Table 1.7.3.

For the entries in the sum of squares column, we may proceed as follows, once the least-squares estimate $\hat{\beta}$ is determined. First, find $SS_r(\beta)$ and compute SS_T using the relationship

$$SS_T = \Sigma\Sigma y_{ij}^2 - \frac{(\Sigma\Sigma y_{ij})^2}{N}. \tag{1.94}$$

Then find SS_e using

$$SS_e = SS_T - SS_r(\beta). \tag{1.94a}$$

Extract the so-called pure error term SS_W, the within sample sum of

TABLE 1.7.3. Lack of Fit Analysis of Variance when Fitting a Straight Line

Source	Degrees of Freedom	Sum of Squares	Expected Mean Square
Regression	1	$SS_r(\beta) = \hat{\beta}^2 S(\dot{x}^2)$	$\sigma^2 + \beta^2 S(\dot{x}^2) + g(\Lambda^2)$
Residual	$N-2$: $m-2$	$SS_L = SS_e - SS_W$	$\sigma^2 + \Lambda^2/(m-2)$
	$N-2$: $N-m$	$SS_W = \Sigma\Sigma(y_{ij} - \bar{y}_{\cdot j})^2$	σ^2
	$N-1$	$SS_T = \Sigma\Sigma(y_{ij} - \bar{y})^2$	

squares, by computing

$$\text{(1.94}b\text{)} \qquad \mathrm{SS}_W = \sum_{j=1}^{m} \left[\sum_{i=1}^{n_j} y_{ij}^2 - \frac{\left(\sum_{i=1}^{n_j} y_{ij}\right)^2}{n_j} \right];$$

that is,

$$\text{(1.94}c\text{)} \qquad \mathrm{SS}_W = \sum_{j=1}^{m} \sum_{i=1}^{n_j} y_{ij}^2 - \sum_{j=1}^{m} \left[\frac{\left(\sum_{i=1}^{n_j} y_{ij}\right)^2}{n_j} \right].$$

Relationship (1.94b) states that SS_W can be computed by finding the sum of squares of deviations $y_{ij} - \bar{y}_{\cdot j}$—that is, deviations of observations in the jth group from their sample mean $\bar{y}_{\cdot j}$, for each $j = 1, \ldots, m$—and then summing these, while (1.94c) states that SS_W can be computed by finding the sum of squares of all observations y_{ij} and subtracting from this the sum of the correction factors of all the m groups, that is, the sum of all the quantities $(\sum_{i=1}^{n_j} y_{ij})^2/n_j$.

If the hypothesis of linearity is rejected, the data support the lack of fit assumption that η may be *any nonlinear function* of x—sinusoidal, exponential, and so on. Now it may be that at the outset, we know that η is (or well approximated by) a polynomial function of x. Then, if the hypothesis of linearity is rejected, a quadratic may be fitted to the data, and a new lack of fit test made to examine the *hypothesis of quadraticness versus cubicness* of the response function. (We will defer a discussion of details until Chapter 3.) In this way, we can find, to the extent that the data indicate, the *polynomial model* that best fits the data.

Now if the test (1.91) accepts linearity, we may still be interested in examining the hypothesis that $\beta = 0$ (response function is a constant function of x) versus the hypothesis that $\beta \neq 0$. To do this we compare $\mathrm{SS}_r(\beta)$ and $\mathrm{SS}_e/(N-2)$, since it may be shown *that if* $\Lambda = 0$ [and we are assuming here that the data indicate this to be so through use of (1.91)] and $\beta = 0$,

$$\text{(1.94}d\text{)} \qquad F_\beta = \frac{\mathrm{SS}_r(\beta)}{\mathrm{SS}_e/(N-2)}$$

is distributed as $F_{1,\,N-2}$ [provided $y_{ij} = N(\alpha + \beta x_j, \sigma^2)$, $\beta = 0$, and so on].

The above discussion has been conducted on the assumption that there are repeat observations at $m \geq 3$ values of the independent variable. However, it is often the case that the experiment was conducted without repeat

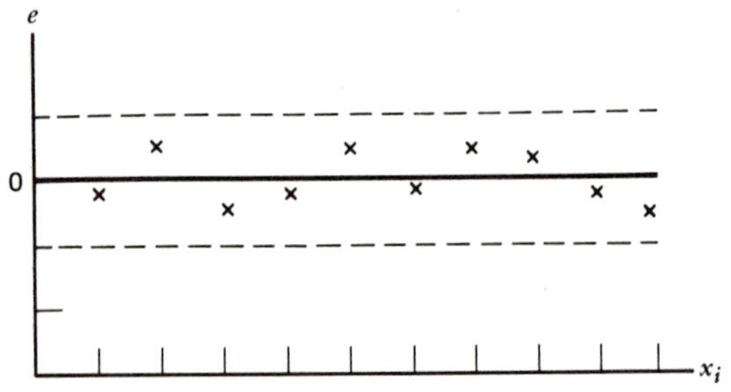

FIGURE 1.7.1. Residuals of observations plotted against x_i.

observations, but inquiries about lack of fit are still of primary consideration. In this case, what is often done is to *examine the residuals* $e_i = y_i - \hat{\eta}_i$. There is a theoretical basis for this, but examining residuals can also be done visually by plotting the e_i values in various ways:

(i) e_i against i;
(ii) e_i against x_i;
(iii) e_i against $\hat{\eta}_i = \hat{\eta}_{x_i}$.

If there is no lack of fit, then these three plots should yield points which hover above and below the $e = 0$ line, as in Figure 1.7.1, where the e_i values seem to fall in a band about $e = 0$.

However, if the residuals do not fall into a band such as in Figure 1.7.1, the pattern exhibited could point to which one (or more) of the assumptions does not hold. For example, if e_i values plotted against x_i exhibit the behavior shown in Figure 1.7.2, where the y_i are regressed on $\alpha + \beta x_i$, then we would suspect the linearity assumption and fit a quadratic ($\gamma_0 + \gamma_1 x_i + \gamma_2 x_i^2$), which we will be able to do after digesting the material of Chapter 3.

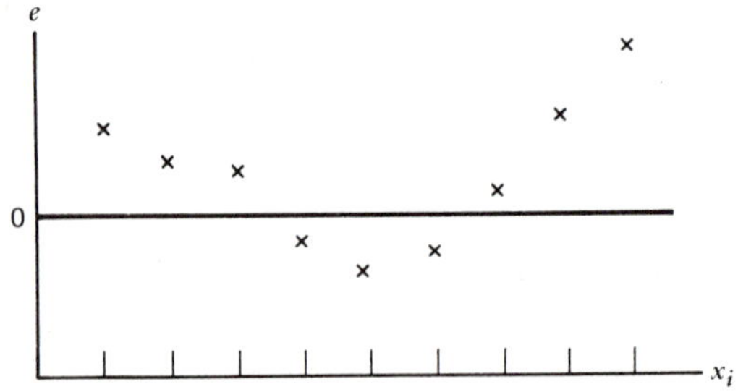

FIGURE 1.7.2. Residuals plotted against x_i exhibiting a quadratic behavior.

For further reading, we recommend the excellent account of the examination of residuals that is given in the book by Draper and Smith (1981).

1.8. PREDICTION

Very often, the objective of an experimental investigation is to gather sample data which will enable the experimenter to predict the outcome of an observed value y of the (dependent) response variable when the independent variable x is set equal to some value of interest. In this situation, we certainly would like to have confidence in the model to which the experimental data will be fitted, and we suppose for the discussion here, that a priori knowledge or "lack of fit" considerations lead us to suppose that the response is linear in x, that is,

$$E(y \mid x) = \alpha + \beta x = \eta_x. \tag{1.95}$$

Suppose further that data (x_i, y_i), $i = 1, \ldots, N$ are available (in fact, it may have been used to test lack of fit, and no lack of fit found) leading to the fitted least-squares line

$$\hat{\eta}_x = \hat{\alpha} + \hat{\beta}x, \tag{1.96}$$

where $(\hat{\alpha}, \hat{\beta})$ are the least-squares estimates of (α, β). Now we wish to predict y, the (future) value of the response when $x = x_0$ is used, where this future y is to be observed independently of the data that led to (1.96). Denote the predicted value by $\tilde{y}$ (we do not use the caret symbol ^ over the letter y since y is a random variable, and not a parameter). The question now is "What is a suitable predictor $\tilde{y}_{x_0}$, based on the sample information $\{(x_i, y_i) \mid i = 1, \ldots, N\}$, of an independent observed response y to be observed at $x = x_0$?"

Intuitively, to predict a y (to be observed at x_0) based on the data (x_i, y_i), $i = 1, \ldots, N$, it seems reasonable to use the predictor

$$\tilde{y}_{x_0} = \hat{\alpha} + \hat{\beta}x_0, \tag{1.97}$$

since we note that the expectation of $\tilde{y}_{x_0}$ given by (1.97) is such that

$$E\big(\tilde{y}_{x_0} \mid (x_i, y_i), i = 1, \ldots, N\big) = E(y \mid x_0), \tag{1.98}$$

since both expectations are equal to η_{x_0}. We will justify the selection of (1.97) as follows. For any predictor $\tilde{y}$ of y, we would like deviations of actual y values at x_0 from predicted values $\tilde{y}_{x_0}$ to be minimum, and we note that a measure of spread of these deviations is, given $x = x_0$,

$$\begin{aligned} &E\left((\tilde{y}_{x_0} - y)^2 \mid (x_i, y_i), i = 1,\ldots,N\right) \\ (1.99) \qquad &= E\left((\tilde{y}_{x_0} - \eta_{x_0})^2 \mid (x_i, y_i), i = 1,\ldots,N\right) \\ &\quad + E\left((y - \eta_{x_0})^2 \mid x_0\right) \end{aligned}$$

The cross-product term vanishes, as is easily verified, since the covariance between y and $\tilde{y}_{x_0}$ is zero by virtue of the assumption that y is to be observed at x_0 and independently of the data (x_i, y_i), $i = 1,\ldots,N$. Hence, if we *restrict ourselves to prediction functions* $\tilde{y}$ *which are linear in the* y's, we know from Corollary 1.2.1.3 of the Gauss theorem that the first term on the right-hand side of (1.99) is minimized if indeed we do use (1.97), and in so doing the "squared error of prediction," given by (1.99), is minimized. In summary, if

$$\begin{aligned} &y_i, i = 1,\ldots,N \quad \text{are} \quad \mathrm{ID}(\alpha + \beta x_i, \sigma^2), \\ (1.100) \qquad &y \text{ is independent of the } y_i, \quad \text{and} \quad D(\alpha + \beta x_0, \sigma^2) \end{aligned}$$

then the predictor of y, $\tilde{y}$, given by (1.97) is a minimum mean square error predictor among all predictors linear in the y_j's, with, using (1.99) and (1.100) along with (1.25a),

$$(1.101) \qquad E\left[(\tilde{y}_{x_0} - y)^2 \mid (x_i, y_i)\right] = \sigma^2\left[1 + \frac{1}{N} + \frac{(x_0 - \bar{x})^2}{S(\dot{x}^2)}\right].$$

If in (1.100), we let $D = N$, that is, if we further *assume normality*, then we immediately have that

$$\begin{aligned} t_{N-2} &= \frac{\tilde{y}_{x_0} - y}{\left[\mathrm{SS}_e/(N-2)\right]^{1/2}\left[1 + 1/N + (x_0 - \bar{x})^2/S(\dot{x}^2)\right]^{1/2}} \\ (1.102) \qquad &= \frac{\tilde{y}_{x_0} - y}{S_{y\cdot x}\left[1 + 1/N + (x_0 - \bar{x})^2/S(\dot{x}^2)\right]^{1/2}} \end{aligned}$$

so that a $(1-\gamma)$ prediction interval for y at x_0 is the interval with endpoints

$$\tilde{y}_{x_0} \pm S_{y\cdot x}\left[1+\frac{1}{N}+\frac{(x_0-\bar{x})^2}{S(\dot{x}^2)}\right]^{1/2} t_{N-2;\,\gamma/2} \tag{1.103}$$

where the point predictor $\tilde{y}_{x_0}$ is given by (1.97). Note then, that if we let x_0 vary, we would obtain a *prediction* band that is wider than the *confidence* band for η_x [see (1.79)].

1.9. AN ILLUSTRATIVE EXAMPLE

Thirty individuals, chosen at random, were given a diet with different (predecided) amounts (x) per day of a riboflavin extract (milligrams per milliliter of solution) and their systolic blood pressure (y) recorded, with the results as given in Table 1.9.1.

Some easy calculations yield the following pertinent information:

$$\begin{gathered} N=30;\quad \Sigma x_i=1354;\quad \Sigma x_i y_i=199{,}576;\quad \Sigma y_i=4276;\\ \Sigma x_i^2=67{,}894;\quad \Sigma y_i^2=624{,}260. \end{gathered} \tag{1.104}$$

This experiment was carried out with the intention of examining the assumption held by heart specialists that the relationship between x and y over the range of x commonly encountered ($15 \le x \le 70$) is linear, and in

TABLE 1.9.1

Patient Number	x	y	Patient Number	x	y	Patient Number	x	y
1	39	144	11	64	162	21	36	136
2	47	220	12	56	150	22	50	142
3	45	138	13	59	140	23	39	120
4	47	145	14	34	110	24	21	120
5	65	162	15	42	128	25	44	160
6	46	142	16	48	130	26	53	158
7	67	170	17	45	135	27	63	144
8	42	124	18	17	114	28	29	130
9	67	158	19	20	116	29	25	125
10	56	154	20	19	124	30	69	175

fact, as the reader will note on examination of the values of x used, several patients were fed a diet containing the same amounts of x so that a "lack of fit" test could be carried out.

To get to the point at which we may examine the question of lack of fit, we first regress y on x, and using (1.104), obtain the normal equations

$$\begin{aligned} 30\hat{\alpha} + 1354\hat{\beta} &= 4276 \\ 1354\hat{\alpha} + 67{,}894\hat{\beta} &= 199{,}576. \end{aligned} \tag{1.105}$$

These yield the solution

$$\hat{\alpha} = 98.71, \qquad \hat{\beta} = 0.97087. \tag{1.105a}$$

Using (1.104) and (1.105a), we may immediately find that

$$S(\dot{x}^2) = \Sigma(x_i - \bar{x})^2 = \Sigma x_i^2 - \frac{(\Sigma x_i)^2}{30} = 6783.47 \tag{1.106}$$

so that the extra regression sum of squares due to β is

$$SS_r(\beta) = \hat{\beta}^2 S(\dot{x}^2) = 6394.02. \tag{1.106a}$$

We also find that

$$S(\dot{y}^2) = \Sigma y_i^2 - \frac{(\Sigma y_i)^2}{30} = 14{,}787.47 \tag{1.107}$$

which yields in turn

$$SS_e = S(\dot{y}^2) - SS_r(\beta) = 8393.45 \tag{1.108}$$

To complete the elements for a "lack of fit" test, we wish to extract the pure error term SS_W. Reorganizing the data of Table 1.9.1, we see that (Table 1.9.2) there are six values of x at which two observations are taken,

TABLE 1.9.2. Observations y Recorded at Values of x for which $n_j \geq 2$

x	39	42	45	47	56	67
y	144;120	124;128	138;135	220;145	150;154	170;158

and in fact the values of x at which this is done are $x = 39, 42, 45, 47, 56$, and 67. At all other values of x, only one observation has been taken so that $m = 24$.

When n_j observations y_{ij} are taken using "condition j", then the within sum of squares is of course

$$(1.109) \qquad \sum_{i=1}^{n_j} (y_{ij} - \bar{y}_{\cdot j})^2, \qquad \bar{y}_{\cdot j} = n_j^{-1} \sum_{i=1}^{n_j} y_{ij}.$$

We remind the reader that when $n_j = 2$, we have

$$(1.109a) \qquad \sum_{i=1}^{2} (y_{ij} - \bar{y}_{\cdot j})^2 = \frac{(y_{1j} - y_{2j})^2}{2}.$$

Hence, it is rapidly seen that for this set of data

$$(1.109b) \quad SS_W = \left[(144 - 120)^2 + \cdots + (170 - 158)^2\right] / 2 = 3193.$$

We are now in a position to construct the relevant analysis of variance table (see Table 1.9.3).

The "lack of fit" sum of squares is of course found by subtraction, since we have

$$(1.110) \qquad SS_L = SS_T - SS_r(\beta) - SS_W$$

yielding the entry for SS_L as given in Table 1.9.3. We have that

$$(1.111) \quad F_L = \frac{5200.45/22}{3193/6} = \frac{236.38}{532.17} = 0.44 < F_{22,6;0.05} = 3.855,$$

that is, we accept the hypothesis that there is no lack of fit when fitting the "straight line" model $\alpha + \beta x$.

TABLE 1.9.3. Analysis of Variance for Data of Table 1.9.1

Source	Degrees of Freedom	Sum of Squares	
β	1	$SS_r(\beta) = 6394.02$	
Lack of fit	22	$SS_L = 5200.45$	$F_L = 0.44$
Pure error	6	$SS_W = 3193$	
Total	29	$SS_T = S(\dot{y}^2) = 14{,}787.47$	

We are now in the position to examine statements about β. The assertion $\beta = 0$ is quickly seen to be rejected at, for example, the 5% level, since, on using (1.94d), we find

$$F_\beta = \frac{6394.02}{8393.45/28} = 21.33 > F_{1,28;0.05} = 4.196. \tag{1.112}$$

It may be that the assertion $\beta = 1$ was of interest if the experimenter were satisfied about linearity in x of the behavior of y. Carrying this out, we of course examine

$$F_\beta^* = \frac{(\hat{\beta} - \beta_0)^2 S(\dot{x}^2)}{SS_e/28}. \tag{1.113}$$

The observed value of F_β^* for this situation is

$$(\text{observed})\ F_\beta^* = \frac{(0.971 - 1)^2(6783.47)}{299.77} = 0.019, \tag{1.113a}$$

and we have that (observed) $F_\beta^* \leq F_{1,28;0.05} = 4.196$, so that we accept the hypothesis that $\beta = 1$.

We may now proceed, since linearity is indicated, to determine the usual entities for inference. For example, a 95% confidence interval for β is $I_{0.95}(\beta)$, where

$$\begin{aligned} I_{0.95}(\beta) &= \left[\hat{\beta} \pm \left(\frac{SS_e}{28}\right)^{1/2}\left(\frac{1}{S(\dot{x}^2)}\right)^{1/2} t_{28;0.025}\right] \\ &= [0.971 \pm 0.4305], \end{aligned} \tag{1.114}$$

that is,

$$I_{0.95}(\beta) = [0.5405, 1.4015]. \tag{1.114a}$$

Similarly, it is easy to see that the 95% confidence band for η_x is found from the limits

$$\hat{\alpha} + \hat{\beta}x \pm \left[\frac{1}{30} + \frac{(x - 45.13)^2}{6783.5}\right]^{1/2}\left(\frac{SS_e}{28}\right)^{1/2} t_{28;0.025} \tag{1.115}$$

and for varying x, the band is as graphed in Figure 1.9.1. The prediction

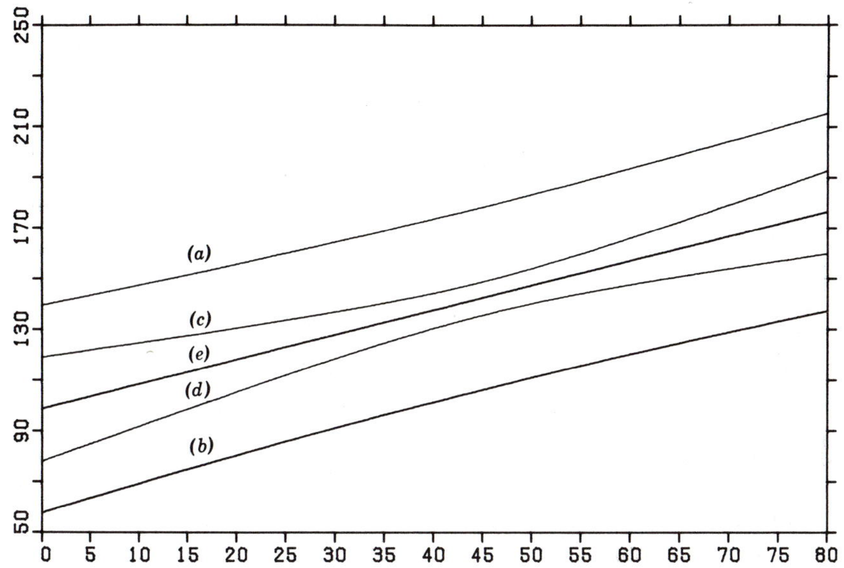

FIGURE 1.9.1. (a and b) The 95% prediction band and (c and d) the 95% confidence band for y and η_x, respectively, based on the data of Table 1.9.1. (e) The least-squares regression line $\eta_x = 98.71 + 0.9709x$.

band of level 0.95 for a future observation y, to be observed independently at x, is also graphed in Figure 1.9.1 and is generated from the limits

$$\text{(1.115}a\text{)} \quad \hat{\alpha} + \hat{\beta}x \pm \left[1 + \frac{1}{30} + \frac{(x - 45.13)^2}{6783.5}\right]^{1/2} \left(\frac{\text{SS}_e}{28}\right)^{1/2} t_{28;\,0.025}.$$

Finally, we may also produce the joint 95% confidence region for (α, β), given by the set of points in the (α, β) plane $C_{0.95}$, where

(1.116)

$$C_{0.95} = \left\{ (\alpha, \beta) \,\middle|\, \begin{array}{r} 30(\alpha - 98.71)^2 + 2(1354)(\alpha - 98.71)(\beta - 0.971) \\ +67{,}894(\beta - 0.971)^2 \le 2\dfrac{\text{SS}_e}{28} F_{2,28;\,0.05} \end{array} \right\}.$$

This is a region interior to the ellipse given by the equation found by using the equality sign in (1.116) and is diagrammed in Figure 1.9.2—the

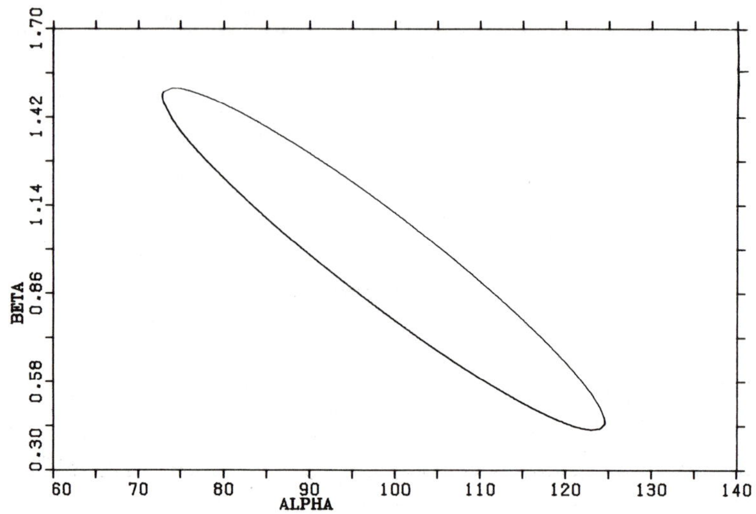

FIGURE 1.9.2. The 95% confidence ellipse for (α, β), based on the data of Table 1.9.1.

center of the ellipse is of course $(\hat{\alpha}, \hat{\beta})$. This problem will be referred to in Chapter 3.

PROBLEMS

1.1. Suppose $\eta_x = E(y \mid x) = \beta x$, and that for a set of N preselected x's, say $x_1, \ldots, x_N$, we observe $y_1, \ldots, y_N$ respectively. Find, by minimizing a suitable sum of squares, the least-squares estimate of β, and hence of η_x. Denoting least-squares estimates by $\hat{\beta}$ and $\hat{\eta}_x = \hat{\beta}x$, show that residuals $e_i = y_i - \hat{\beta}x_i$ do not sum to zero.

1.2. Verify (1.17a).

1.3. Verify properties (i)–(iii) of (1.22c).

1.4. (a) (i) Deduce (1.24a). (ii) Deduce (1.25a). (iii) Verify (1.35a) and hence (1.36). (iv) Deduce (1.40b) and (1.40c).
Hint: For the latter write $\Sigma\hat{\eta}_u^2$ as $\Sigma_{u=1}^N(\hat{\alpha} + \hat{\beta}x_u)^2$ and expand the square, which will show that $\Sigma\hat{\eta}_u^2$ has the form of the right-hand side of (1.40b). (v) Derive (1.52) using the model (1.50). Show that the covariance of $\bar{y} = \hat{\phi}$ and $\hat{\beta}$ is zero.

(b) A breakdown leading to (1.40a) may be derived directly, using (1.15a), which states $\Sigma_{i=1}^N e_i = \Sigma_{i=1}^N(y_i - \hat{\alpha} - \hat{\beta}x_i) = 0$ and

$\Sigma_{i=1}^{N} e_i x_i = 0$, and (1.20$c$), which states $SS_e = \Sigma e_i^2 = \Sigma(y_i - \hat{\eta}_i)^2$, with $\hat{\eta} = \hat{\alpha} + \hat{\beta}x_i$, by proceeding as follows: Write Σy_i^2 as

$$\begin{aligned}\Sigma y_i^2 &= \Sigma[(y_i - \hat{\eta}_i) + \hat{\eta}_i]^2 \\ &= \Sigma(y_i - \hat{\eta}_i)^2 + \Sigma\hat{\eta}_i^2 + 2\Sigma\hat{\eta}_i(y_i - \hat{\eta}_i) \\ &= SS_e + SS_r.\end{aligned}$$

Show that $\Sigma\hat{\eta}_i(y_i - \hat{\eta}_i) = 0$. Thus $\Sigma y_i^2 = \Sigma(y_i - \hat{\eta}_i)^2 + \Sigma\hat{\eta}_i^2$; this equation is often called the *fundamental identity*, which states that the sum of squares of the y_i values is made up of SS_e, which may be viewed as a sum of squares of the y_i values *about* the regresssion line $\hat{\eta} = \hat{\alpha} + \hat{\beta}x$, and SS_r, a part due to regression, where $SS_r = \Sigma\hat{\eta}_i^2$.

1.5. (a) Prove the breakdown (1.56).

(b) Derive the expected mean squares tabulated in Table 1.5.1 and show that the sum of the first two entries is as in (1.42a).

(c) Using parts (a) and (b) and the identity (1.56), show that $E\Sigma(y_i - \bar{y})^2 = (N-1)\sigma^2 + \beta^2 S(\dot{x}^2)$.

1.6. (a) Substantiate that the test procedure of (1.67) is of significance level γ.

(b) Explain why (1.68d) is the regression sum of squares found when regressing y_i^* on x_i using the model (1.68b).

(c) Prove (1.68f).

1.7. (a) Under the conditions of Corollary 1.6.1.2, derive (1.72) and (1.72a), thus showing the procedures (1.75a) and (1.75b) are of level γ.

(b) Referring to (1.75b), show that $(\hat{\beta} - \beta_0)^2 S(\dot{x}^2)$ is the extra regression sum of squares due to $\beta^{**} = \beta - \beta_0$ when regressing y_i^{**} on x_i, with $y_i^{**} = y_i - \beta_0 x_i$, and $E(y_i^{**} \mid x_i) = \alpha + \beta^{**}x_i$ [equivalent to $E(y_i \mid x_i) = \alpha + \beta x_i$]. Note that $\beta = \beta_0$ is equivalent to $\beta^{**} = 0$, and so on.

(c) Show that $\hat{\beta}^{**} = \hat{\beta} - \beta_0$.

1.8. Derive (1.76).

1.9. Use (1.77) to derive (a) a test procedure for the hypothesis that $\eta_{x_0} = E(y \mid x_0) = \delta$, where δ is a specified number, and (b) a confidence interval for η_{x_0}, at level $1 - \gamma$.

TABLE P1.10

x (gal/min)	2	4	6	8	10	12	14
y (% waste removed)	24.3	19.7	17.8	14.0	12.3	7.2	5.5

1.10. A study was instituted to determine the percent of waste solids removed in a filtration system as a function of the flow rate (x) of the effluent being fed into the system. The x values used and the corresponding observed percent of waste solid removed (y) are shown in Table P1.10. We make the usual normality and independence assumptions.

(a) Fit the model $E(y \mid x) = \alpha + \beta x = \eta_x$. Examine the residuals and comment on the assumption of linearity (in x).

(b) (i) Find a 95% confidence interval for α; (ii) find a 95% confidence interval for β.

(c) Determine the equations for the 95% confidence belt for η_x, and graph the belt as well as the fitted line as a function of x.

(d) Find the 95% (simultaneous) confidence region for (α, β).

(e) Find a 95% prediction belt for a future observation to be "taken at x" and graph on the same sheet of paper used in part (c).

(f) Using (d), is the postulated pair of values $(\alpha, \beta) = (26, -1.7)$ tenable?

(g) (i) Using part (i) of (b), is the postulated value $\alpha = 26$ tenable? (ii) Using part (ii) of (b), is the postulated value $\beta = -1.7$ tenable?

(h) Comment on (e) and (f).

1.11. Justify (1.84*a*)–(1.84*d*).

1.12. Table P1.12 gives observations (y) on the yield of a certain by-product (percent volume of fixed volume batches of the main product) when certain temperatures x (°C) are used in the process.

(a) Carry out the instructions of parts (a)–(e) of Problem 1.10.

(b) Note that there are repeat observations at certain values of x. Exploit this and perform a lack of fit test when a linear model is fitted. Compare the results with the analysis of residuals performed in part (a).

TABLE P1.12

Observation Number	1	2	3	4	5	6	7	8	9	10	11	12	13	14	15
x	22	30	28	30	48	29	50	15	50	39	47	39	42	17	25
y	72	81	69	73	81	71	85	62	79	77	81	75	70	55	64

Observation Number	16	17	18	19	20	21	22	23	24	25	26	27	28	29	30
x	31	28	0	3	12	19	33	22	4	27	36	46	12	8	52
y	65	62	57	58	62	68	71	60	60	80	79	72	65	63	87

1.13. The yields y of a certain by-product obtained when using an amount x of catalyst were observed for 13 values of x, with the results shown in Table P1.13.

(a) Exploit the fact that there are repeat observations and perform a lack of fit test, if a linear model is fitted.

(b) If the hypothesis of "no lack of fit" is accepted, carry out the instructions of parts (a)–(e) of Problem 1.10.

(c) Examine the hypothesis $(\alpha, \beta) = (130, -25)$.

TABLE P1.13

Observation Number	1	2	3	4	5	6	7
x	0.01	0.48	0.71	0.95	1.19	0.01	0.48
y	127.6	124.0	110.8	103.9	101.5	130.1	122.0

Observation Number	8	9	10	11	12	13
x	1.44	0.71	1.96	0.01	1.44	1.96
y	92.3	113.1	83.7	128.0	91.4	86.2

1.14. Table P1.14 gives the 12 results of measuring the thickness (y) of silver film deposited when an amount (x) of a certain acid mix is used in the process. (The values of x were predetermined, and the results shown in Table P1.14 are given in the order obtained.) Carry out the instructions (a) and (b) of Problem 1.13.

TABLE P1.14

x	4	6	2	5	7	6	3	8	5	3	1	5
y	197	272	100	228	327	279	148	377	238	142	66	239

1.15. Repeat the instructions of Problem 1.14 for the set of data given in Table P1.15 (x, time in hours; y, observed concentration of solution in 10^{-3} mg/ml.) Carry out the instructions (a) and (b) of Problem 1.13.

TABLE P1.15

x	6	6	6	8	8	8	10	10	10
y	29	32	27	79	72	88	181	165	201
x	12	12	12	14	14	14	16	16	16
y	425	384	472	1130	1020	1249	2812	2465	3099

1.16. The set of data in Table P1.16 was "fitted to a straight line" and residuals plotted against (i) the corresponding $\hat{\eta}$'s and (ii) the corresponding x's. After examination of these plots, it was decided to do a lack of fit test by taking advantage of the "design of the experiment," which for this set, called for three observations to be taken at each of the six values of x used.

(a) Regress y on x and plot the residuals against η's and x's. Comment on the advisability of proceeding with a lack of fit test.

(b) Perform the lack of fit test.

(c) Save your calculations—this problem is referred to in Chapter 3, Problem 3.58.

TABLE P1.16

x	5	5	5	8	8	8	11	11	11
y	115.42	111.70	109.88	260.19	261.14	259.34	478.46	484.24	478.01
x	14	14	14	17	17	17	20	20	20
y	771.69	770.75	773.76	1137.36	1137.28	1135.79	1572.79	1570.73	1574.13

APPENDIX A1.1.

In this chapter, we have developed a formula for the regression line of y on x, using calculus, This of course can be done purely algebraically.

First, suppose we have N independent observations $(w_1, \ldots, w_N)$ and we wish to minimize the quantity $V(k)$ with respect to k, where

$$V(k) = \sum_{i=1}^{N} (w_i - k)^2. \tag{A1.1.1}$$

Then, V is minimized over k at the value $k = \bar{w}$. We may write

$$V(k) = \sum_{i=1}^{N} [(w_i - \bar{w}) + (\bar{w} - k)]^2,$$

$$= \sum_{i=1}^{N} \left[(w_i - \bar{w})^2 + (\bar{w} - k)^2 + 2(\bar{w} - k)(w_i - \bar{w})\right] \tag{A1.1.2}$$

$$= \sum_{i=1}^{N} (w_i - \bar{w})^2 + N(\bar{w} - k)^2$$

since

$$\sum_{i=1}^{N} (w_i - \bar{w}) = \Sigma w_i - N\bar{w} = N\bar{w} - N\bar{w} = 0.$$

Now, $V(k)$ is the sum of two non-negative quantities and obviously has its minimum for fixed $(w_1, \ldots, w_N)$ at $k = \bar{w}$.

The regression problem requires the values of α and β which minimize $Q(\alpha, \beta)$, defined in (1.14) as

$$Q(\alpha, \beta) = \sum_{i=1}^{N} (y_i - \alpha - \beta x_i)^2, \tag{A1.1.3}$$

where $x_1, \ldots, x_N$ are not all equal. If we let $y_i - \beta x_i = w_i$, we have $\bar{y} - \beta\bar{x} = \bar{w}$ and, using (A1.1.2), we have

$$Q(\alpha, \beta) = \sum_{i=1}^{N} (w_i - \alpha)^2 = \sum_{i=1}^{N} (w_i - \bar{w})^2 + N(\bar{w} - \alpha)^2. \tag{A1.1.4}$$

Of course, this is minimized for any fixed value of β by choosing

(A1.1.5) $$\alpha = \overline{w} = \bar{y} - \beta\bar{x}.$$

That is to say, for any value of β, $Q(\alpha, \beta)$ is minimized with respect to α by choosing α to be $\bar{y} - \beta\bar{x}$, in which case $Q(\alpha, \beta)$ has the value

(A1.1.6) $$\begin{aligned} Q(\bar{y} - \beta\bar{x}, \beta) &= \sum_{i=1}^{N} [y_i - (\bar{y} - \beta\bar{x}) - \beta x_i]^2 \\ &= \sum_{i=1}^{N} [(y_i - \bar{y}) - \beta(x_i - \bar{x})]^2. \end{aligned}$$

Note that (A1.1.6) is a function of β alone, which we will denote by $Q(\beta)$. To find the value of β which minimizes $Q(\beta)$ for fixed values of $x_1, \ldots, x_N$ not all equal, we write

(A1.1.7) $$\begin{aligned} Q(\beta) &= A\beta^2 - 2B\beta + C \\ &= A\left[\left(\beta - \frac{B}{A}\right)^2\right] + \frac{AC - B^2}{A} \end{aligned}$$

where

$$A = \sum_{i=1}^{N} (x_i - \bar{x})^2 > 0, \quad B = \Sigma(x_i - \bar{x})(y_i - \bar{y}), \quad \text{and} \quad C = \Sigma(y_i - \bar{y})^2.$$

Hence, $Q(\beta)$ is minimized with respect to β if we choose

(A1.1.8) $$\beta = \frac{B}{A} = \frac{\Sigma(x_i - \bar{x})(y_i - \bar{y})}{\Sigma(x_i - \bar{x})^2}.$$

That is, $Q(\alpha, \beta)$ is minimized if we choose α and β to be, for example, $\hat{\alpha}$ and $\hat{\beta}$, respectively, where

(A1.1.9) $$\begin{aligned} \hat{\alpha} &= \bar{y} - \hat{\beta}\bar{x} \\ \hat{\beta} &= \frac{\Sigma(x_i - \bar{x})(y_i - \bar{y})}{\Sigma(x_i - \bar{x})^2}. \end{aligned}$$

We note at this point that (1.56) may be obtained directly here, since the minimum of $Q(\alpha, \beta)$ is $Q(\hat{\alpha}, \hat{\beta})$, and from (A1.1.7) this is

$$Q(\hat{\alpha}, \hat{\beta}) = C - \frac{B^2}{A}$$

$$= \Sigma(y_i - \bar{y})^2 - \frac{[\Sigma(x_i - \bar{x})(y_i - \bar{y})]^2}{\Sigma(x_i - \bar{x})^2} \tag{A1.1.10}$$

$$= \Sigma(y_i - \bar{y})^2 - \hat{\beta}^2\Sigma(x_i - \bar{x})^2.$$

We note from (1.56) that $Q(\hat{\alpha}, \hat{\beta}) = \mathrm{SS}_e$.

2 Sampling from the Multivariate Normal

In subsequent chapters, we will be discussing the linear univariate model in its full generality, and at various points in the discussion, it will be necessary to have results available that allow us to state whether certain statistics are or are not independent, what the distributions of certain statistics are, and so on. In fact, the latter will be done under the assumption of normality, and it is for this reason that we now discuss sampling from the multivariate normal distribution. (Univariate normal theory will be used and assumed known.) The results of this chapter are used throughout the remaining chapters of this book. For a summary of the matrix results used, see Appendixes A2.1 and A2.2. (For further reading on matrices, the interested reader is advised to consult Searle (1966).)

2.1. THE MULTIVARIATE NORMAL DISTRIBUTION

We begin this section with a formal definition.

Definition 2.1. *The $m \times 1$ random vector $\mathbf{y}$ is said to be normally distributed if its probability density function $p_{\mathbf{y}}(\mathbf{y})$ is of the form*

$$(2.1) \quad p_{\mathbf{y}}(\mathbf{y}) = p_{\mathbf{y}}(y_1,\ldots,y_m) = \frac{|\Sigma^{-1}|^{1/2}}{(2\pi)^{m/2}} \exp\left[-\tfrac{1}{2}(\mathbf{y}-\boldsymbol{\mu})'\Sigma^{-1}(\mathbf{y}-\boldsymbol{\mu})\right]$$

for $(y_1,\ldots,y_m) \in R^m$, where the matrix $\Sigma^{-1} = (\sigma^{ij})$ is a $m \times m$ positive definite matrix of constants, with $\sigma^{ij} = \sigma^{ji}$, and $\boldsymbol{\mu} = (\mu_1,\ldots,\mu_m)'$ is such that $-\infty < \mu_j < \infty, j = 1,\ldots,m$. ($R^m$ is standard notation for Euclidean space of m dimensions.)

To prove that (2.1) is a density function, we should show

(2.2*a*) $p_{\mathbf{y}}(\mathbf{y}) = p_{\mathbf{y}}(y_1,\ldots,y_m) \geq 0$ for all points $(y_1,\ldots,y_m) \in R^m$;

(2.2*b*)
$$\int_{R^m} p_{\mathbf{y}}(\mathbf{y})\, d\mathbf{y} = \int_{R^m} p_{\mathbf{y}}(y_1,\ldots,y_m)\, dy_1 \cdots dy_m = 1.$$

The property (2.2*a*) obviously holds. To show (2.2*b*), consider the transformation

(2.3)
$$\mathbf{y} = \mathbf{v} + \boldsymbol{\mu} = I\mathbf{v} + \boldsymbol{\mu}$$

where I is the identity matrix of order m. The Jacobian J of this transformation is the determinant of I, and this is clearly 1. Hence we have

(2.3*a*)
$$p_{\mathbf{v}}(\mathbf{v}) = \frac{|\Sigma^{-1}|^{1/2}}{(2\pi)^{m/2}} \exp\left(-\tfrac{1}{2}\mathbf{v}'\Sigma^{-1}\mathbf{v}\right).$$

Now since Σ^{-1} is assumed positive definite and symmetric, there exists an orthogonal matrix P such that

(2.4)
$$P'\Sigma^{-1}P = D \quad \text{or} \quad \Sigma^{-1} = PDP'$$

where D is a diagonal matrix

(2.4*a*)
$$D = \begin{pmatrix} \lambda_1 & & \bigcirc \\ & \ddots & \\ \bigcirc & & \lambda_m \end{pmatrix}, \qquad \lambda_j > 0$$

with λ_j a characteristic root of Σ^{-1}. We note that

(2.4*b*)
$$|D| = |\Sigma^{-1}| = \prod_{j=1}^{m} \lambda_j$$

since $|P'| = |P| = \pm 1$, because P is orthogonal.

We then consider the so-called *canonical* transformation

(2.5)
$$\mathbf{z} = P'\mathbf{v} = P'(\mathbf{y} - \boldsymbol{\mu}) \quad \text{or} \quad \mathbf{v} = P\mathbf{z}.$$

The Jacobian J of this transformation is clearly $|P| = \pm 1$, so that the absolute value of J is $|J| = 1$. This means that the density of $\mathbf{z}$ is

$$p_{\mathbf{z}}(\mathbf{z}) = \frac{|\Sigma^{-1}|^{1/2}}{(2\pi)^{m/2}} \exp(-\tfrac{1}{2}\mathbf{z}'P'\Sigma^{-1}P\mathbf{z}) \tag{2.6}$$

or, using (2.4) and (2.4b),

$$p_{\mathbf{z}}(\mathbf{z}) = \frac{(\prod_{j=1}^{m} \lambda_j)^{1/2}}{(2\pi)^{m/2}} \exp(-\tfrac{1}{2}\mathbf{z}'D\mathbf{z}). \tag{2.6a}$$

Since D is diagonal, so that $\mathbf{z}'D\mathbf{z} = \Sigma_{j=1}^{m} \lambda_j z_j^2$, we have

$$p_{\mathbf{z}}(\mathbf{z}) = \prod_{j=1}^{m} \frac{\lambda_j^{1/2}}{\sqrt{2\pi}} \exp\left(-\frac{\lambda_j z_j^2}{2}\right), \tag{2.6b}$$

which is to say, $p_{\mathbf{z}}(\mathbf{z})$ is the product of densities of random variables z_j which are distributed independently as univariate normal variables, *mean* 0, *and variance* $1/\lambda_j$. We thus have that

$$\int_{R^m} p_{\mathbf{y}}(\mathbf{y})\, d\mathbf{y} = \prod_{j=1}^{m} \int_{-\infty}^{\infty} \left(\frac{\lambda_j}{2\pi}\right)^{1/2} \left[\exp\left(-\frac{\lambda_j z_j^2}{2}\right)\right] dz_j = 1, \tag{2.7}$$

since the area under a univariate normal density function is 1. Thus, (2.2) holds and $p_{\mathbf{y}}(\mathbf{y})$ is indeed a probability density function (PDF).

We note in passing that the above gives rise to the identity

$$\begin{aligned} \int_{R^m} \exp[-\tfrac{1}{2}(\mathbf{y}-\boldsymbol{\mu})'\Sigma^{-1}(\mathbf{y}-\boldsymbol{\mu})]\, d\mathbf{y} &= (2\pi)^{m/2}|\Sigma^{-1}|^{-1/2} \\ &= (2\pi)^{m/2}|\Sigma|^{1/2}. \end{aligned} \tag{2.8}$$

Further properties of the multivariate normal are contained in the following theorem.

Theorem 2.1.1. *Suppose* $\mathbf{y}$ *is a random* $m \times 1$ *vector having PDF given by* (2.1). *Then the mean vector* $E(\mathbf{y})$ *of* $\mathbf{y}$ *is such that*

$$E(\mathbf{y}) = \boldsymbol{\mu} \tag{2.9}$$

and the dispersion matrix (i.e., the variance–covariance matrix) V of $\mathbf{y}$ *is*

$$V(\mathbf{y}) = \Sigma. \tag{2.9a}$$

Proof. Recall that in proving $p_{\mathbf{y}}(\mathbf{y})$ of (2.1) is indeed a density, we found that $\mathbf{z}$ is such that $E(\mathbf{z}) = \mathbf{0}$. That is, since $\mathbf{z} = P'(\mathbf{y} - \boldsymbol{\mu})$, or $\mathbf{y} - \boldsymbol{\mu} = P\mathbf{z}$,

$$E(\mathbf{y} - \boldsymbol{\mu}) = E[P\mathbf{z}] = P\mathbf{0} = \mathbf{0} \tag{2.10}$$

or

$$E(\mathbf{y}) = \boldsymbol{\mu}. \tag{2.10a}$$

Furthermore, from (2.6b), we found that the z_j are independent, with $V(z_j) = \lambda_j^{-1}$, so that

$$V(\mathbf{z}) = D^{-1} = \begin{pmatrix} \lambda_1^{-1} & & \\ & \ddots & \\ & & \lambda_m^{-1} \end{pmatrix}. \tag{2.11}$$

That is, since $\mathbf{y} - \boldsymbol{\mu} = P\mathbf{z}$,

$$\begin{aligned} V(\mathbf{y}) &= V(P\mathbf{z}) = PV(\mathbf{z})P' \\ &= PD^{-1}P'. \end{aligned} \tag{2.11a}$$

But from (2.4) we have that

$$D^{-1} = P^{-1}\Sigma(P')^{-1}, \tag{2.11b}$$

so that

$$\begin{aligned} V(\mathbf{y}) &= P\left[P^{-1}\Sigma(P')^{-1}\right]P' \\ &= \Sigma. \end{aligned} \tag{2.11c}$$

In view of this result [which in words states that if a random vector variable $\mathbf{y}$ has PDF as given in (2.1), then the constants $\boldsymbol{\mu}$ of the quadratic form in the exponent are the mean values of $\mathbf{y}$, and the inverse of the matrix of the quadratic form is the dispersion matrix of $\mathbf{y}$], we denote $\mathbf{y}$ as

$$\mathbf{y} = N(\boldsymbol{\mu}, \Sigma). \tag{2.12}$$

Theorem 2.1.1 has a useful alternative proof. Suppose we calculate the characteristic function of $\mathbf{y}$, say $\phi_{\mathbf{y}}(\mathbf{t})$, then we have ($i^2 = -1$)

$$\begin{aligned}\phi_{\mathbf{y}}(\mathbf{t}) &= E[\exp(i\mathbf{t}'\mathbf{y})] \\ &= \int_{R^m} \exp(i\mathbf{t}'\mathbf{y}) \frac{|\Sigma^{-1}|^{1/2}}{(2\pi)^{m/2}} \exp[-\tfrac{1}{2}(\mathbf{y}-\boldsymbol{\mu})'\Sigma^{-1}(\mathbf{y}-\boldsymbol{\mu})]\, d\mathbf{y}\end{aligned} \tag{2.13}$$

where $\mathbf{t} = (t_1, \ldots, t_m)'$. Now suppose we make the transformation (2.5), that is, $\mathbf{z} = P'(\mathbf{y} - \boldsymbol{\mu})$, where P is orthogonal and such that (2.4) holds. Then, from (2.4b) and (2.6b) we clearly have

$$\phi_{\mathbf{y}}(\mathbf{t}) = \exp(i\mathbf{t}'\boldsymbol{\mu}) \int_{R^m} \exp(i\mathbf{t}'P\mathbf{z}) \prod_{j=1}^{m} \left(\frac{\lambda_j}{2\pi}\right)^{1/2} \exp\left(-\frac{\lambda_j z_j^2}{2}\right) \pi\, dz_j. \tag{2.13a}$$

($\pi dz_j = dz_1 \cdots dz_m$). Now write $\mathbf{u} = P'\mathbf{t}$, so that

$$\begin{aligned}\phi_{\mathbf{y}}(\mathbf{t}) &= \exp(i\mathbf{t}'\boldsymbol{\mu}) \int_{R^m} \exp(i\mathbf{u}'\mathbf{z}) \prod_{j=1}^{m} \left(\frac{\lambda_j}{2\pi}\right)^{1/2} \exp\left(-\frac{\lambda_j z_j^2}{2}\right) \pi\, dz_j \\ &= \exp(i\mathbf{t}'\boldsymbol{\mu}) \prod_{j=1}^{m} \int_{-\infty}^{\infty} e^{iu_j z_j} \left(\frac{\lambda_j}{2\pi}\right)^{1/2} \exp\left(-\frac{\lambda_j z_j^2}{2}\right) dz_j\end{aligned} \tag{2.13b}$$

and the jth integral in the above product is the characteristic function of a $N(0, 1/\lambda_j)$ variable, well known to be $\exp(-u_j^2/2\lambda_j)$, so that

$$\begin{aligned}\phi_{\mathbf{y}}(\mathbf{t}) &= \exp(i\mathbf{t}'\boldsymbol{\mu}) \prod_{j=1}^{m} \exp\left(-\frac{1}{2}\frac{u_j^2}{\lambda_j}\right) \\ &= \exp(i\mathbf{t}'\boldsymbol{\mu})\exp(-\tfrac{1}{2}\mathbf{u}'D^{-1}\mathbf{u}) \\ &= \exp(i\mathbf{t}'\boldsymbol{\mu})\exp(-\tfrac{1}{2}\mathbf{t}'[PD^{-1}P']\mathbf{t}).\end{aligned} \tag{2.13c}$$

Now from (2.11b) we have

$$PD^{-1}P' = P\left[P^{-1}\Sigma(P')^{-1}\right]P' = \Sigma, \tag{2.14}$$

so that

$$\phi_{\mathbf{y}}(\mathbf{t}) = \exp(i\mathbf{t}'\boldsymbol{\mu} - \tfrac{1}{2}\mathbf{t}'\Sigma\mathbf{t}). \tag{2.15}$$

Differentiating once with respect to the t_j, setting $\mathbf{t} = \mathbf{0}$, and multiplying by

$-i$, yields (2.10a). Differentiating (2.15) appropriately twice, setting $\mathbf{t} = \mathbf{0}$, and multiplying by $(-i)^2$, leads to (2.11c). See Problem 2.2.

From this characteristic function, we see immediately that if the y_j are *uncorrelated* ($\sigma_{rs} = 0$, all $r \neq s$),

$$\phi_{\mathbf{y}}(\mathbf{t}) = \prod_{j=1}^{m} \exp\left(it_j\mu_j - \tfrac{1}{2}\sigma_j^2 t_j^2\right), \tag{2.16}$$

which is to say that if the y_j's are multivariate normally distributed and are *uncorrelated*, they are *statistically independent*—a property unique to the multivariate normal distribution.

The canonical transformation (2.5)

$$\mathbf{z} = P'(\mathbf{y} - \boldsymbol{\mu}), \qquad P'\Sigma^{-1}P = D \tag{2.17}$$

is useful in many other ways. Examining (2.1), we see that the normal density is a function of the quadratic form

$$Q = (\mathbf{y} - \boldsymbol{\mu})'\Sigma^{-1}(\mathbf{y} - \boldsymbol{\mu}) \tag{2.18}$$

and is constant along the ellipsoids $Q = q$. Another way of saying this is that the contours of the PDF of $\mathbf{y}$ are ellipsoids in m dimensions, centered at the point $\boldsymbol{\mu}' = (\mu_1, \ldots, \mu_m)$. Now the canonical transformation takes the quadratic form (2.18) in y_j into

$$Q^* = \sum_{j=1}^{m} \lambda_j z_j^2 \tag{2.18a}$$

that is, $p_{\mathbf{y}}(\mathbf{y})$ goes into

$$\begin{aligned} p_{\mathbf{z}}(\mathbf{z}) &= \frac{\left(\prod \lambda_j\right)^{1/2}}{(2\pi)^{m/2}} \exp\left(-\frac{1}{2}\sum_{j=1}^{m} \lambda_j z_j^2\right) \\ &= \prod_{j=1}^{m} \left(\frac{\lambda_j}{2\pi}\right)^{1/2} \exp\left(-\frac{\lambda_j z_j^2}{2}\right), \end{aligned} \tag{2.18b}$$

that is, into a product of independent $N(0, 1/\lambda_j)$ variables. The contours of equidensity of (2.18b) are ellipsoids whose axes correspond to the principal axes of the ellipsoids $Q = q$ and whose origin is the point $(\mu_1, \ldots, \mu_m)$ in the

y system. Furthermore,

$$Q^* = q^* \tag{2.19}$$

is an ellipsoid in m dimensions whose principal axes correspond to the z_j axes in R^m and have semilength proportional to $(1/\lambda_j)^{1/2}$, where the λ_j are characteristic roots of Σ^{-1}, or λ_j^{-1} are characteristic roots of Σ. (See Problems 2.1 and 2.9 as examples.)

The characteristic function (2.15) can be used to find marginal distributions of the m-variate normal distribution. To this end, let

$$\mathbf{y}' = (\mathbf{y}_1' \mid \mathbf{y}_2'), \tag{2.20}$$

where $\mathbf{y}_1$ is $m_1 \times 1$, $\mathbf{y}_2$ is $m_2 \times 1$, $m_1 + m_2 = m$, and similarly, let

$$\boldsymbol{\mu}' = (\boldsymbol{\mu}_1' \mid \boldsymbol{\mu}_2') \quad \text{and} \quad \mathbf{t}' = (\mathbf{t}_1' \mid \mathbf{t}_2'). \tag{2.20a}$$

Furthermore, partition Σ in similar fashion,

$$\Sigma = \left(\begin{array}{c|c} \Sigma_{11} & \Sigma_{12} \\ \hline \Sigma_{21} & \Sigma_{22} \end{array}\right) \tag{2.20b}$$

where Σ_{11} is $m_1 \times m_1$, Σ_{12} is $m_1 \times m_2$, $\Sigma_{21} = \Sigma_{12}'$ is $m_2 \times m_1$, and Σ_{22} is $m_2 \times m_2$. Now

(2.21)

$$\phi_{\mathbf{y}_1,\mathbf{y}_2}(\mathbf{t}_1', \mathbf{t}_2') = \exp\left[i(\mathbf{t}_1'\boldsymbol{\mu}_1 + \mathbf{t}_2'\boldsymbol{\mu}_2) - \tfrac{1}{2}(\mathbf{t}_1' \mid \mathbf{t}_2')\left(\begin{array}{c|c} \Sigma_{11} & \Sigma_{12} \\ \hline \Sigma_{21} & \Sigma_{22} \end{array}\right)\left(\begin{array}{c} \mathbf{t}_1 \\ \hline \mathbf{t}_2 \end{array}\right)\right]$$

so that the characteristic function of $\mathbf{y}_1$ is

$$\phi_{\mathbf{y}_1}(\mathbf{t}_1) = \phi_{\mathbf{y}_1,\mathbf{y}_2}(\mathbf{t}_1', \mathbf{0}') = \exp\left[i\mathbf{t}_1'\boldsymbol{\mu}_1 - \tfrac{1}{2}\mathbf{t}_1'\Sigma_{11}\mathbf{t}_1\right], \tag{2.22}$$

which is to say, $\mathbf{y}_1 = N(\boldsymbol{\mu}_1, \Sigma_{11})$.

Now suppose we are interested in the conditional distribution of $\mathbf{y}_2$, given $\mathbf{y}_1$. We need some relationships between partitions of a matrix and its inverse. Assume we have, in addition to (2.20b), a similarly partitioned inverse of Σ^{-1},

(2.23) $$\Sigma^{-1} = \left(\begin{array}{c|c} C_{11} & C_{12} \\ \hline C_{21} & C_{22} \end{array}\right), \qquad C_{11} \text{ is } m_1 \times m_1, \text{ etc.}$$

Then [see Problem 2.3(a), and Section 14 of Appendix A2.1], it can be shown that

(2.24)
$$\begin{aligned}
\Sigma_{11} &= \left(C_{11} - C_{12}C_{22}^{-1}C_{21}\right)^{-1} \\
\Sigma_{12} &= \left(C_{12}C_{22}^{-1}C_{21} - C_{11}\right)^{-1}C_{12}C_{22}^{-1} \\
\Sigma_{21} &= \left(C_{21}C_{11}^{-1}C_{12} - C_{22}\right)^{-1}C_{21}C_{11}^{-1} \\
\Sigma_{22} &= \left(C_{22} - C_{21}C_{11}^{-1}C_{12}\right)^{-1}
\end{aligned}$$

and, similarly,

(2.24a)
$$\begin{aligned}
C_{11} &= \left(\Sigma_{11} - \Sigma_{12}\Sigma_{22}^{-1}\Sigma_{21}\right)^{-1} \\
C_{12} &= \left(\Sigma_{12}\Sigma_{22}^{-1}\Sigma_{21} - \Sigma_{11}\right)^{-1}\Sigma_{12}\Sigma_{22}^{-1} \\
C_{21} &= \left(\Sigma_{21}\Sigma_{11}^{-1}\Sigma_{12} - \Sigma_{22}\right)^{-1}\Sigma_{21}\Sigma_{11}^{-1} \\
C_{22} &= \left(\Sigma_{22} - \Sigma_{21}\Sigma_{11}^{-1}\Sigma_{12}\right)^{-1}.
\end{aligned}$$

Computationally, the relationships (2.24) can be used to find the Σ_{ij} partitioning Σ, given the C_{ij} partitioning Σ^{-1}, while the relationships (2.24a) can be used to find the C_{ij} of Σ^{-1}, given the Σ_{ij} of Σ, and so on.

Now from the definition of the conditional distribution of $\mathbf{y}_2$, given $\mathbf{y}_1$, we have

(2.25) $$p(\mathbf{y}_2 | \mathbf{y}_1) = \frac{p_{\mathbf{y}}(\mathbf{y})}{p_{\mathbf{y}_1}(\mathbf{y}_1)}.$$

Using (2.1) and (2.22), we have

$$p(\mathbf{y}_2|\mathbf{y}_1) = \frac{|\Sigma^{-1}|^{1/2}}{(2\pi)^{m_2/2}|\Sigma_{11}^{-1}|^{1/2}} \exp\left(-\frac{Q}{2}\right), \tag{2.25a}$$

where

$$Q = (\mathbf{y} - \boldsymbol{\mu})'\Sigma^{-1}(\mathbf{y} - \boldsymbol{\mu}) - (\mathbf{y}_1 - \boldsymbol{\mu}_1)'\Sigma_{11}^{-1}(\mathbf{y}_1 - \boldsymbol{\mu}_1). \tag{2.26}$$

Now in view of (2.20*b*), (2.23), (2.24), and (2.24*a*), we may write

(2.26*a*)

$$Q = [(\mathbf{y}_1 - \boldsymbol{\mu}_1)', (\mathbf{y}_2 - \boldsymbol{\mu}_2)'] \left(\begin{array}{c|c} C_{11} & C_{12} \\ \hline C_{21} & C_{22} \end{array}\right) [(\mathbf{y}_1 - \boldsymbol{\mu}_1)', (\mathbf{y}_2 - \boldsymbol{\mu}_2)']'$$

$$- (\mathbf{y}_1 - \boldsymbol{\mu}_1)'\left[C_{11} - C_{12}C_{22}^{-1}C_{21}\right](\mathbf{y}_1 - \boldsymbol{\mu}_1).$$

Using the fact that $C_{12}' = C_{21}$, we have

$$\begin{aligned} Q = {} & (\mathbf{y}_2 - \boldsymbol{\mu}_2)'C_{22}(\mathbf{y}_2 - \boldsymbol{\mu}_2) + 2(\mathbf{y}_2 - \boldsymbol{\mu}_2)'C_{21}(\mathbf{y}_1 - \boldsymbol{\mu}_1) \\ & + (\mathbf{y}_1 - \boldsymbol{\mu}_1)'C_{12}C_{22}^{-1}C_{21}(\mathbf{y}_1 - \boldsymbol{\mu}_1). \end{aligned} \tag{2.26b}$$

Completing the square, we find

(2.26*c*)

$$Q = \left[(\mathbf{y}_2 - \boldsymbol{\mu}_2) + C_{22}^{-1}C_{21}(\mathbf{y}_1 - \boldsymbol{\mu}_1)\right]'C_{22}\left[(\mathbf{y}_2 - \boldsymbol{\mu}_2) + C_{22}^{-1}C_{21}(\mathbf{y}_1 - \boldsymbol{\mu}_1)\right].$$

The function

$$\boldsymbol{\mu}_{2\cdot 1} = \boldsymbol{\mu}_2 - C_{22}^{-1}C_{21}(\mathbf{y}_1 - \boldsymbol{\mu}_1) \tag{2.26d}$$

is called the *regression function* of $\mathbf{y}_2$ on $\mathbf{y}_1$ (recall that we are given $\mathbf{y}_1$), and the matrix [see relationships (2.24*a*) and Problem 2.3(ii)]

$$-C_{22}^{-1}C_{21} = \Sigma_{21}\Sigma_{11}^{-1} \tag{2.26e}$$

is called the *matrix of regression coefficients* of $\mathbf{y}_2$ on $\mathbf{y}_1$. From (2.25*a*) and

(2.26e) we have

$$(2.27)\quad p(\mathbf{y}_2|\mathbf{y}_1) = \frac{|\Sigma^{-1}|^{1/2}}{(2\pi)^{m_2/2}|\Sigma_{11}^{-1}|^{1/2}}\exp\{-\tfrac{1}{2}[\mathbf{y}_2 - \boldsymbol{\mu}_{2\cdot 1}]'C_{22}[\mathbf{y}_2 - \boldsymbol{\mu}_{2\cdot 1}]\},$$

and it remains to show that

$$(2.27a)\qquad \frac{|\Sigma^{-1}|}{|\Sigma_{11}^{-1}|} = |C_{22}|.$$

Now the left-hand side of (2.27a) is

$$(2.27b)\qquad |\Sigma^{-1}||\Sigma_{11}| = \left\{\det\begin{pmatrix} C_{11} & C_{12} \\ C_{21} & C_{22}\end{pmatrix}\right\}\left\{\det\begin{pmatrix} \Sigma_{11} & O \\ \Sigma_{21} & I_{m_2}\end{pmatrix}\right\}$$

where I_{m_2} denotes the identity matrix of order m_2. Equation (2.27b) is easily verified by expanding along the last m_2 columns of the matrix mentioned on the extreme right-hand side, and we then have, using the identity that $(\det A)(\det B) = \det(AB)$,

$$(2.27c)\qquad |\Sigma^{-1}||\Sigma_{11}| = \det\left\{\begin{matrix} C_{11}\Sigma_{11} + C_{12}\Sigma_{21} & C_{12} \\ C_{21}\Sigma_{11} + C_{22}\Sigma_{21} & C_{22}\end{matrix}\right\}.$$

But from Problem 2.2, which may be used to derive (2.24) and (2.24a), we have

$$(2.27d)\qquad \begin{aligned} C_{11}\Sigma_{11} + C_{12}\Sigma_{21} &= I_{m_1} \\ C_{21}\Sigma_{11} + C_{22}\Sigma_{21} &= O. \end{aligned}$$

Hence

$$(2.27e)\qquad |\Sigma^{-1}||\Sigma_{11}| = \det\begin{pmatrix} I_{m_1} & C_{12} \\ O & C_{22}\end{pmatrix} = \det C_{22}.$$

We usually denote C_{22} by $\Sigma_{2\cdot 1}^{-1}$, that is,

$$(2.28)\qquad C_{22} = (\Sigma_{22} - \Sigma_{21}\Sigma_{11}^{-1}\Sigma_{12})^{-1} = \Sigma_{2\cdot 1}^{-1}$$

and in view of (2.27) and (2.28) we have that the *conditional distribution of* $\mathbf{y}_2$

given $\mathbf{y}_1$ *is such that*

$$\mathbf{y}_2|\mathbf{y}_1 = N(\boldsymbol{\mu}_{2\cdot 1}, \Sigma_{2\cdot 1}). \tag{2.29}$$

EXAMPLE 2.1.1

Suppose $m = 2$, $m_1 = m_2 = 1$, so that

$$\mathbf{y} = (y_1 \mid y_2)', \qquad \boldsymbol{\mu} = (\mu_1 \mid \mu_2)'$$

$$\Sigma = \left(\begin{array}{c|c} \sigma_1^2 & \rho\sigma_1\sigma_2 \\ \hline \rho\sigma_1\sigma_2 & \sigma_2^2 \end{array}\right), \qquad \Sigma^{-1} = \frac{1}{\Delta}\left(\begin{array}{c|c} \sigma_2^2 & -\rho\sigma_1\sigma_2 \\ \hline -\rho\sigma_1\sigma_2 & \sigma_1^2 \end{array}\right) \tag{2.30}$$

$$\Delta = \sigma_1^2\sigma_2^2(1-\rho^2).$$

Of course, we have

$$\begin{aligned} p_{\mathbf{y}}(\mathbf{y}) = \frac{1}{2\pi\left[\sigma_1^2\sigma_2^2(1-\rho^2)\right]^{1/2}} \exp\left\{-\frac{1}{2(1-\rho^2)}\left[\frac{(y_1-\mu_1)^2}{\sigma_1^2}\right.\right. \\ \left.\left. -2\frac{\rho}{\sigma_1\sigma_2}(y_1-\mu_1)(y_2-\mu_2) + \frac{(y_2-\mu_2)^2}{\sigma_2^2}\right]\right\}. \end{aligned} \tag{2.31}$$

Suppose we now wish to determine the conditional distribution of y_2, given y_1. We have

$$\Sigma_{21}\Sigma_{11}^{-1} = \frac{\rho\sigma_1\sigma_2}{\sigma_1^2} = \rho\frac{\sigma_2}{\sigma_1} \tag{2.32}$$

because $\Sigma_{21} = \rho\sigma_1\sigma_2$, $\Sigma_{11} = \sigma_1^2$, and $\Sigma_{11}^{-1} = 1/\sigma_1^2$. Furthermore,

$$\begin{aligned} \Sigma_{2\cdot 1} &= \Sigma_{22} - \Sigma_{21}\Sigma_{11}^{-1}\Sigma_{12} = \sigma_2^2 - \frac{\rho^2\sigma_1^2\sigma_2^2}{\sigma_1^2} \\ &= \sigma_2^2(1-\rho^2), \end{aligned} \tag{2.32a}$$

$$\begin{aligned} \boldsymbol{\mu}_{2\cdot 1} = \mu_{2\cdot 1} &= \mu_2 + \Sigma_{21}\Sigma_{11}^{-1}(y_1 - \mu_1) \\ &= \mu_2 + \rho\frac{\sigma_2}{\sigma_1}(y_1-\mu_1). \end{aligned} \tag{2.32b}$$

Note that from (2.30), C_{22} is given by

$$C_{22} = \frac{\sigma_1^2}{\Delta} = \left[\sigma_2^2(1-\rho^2)\right]^{-1} \tag{2.32c}$$

so that

$$C_{22}^{-1} = \sigma_2^2(1-\rho^2), \tag{2.32d}$$

providing a check on the result (2.32a)—see (2.28). Thus,

(2.33)

$$p(y_2|y_1) = \frac{1}{\sqrt{2\pi}\left[\sigma_2^2(1-\rho^2)\right]^{1/2}}\exp\left[-\frac{1}{2\sigma_2^2(1-\rho^2)}(y_2-\alpha-\beta y_1)^2\right]$$

where

$$\alpha = \mu_2 - \beta\mu_1, \qquad \beta = \rho\frac{\sigma_2}{\sigma_1}. \tag{2.33a}$$

So far in our discussion of the multivariate normal, we have assumed that the dispersion matrix of the vector random variable $\mathbf{y}$ is positive definite. We now consider the following situation—one that leads to a dispersion matrix of a random variable which is positive semidefinite.

Suppose $\boldsymbol{\omega} = N(\boldsymbol{\eta}, V)$, where V, the dispersion matrix of $\boldsymbol{\omega}$, is positive definite. Let $\boldsymbol{\omega}$ be $m_1 \times 1$, so that $\boldsymbol{\eta}$ is $m_1 \times 1$ and V is $m_1 \times m_1$. Now suppose we let

$$\mathbf{y} = A\boldsymbol{\omega} \tag{2.34}$$

where A is $m_2 \times m_1$, so that $\mathbf{y}$ is $m_2 \times 1$, where $m_2 \leq m_1$. Then, for $\mathbf{t} = (t_1, \ldots, t_{m_2})'$, we know that the characteristic function of $\mathbf{y}$ is

$$\phi_{\mathbf{y}}(\mathbf{t}) = \phi_{\omega}(A'\mathbf{t}) = \exp(i\mathbf{t}'A\boldsymbol{\eta} - \tfrac{1}{2}\mathbf{t}'AVA'\mathbf{t}), \tag{2.35}$$

that is to say, $\mathbf{y}$ is *normally distributed* with

$$E(\mathbf{y}) = \boldsymbol{\mu}_{\mathbf{y}} = A\boldsymbol{\eta}, \qquad V(\mathbf{y}) = \Sigma_{\mathbf{y}} = AVA' \tag{2.35a}$$

and $\boldsymbol{\mu}_{\mathbf{y}}$ is $m_2 \times 1$, and Σ_y is $m_2 \times m_2$. Now we know that we may write V (recall that V is symmetric and assumed positive definite) as $V = M'M$, so that

$$\Sigma_{\mathbf{y}} = (AM'MA') = (MA')'(MA') \tag{2.35b}$$

so that $\Sigma_{\mathbf{y}}$ is symmetric and positive semidefinite at least. Two cases may arise.

CASE 1. The rank of A, denoted by $r(A)$, is

$$r(A) = m_2. \tag{2.36}$$

This means that A is of full rank so that $\mathbf{y} = A\boldsymbol{\omega}$ constitutes a nonsingular transformation of $\boldsymbol{\omega}$, and furthermore, the dispersion matrix of $\mathbf{y}$, $\Sigma_{\mathbf{y}} = AVA'$ is positive definite, since

$$r(\Sigma_{\mathbf{y}}) = r(AM')(MA') = r(AM') = r(A) = m_2, \tag{2.37}$$

that is, the $m_2 \times m_2$ matrix $\Sigma_{\mathbf{y}}$ is of full rank. Hence $\Sigma_{\mathbf{y}}$ possesses an inverse, and the PDF of $\mathbf{y}$ may be written explicitly. We note in passing the case $m_2 = 1$: Then $A = \mathbf{a}'$ is a nontrivial $1 \times m_1$ row vector. This in turn implies that $y = \mathbf{a}'\boldsymbol{\omega}$, a *linear combination of normal random variables*, and a linear combination of normal variables is itself *normally distributed*, since (2.35) takes the form

$$\phi_y(t) = \exp\left[i(\mathbf{a}'\boldsymbol{\eta})t - \tfrac{1}{2}(\mathbf{a}'V\mathbf{a})t^2\right] \tag{2.38}$$

which is the characteristic function of the $N(\mathbf{a}'\boldsymbol{\eta}, \mathbf{a}'V\mathbf{a})$ distribution ($\mathbf{a}'V\mathbf{a} > 0$ since $\mathbf{a} \neq \mathbf{0}$).

CASE 2. $r(A) = r < m_2$. In this case, we find that $r(\Sigma_{\mathbf{y}}) = r(A) = r < m_2$, which means that $\Sigma_{\mathbf{y}}$ is positive semidefinite and does not possess an inverse, so that we can not write down a density function for $\mathbf{y}$ explicitly. Now the characteristic function (2.35) still exists and may be used in the usual way to determine joint moments. What has happened is this: $\mathbf{y} = A\boldsymbol{\omega}$ is a singular transformation of $\boldsymbol{\omega}$, and because $r(A) = r < m_2$, *only r of the m_2 rows of A are linearly independent*, and the rest of the m_2 rows are linear combinations of these, which means that $m_2 - r$ of the variables $y_1, \ldots, y_{m_2}$ can be expressed as linear functions of r of the $y_1, \ldots, y_{m_2}$ variables which are linearly independent. Because of this, we say that $\mathbf{y}$, which is $m_2 \times 1$ has a *nonsingular normal distribution in $r < m_2$ dimensions*, and a *singular normal distribution* in m_2 dimensions. It is nonsingular in r dimensions, and not m_2, since $m_2 - r$ variables can be expressed as linear functions of r variables, so that *the distribution is concentrated*, that is, defined, in Euclidean space of dimension r, R^r, which is contained in R^{m_2}. In fact, $\mathbf{y}$ is distributed entirely in

$$\{A\mathbf{x}|\mathbf{x} \in R^{m_1}\}, \tag{2.38a}$$

which is an r-dimensional subspace of R^{m_2}, called the column space, or image space of A. (See Section 2 of Appendix A2.2 for further discussion of column or image spaces.) A simple example will clarify ideas here.

EXAMPLE 2.1.2

Suppose $\boldsymbol{\omega} = N(\mathbf{0}, I_2)$, that is, $\boldsymbol{\omega} = (\omega_1, \omega_2)'$ constitutes a random sample of two independent observations on $\omega = N(0, 1)$. Consider

$$\mathbf{y} = \frac{1}{2}\begin{pmatrix} 1 & -1 \\ -1 & 1 \end{pmatrix}\begin{pmatrix} \omega_1 \\ \omega_2 \end{pmatrix} = \begin{pmatrix} \omega_1 - \bar{\omega} \\ \omega_2 - \bar{\omega} \end{pmatrix}, \qquad \bar{\omega} = \frac{(\omega_1 + \omega_2)}{2}. \tag{2.39}$$

Here $m_1 = m_2 = 2$, and

$$A = \frac{1}{2}\begin{pmatrix} 1 & -1 \\ -1 & 1 \end{pmatrix}, \qquad \boldsymbol{\eta} = (0, 0)'.$$

Using (2.35) with $\mathbf{t} = (t_1, t_2)'$

$$\begin{aligned} \phi_{\mathbf{y}}(\mathbf{t}) &= \exp\left\{it'\left[\frac{1}{2}\begin{pmatrix} 1 & -1 \\ -1 & 1 \end{pmatrix}\begin{pmatrix} 0 \\ 0 \end{pmatrix}\right]\right. \\ &\qquad \left. -\frac{1}{2}\mathbf{t}'\left[\frac{1}{4}\begin{pmatrix} 1 & -1 \\ -1 & 1 \end{pmatrix} I \begin{pmatrix} 1 & -1 \\ -1 & 1 \end{pmatrix}\right]\mathbf{t}\right\} \\ &= \exp\left\{it'\mathbf{0} - \frac{1}{2}\mathbf{t}'\left[\frac{1}{2}\begin{pmatrix} 1 & -1 \\ -1 & 1 \end{pmatrix}\right]\mathbf{t}\right\} \end{aligned} \tag{2.40}$$

so that $E(y_1) = E(y_2) = 0$, $\sigma_{y_1}^2 = \sigma_{y_2}^2 = \frac{1}{2}$, and $\sigma_{y_1 y_2} = -\frac{1}{2}$. That is,

$$\boldsymbol{\mu}_{\mathbf{y}} = (0, 0)', \qquad \Sigma_{\mathbf{y}} = \frac{1}{2}\begin{bmatrix} 1 & -1 \\ -1 & 1 \end{bmatrix}, \tag{2.41}$$

and

$$\det \Sigma_{\mathbf{y}} = \tfrac{1}{4}(1 - 1) = 0, \tag{2.41a}$$

and clearly $r(\Sigma_{\mathbf{y}}) = 1 < 2$. This is because

$$y_1 + y_2 = (\omega_1 - \bar{\omega}) + (\omega_2 - \bar{\omega}) = \omega_1 + \omega_2 - 2\bar{\omega} = 0; \tag{2.42}$$

that is, $\mathbf{y}$ is distributed entirely along the one-dimensional subspace given by

$y_1 + y_2 = 0$. (Note that A is the matrix which orthogonally projects R^2 onto this subspace; see Section 2 of Appendix A2.2.)

2.2. DISTRIBUTION OF QUADRATIC FORMS

As we have indicated before, we will impose the condition of normality at various points in subsequent chapters and derive the distribution of certain statistics. The reader will recall that in Chapter 1, we assumed the condition of normality and when working with the analysis of variance tables, for example, we found ourselves involved with statistics such as the sum of squares due to regression, error, and so on, and these of course are quadratic in the observations y_i. We will be interested in the distributions of more general quadratic forms in the y_i and/or in variables $(y_i - E(y_i)) = (y_i - \mu_i)$.

As a start, we have the following important theorem.

Theorem 2.2.1. *Suppose a $m \times 1$ vector of random variables $\mathbf{y}$ is such that $\mathbf{y} = N(\boldsymbol{\mu}, \Sigma)$, with Σ positive definite. Consider the (central) quadratic form Q in the variables y_i centered about their mean μ_i, where*

$$Q = (\mathbf{y} - \boldsymbol{\mu})'G(\mathbf{y} - \boldsymbol{\mu}), \tag{2.43}$$

and where G is an arbitrary real symmetric matrix. Then the random variable Q is distributed as a linear combination of m independent chi-square random variables, each of one degree of freedom, where the weights, λ_j, are characteristic roots of ΣG or $G\Sigma$, or writing $\Sigma = PP'$, roots of $P'GP$. That is,

$$Q = \lambda_1 \chi_1^2(1) + \lambda_2 \chi_1^2(2) + \cdots + \lambda_m \chi_1^2(m) \tag{2.44}$$

where $\chi_1^2(j), j = 1, \ldots, m$, are independent chi-square variables, each with one degree of freedom.

Proof. We consider the characteristic function of Q,

$$\phi_Q(t) = E(e^{itQ})$$

$$= \frac{|\Sigma^{-1}|^{1/2}}{(2\pi)^{m/2}} \int_{R^m} \exp[-\tfrac{1}{2}(\mathbf{y} - \boldsymbol{\mu})'[\Sigma^{-1} - 2itG](\mathbf{y} - \boldsymbol{\mu})]\, d\mathbf{y} \tag{2.45}$$

and using (2.8), we find

$$\phi_Q(t) = \frac{|\Sigma^{-1}|^{1/2}}{(2\pi)^{m/2}} \frac{(2\pi)^{m/2}}{|\Sigma^{-1} - 2itG|^{1/2}}, \tag{2.45a}$$

or,

$$\phi_Q(t) = \frac{1}{|I - 2itG\Sigma|^{1/2}}. \tag{2.45b}$$

Note that, in general, ΣG or $G\Sigma$ need not be symmetric. Now since Σ is positive definite, we know that there exists a nonsingular square matrix P such that $\Sigma = PP'$. Thus,

$$|I - 2itG\Sigma| = |I - 2itGPP'| \tag{2.46}$$

and using the identity that $|I - AB| = |I - BA|$ [see (A2.1.18s) of Appendix A2.1] we have that

$$|I - 2itG\Sigma| = |I - 2itP'GP|. \tag{2.46a}$$

Now $P'GP$ is symmetric, and thus there exists an orthogonal matrix U, such that

$$U'(P'GP)U = D = \begin{pmatrix} \lambda_1 & & \\ & \ddots & \\ & & \lambda_m \end{pmatrix}, \tag{2.47}$$

where the λ_j are the roots of $P'GP$. From (2.47), of course, we obtain

$$P'GP = UDU' \tag{2.47a}$$

so that

$$\begin{aligned} |I - 2itG\Sigma| &= |I - 2itUDU'| \\ &= |I - 2itU'UD| \end{aligned} \tag{2.48}$$

and since U is orthogonal, we have

$$|I - 2itG\Sigma| = |I - 2itD|. \tag{2.48a}$$

But D is diagonal, so that

$$|I - 2itG\Sigma| = |I - 2itP'GP| = |I - 2it\Sigma G|$$

$$= \prod_{j=1}^{m} (1 - 2it\lambda_j). \tag{2.48b}$$

Using the result

$$\phi_{\chi_1^2}(t) = (1 - 2it)^{-1/2} \tag{2.48c}$$

(see Problem 2.4), we have

$$\phi_Q(t) = \prod_{j=1}^{m} (1 - 2it\lambda_j)^{-1/2} = \prod_{j=1}^{m} \phi_{\lambda_j \chi_1^2(j)}(t), \tag{2.49}$$

that is,

$$\phi_Q(t) = \phi_{\Sigma \lambda_j \chi_1^2(j)}(t). \tag{2.49a}$$

From the uniqueness theorem of characteristic functions we have

$$Q = \sum_{j=1}^{m} \lambda_j \chi_1^2(j), \tag{2.50}$$

where the $\chi_1^2(j)$, $j = 1, \ldots, m$, are m independent chi-square one variables, and the λ_j are roots of $P'GP$. It is easily established that the λ_j are also the roots of $G\Sigma$ or ΣG, and so the theorem is proved.

We proceed to use Theorem 2.2.1, and it will be necessary to use the concept of idempotency. The reader not familiar with idempotent matrices should read Appendix A2.2 at this point.

***Theorem* 2.2.2.** *Suppose* $\mathbf{y}$ *is* $m \times 1$, $\mathbf{y} = N(\boldsymbol{\mu}, \Sigma)$, *where* Σ *is positive definite, so that* $\Sigma = PP'$, P *nonsingular. Consider the central quadratic form* Q *given by* (2.43).

Then, a necessary and sufficient condition that Q *be distributed as a* χ_r^2 *variable is that the matrix* $P'GP$ *or* ΣG *or* $G\Sigma$ *be idempotent of rank* r.

Proof. We have seen that

$$Q = \lambda_1 \chi_1^2(1) + \cdots + \lambda_j \chi_1^2(j) + \cdots + \lambda_m \chi_1^2(m), \tag{2.51}$$

where the $\chi_1^2(j)$ are m independent chi-square variables each with one degree of freedom, and the λ_j are roots of the symmetric matrix $P'GP$, or the matrix ΣG, or the matrix $G\Sigma$.

(i) *Necessity.* If $Q = \chi_r^2$, then from properties of chi-square variables, we know that

$$Q = \chi_1^2(1) + \cdots + \chi_1^2(r) + 0\chi_1^2(r+1) + \cdots + 0\chi_1^2(m) \tag{2.52}$$

which is to say that, using (2.51),

$$\lambda_1 = \cdots = \lambda_r = 1, \qquad \lambda_{r+1} = \cdots = \lambda_m = 0. \tag{2.52a}$$

(Of course the r ones and $m - r$ zeros could occur in some other order.) Since $P'GP$ is symmetric, we have that $P'GP$ is *idempotent of rank r.*

Now $P'GP$ idempotent implies

$$P'G(PP')GP = P'GP. \tag{2.53}$$

Pre-multiplying by P and post-multiplying by P^{-1}, we have

$$\Sigma G \Sigma G = \Sigma G, \tag{2.53a}$$

that is, ΣG is idempotent. Furthermore, the rank of ΣG is

$$r(\Sigma G) = r(PP'G) = r(P'GP) = r \tag{2.53b}$$

so that ΣG is idempotent of rank r. Similarly, $G\Sigma$ is idempotent of rank r and the necessity part is proven.

(ii) *Sufficiency.* If $P'GP$ or ΣG or $G\Sigma$ is idempotent of rank r, then

$$\lambda_1 = \cdots = \lambda_r = 1, \qquad \lambda_{r+1} = \cdots = \lambda_m = 0, \tag{2.53c}$$

so that on using (2.51)

$$Q = \chi_1^2(1) + \cdots + \chi_1^2(r) + 0\chi_1^2(r+1) + \cdots + 0\chi_1^2(m), \tag{2.53d}$$

that is,

$$Q = \chi_r^2 \tag{2.53e}$$

and the sufficiency part is proved.

An important fact which we state as a corollary and ask the reader to verify is as follows:

Corollary 2.2.2.1. *Suppose* $\mathbf{y} = N(\boldsymbol{\mu}, \sigma^2 I)$, *and consider*

$$Q = (\mathbf{y} - \boldsymbol{\mu})'G(\mathbf{y} - \boldsymbol{\mu}), \tag{2.54}$$

where G is symmetric. Then, a necessary and sufficient condition that Q/σ^2 *be distributed as* χ^2_r *is that G be idempotent of rank r.*

We turn now to some examples.

EXAMPLE 2.2.1

Suppose we have a $m \times 1$ vector random variable $\mathbf{y} = N(\boldsymbol{\mu}, \Sigma)$, and consider

$$Q = (\mathbf{y} - \boldsymbol{\mu})'\Sigma^{-1}(\mathbf{y} - \boldsymbol{\mu}). \tag{2.55}$$

In the notation of Theorems 2.2.1 and 2.2.2, $G = \Sigma^{-1}$ so that $\Sigma G = I_m$, which of course is idempotent, with all roots equal to 1. Hence, since $r(I_m) = \operatorname{tr}(I_m) = m$, we have that the quadratic form Q given in (2.55) is distributed as χ^2_m. The reader is no doubt familiar with the result for $m = 1$. The exponent, apart from the factor $-\frac{1}{2}$, in the PDF of a univariate normal variable $y = N(\mu, \sigma^2)$ is $[(y - \mu)/\sigma]^2 = (y - \mu)\sigma^{-2}(y - \mu)$, and as is well known, this has the χ^2_1 distribution. The result in this example states that for any m-variate normal, with dispersion matrix Σ positive definite, we have that the quadratic form present in the exponent of its PDF is distributed as a chi-square variable, with degrees of freedom equal to the order of the vector random variable $\mathbf{y}$, which in this example is m.

EXAMPLE 2.2.2

Suppose again that $\mathbf{y}$ is $m \times 1$, $\mathbf{y} = N(\boldsymbol{\mu}, \Sigma)$, where

$$\Sigma = \sigma^2 \begin{bmatrix} 1 & \rho & \rho & & \cdots & & \rho \\ \rho & 1 & \rho & & \cdots & & \rho \\ \rho & \rho & 1 & & \cdots & & \rho \\ \vdots & \vdots & \vdots & & \ddots & & \vdots \\ \rho & \rho & \rho & \cdots & \rho & \cdots & 1 \end{bmatrix} \tag{2.56}$$

and

$$\boldsymbol{\mu} = \mu \mathbf{1}_m \tag{2.56a}$$

where $\mathbf{1}_m$ is a $m \times 1$ vector each of whose components are equal to 1, that is, $\mathbf{1}_m = (1, \ldots, 1)'$. Hence, the y_j have common mean, μ, common variance, σ^2, and any two of the y, for example, y_r and y_s, $r \neq s$, are equicorrelated, with correlation coefficient ρ. Note that if $\rho = 0$, we are saying that the y_j are IID $N(\mu, \sigma^2)$ (recall that IID means identically and independently distributed) and hence are m independent observations on $y = N(\mu, \sigma^2)$.

Now consider

$$Q = \sum_{i=1}^{m} (y_i - \bar{y})^2 = (m-1)S_y^2, \tag{2.57}$$

where S_y^2 denotes the sample variance. Using

$$Q = \sum_{i=1}^{m} y_i^2 - \frac{1}{m}\left(\sum_{i=1}^{m} y_i\right)^2, \tag{2.57a}$$

we very quickly see that

$$Q = \mathbf{y}'\left(I_m - \frac{1}{m}\mathbf{1}_m\mathbf{1}'_m\right)\mathbf{y}. \tag{2.57b}$$

As it stands, Q does not resemble a quadratic form in $y_j - \mu$, that is, in $(\mathbf{y} - \mu\mathbf{1}_m)$. But note that ($\mathbf{0}_m$ denotes a $m \times 1$ vector of zeros)

$$\begin{aligned}\mathbf{1}'_m\left(I_m - \frac{1}{m}\mathbf{1}_m\mathbf{1}'_m\right) &= \mathbf{1}'_m - \frac{1}{m}(\mathbf{1}'_m\mathbf{1}_m)\mathbf{1}'_m = \mathbf{1}'_m - \mathbf{1}'_m \\ &= \mathbf{0}'_m,\end{aligned} \tag{2.57c}$$

since $\mathbf{1}'_m\mathbf{1}_m = 1^2 + \cdots + 1^2 = m$, and similarly we have that $[I_m - (1/m)\mathbf{1}_m\mathbf{1}'_m]\mathbf{1}_m = \mathbf{0}_m$. Hence

$$Q = (\mathbf{y} - \boldsymbol{\mu})'M(\mathbf{y} - \boldsymbol{\mu}), \tag{2.58}$$

where $M = I_m - (1/m)\mathbf{1}_m\mathbf{1}'_m$ and $\boldsymbol{\mu} = \mu\mathbf{1}_m$. Also, the dispersion matrix Σ, given by (2.56), may be written as

$$\Sigma = \sigma^2[(1-\rho)I_m + \rho\mathbf{1}_m\mathbf{1}'_m]. \tag{2.59}$$

Consider now

$$Q' = \frac{Q}{\sigma^2(1-\rho)} = (\mathbf{y} - \boldsymbol{\mu})'\left[\frac{M}{\sigma^2(1-\rho)}\right](\mathbf{y} - \boldsymbol{\mu}). \tag{2.60}$$

Q' is a central quadratic form with $G = M/\sigma^2(1-\rho)$. Hence,

(2.61) $$\Sigma G = \frac{1}{1-\rho}[(1-\rho)I_m + \rho \mathbf{1}_m\mathbf{1}'_m]\left(I_m - \frac{1}{m}\mathbf{1}_m\mathbf{1}'_m\right)$$

and

(2.61a) $$\Sigma G = I_m - \frac{1}{m}\mathbf{1}_m\mathbf{1}'_m = M.$$

But

$$MM = \left(I_m - \frac{1}{m}\mathbf{1}_m\mathbf{1}'_m\right)\left(I_m - \frac{1}{m}\mathbf{1}_m\mathbf{1}'_m\right)$$

(2.62) $$= I_m - \frac{1}{m}\mathbf{1}_m\mathbf{1}'_m - \frac{1}{m}\mathbf{1}_m\mathbf{1}'_m + \frac{1}{m^2}\mathbf{1}_m(\mathbf{1}'_m\mathbf{1}_m)\mathbf{1}'_m$$

$$= I_m - \frac{2}{m}\mathbf{1}_m\mathbf{1}'_m + \frac{1}{m}\mathbf{1}_m\mathbf{1}'_m$$

$$= M,$$

that is, $M = I_m - (1/m)\mathbf{1}_m\mathbf{1}'_m$ is *idempotent*, and hence the rank of M, $r(M)$, is (tr stands for trace)

(2.62a) $$r(M) = \operatorname{tr}\left(I_m - \frac{1}{m}\mathbf{1}_m\mathbf{1}'_m\right) = \operatorname{tr}(I_m) - \frac{1}{m}\operatorname{tr}(\mathbf{1}_m\mathbf{1}'_m).$$

Using the result that $\operatorname{tr}(AB) = \operatorname{tr}(BA)$, we have

(2.62b) $$r(M) = m - \frac{1}{m}\operatorname{tr}\mathbf{1}'_m\mathbf{1}_m = m - \frac{1}{m}m$$
$$= m - 1.$$

From Theorem 2.2.2,

(2.62c) $$Q' = \chi^2_{m-1},$$

or

(2.63) $$Q = (1-\rho)\sigma^2\chi^2_{m-1}.$$

We note in passing that if $\rho = 0$, $Q = \Sigma_{i=1}^m (y_i - \bar{y})^2$—which may be regarded as a sample sum of squares of deviations based on m independent observations y_j—is distributed as $\sigma^2 \chi^2_{m-1}$.

We now are in a position to state a very important theorem due to C. C. Craig (1943). We will use this theorem extensively in proving the independence of certain quadratic forms in analyzing the linear model, when we assume normality for the distribution of observed values. Typical quadratic forms include the residual sum of squares and the regression sum of squares.

Theorem 2.2.3. *(Craig's Theorem). Suppose* $\mathbf{y} = N(\boldsymbol{\mu}, \Sigma)$ *is* $m \times 1$, *where* Σ *is positive definite. Consider the two (central) quadratic forms*

$$Q_j = (\mathbf{y} - \boldsymbol{\mu})' A_j (\mathbf{y} - \boldsymbol{\mu}), \qquad j = 1, 2, \tag{2.64}$$

where, as in (2.43), A_1 *and* A_2 *are arbitrary real symmetric matrices.*

Then Q_1 *and* Q_2 *are statistically independent if and only if*

$$A_1 \Sigma A_2 = \bigcirc \tag{2.65}$$

where $\bigcirc$ *stands for a* $m \times m$ *matrix of zeros.*

Proof. [The sufficiency part of this proof turns out to be quite simple, in contrast to the necessity part. The proof of the necessity part is due to Lancaster (1954)].

Recall that a necessary and sufficient condition that two random variables Q_j be statistically independent is that the joint characteristic function factors into a product of the characteristic functions of the marginals, that is,

$$\phi_{\mathbf{Q}}(\mathbf{t}) = \phi_{Q_1, Q_2}(t_1, t_2) = \phi_{Q_1}(t_1)\phi_{Q_2}(t_2), \tag{2.66}$$

where ϕ_{Q_1} and ϕ_{Q_2} are the characteristic functions of Q_1 and Q_2, respectively.

Here, Q_j are central quadratic forms, and we know [see (2.45*b*)] that

$$\phi_{Q_j}(t_j) = |I_m - 2it_j \Sigma A_j|^{-1/2}. \tag{2.66a}$$

Using the identity (2.8), it is easily verified (see Problem 2.5) that

$$\phi_{\mathbf{Q}}(\mathbf{t}) = |I_m - 2it_1 \Sigma A_1 - 2it_2 \Sigma A_2|^{-1/2}. \tag{2.66b}$$

We also note that

(2.66c)

$$\phi_{Q_1}(t_1)\phi_{Q_2}(t_2) = |I_m - 2it_1\Sigma A_1 - 2it_2\Sigma A_2 - 4t_1t_2\Sigma A_1\Sigma A_2|^{-1/2}.$$

(i) *Sufficiency.* If $A_1\Sigma A_2 = \bigcirc$, then

(2.67)
$$\begin{aligned}\phi_{Q_1}(t_1)\phi_{Q_2}(t_2) &= |I_m - 2it_1\Sigma A_1 - 2it_2\Sigma A_2|^{-1/2}\\ &= \phi_{Q_1,Q_2}(t_1,t_2),\end{aligned}$$

from (2.66b). That is, the joint characteristic function factors so that Q_1, Q_2 are independent, and the sufficiency part is proved.

(ii) *Necessity.* We assume in this part that Q_1 and Q_2 are independent, that is,

(2.68) $$|I_m - 2it_1\Sigma A_1 - 2it_2\Sigma A_2| = |I_m - 2it_1\Sigma A_1||I_m - 2it_2\Sigma A_2|$$

and we wish to show that this implies that $A_1\Sigma A_2 = \bigcirc$.

Suppose we set

(2.69) $$2it_j = \frac{\tau_j}{\lambda}, \qquad j = 1,2.$$

Since Σ is assumed positive definite, there exists a $m \times m$ nonsingular matrix P such that $\Sigma = PP'$. Suppose then that we write

(2.69a) $$H_j = P'A_jP.$$

We have, on multiplying by $|P^{-1}| \times |P| = 1$ where needed, that [see (2.68)]

(2.70) $$\left|I_m - \frac{\tau_1}{\lambda}H_1 - \frac{\tau_2}{\lambda}H_2\right| = \left|I_m - \frac{\tau_1}{\lambda}H_1\right| \times \left|I_m - \frac{\tau_2}{\lambda}H_2\right|$$

or, multiplying by λ^{2m}, that

(2.70a) $$\lambda^m|\lambda I_m - \tau_1H_1 - \tau_2H_2| = |\lambda I_m - \tau_1H_1| \times |\lambda I_m - \tau_2H_2|.$$

Since $H_1 = P'A_1P$ is symmetric, there exists an orthogonal matrix that diagonalizes H_1; that is, we have

$$U'H_1U = C = \begin{bmatrix} c_1 & & & & & \\ & \ddots & & & & \\ & & c_r & & & \\ & & & 0 & & \\ & & & & \ddots & \\ & & & & & 0 \end{bmatrix}, \tag{2.71}$$

where $r = r(H_1) = r(A_1)$. We also let

$$G = U'H_2U. \tag{2.71a}$$

Hence, we have from (2.70a) that

$$\lambda^m |\lambda I_m - \tau_1 C - \tau_2 G| = |\lambda I_m - \tau_1 C| \times |\lambda I_m - \tau_2 G| \tag{2.71b}$$

We now partition C and G similarly and write

$$C = \left(\begin{array}{c|c} C_{11} & \bigcirc_{12} \\ \hline \bigcirc_{21} & \bigcirc_{22} \end{array}\right), \qquad G = \left(\begin{array}{c|c} G_{11} & G_{12} \\ \hline G_{21} & G_{22} \end{array}\right) \tag{2.72}$$

where G_{11} and C_{11} are $r \times r$, G_{12} and $\bigcirc_{12}$ are of order $r \times (n-r)$, and so on, and

$$C_{11} = \begin{pmatrix} c_1 & & \bigcirc \\ & \ddots & \\ \bigcirc & & c_r \end{pmatrix} \tag{2.72a}$$

(The $\bigcirc_{ij}$ submatrices have only zero elements.) The symmetric matrix whose determinant is mentioned on the left-hand side of (2.71b) is of the form

(2.73)

$$\left[\begin{array}{cccc|ccc} \lambda - \tau_1 c_1 - \tau_2 g_{11} & -\tau_2 g_{12} & \cdots & -\tau_2 g_{1r} & & & \\ -\tau_2 g_{12} & \lambda - \tau_1 c_2 - \tau_2 g_{22} & & \vdots & & -\tau_2 G_{12} & \\ \vdots & & \ddots & -\tau_2 g_{r-1,r} & & & \\ -\tau_2 g_{1r} & \cdots & & \lambda - \tau_1 c_r - \tau_2 g_{rr} & & & \\ \hline & & & & \lambda - \tau_2 g_{r+1,r+1} & \cdots & -\tau_2 g_{r+1,m} \\ & -\tau_2 G_{21} & & & \vdots & \ddots & \vdots \\ & & & & -\tau_2 g_{r+1,m} & \cdots & \lambda - \tau_2 g_{mm} \end{array}\right]$$

which we denote as

$$E = \left(\begin{array}{c|c} E_{11} & E_{12} \\ \hline E_{21} & E_{22} \end{array}\right) \tag{2.74}$$

where E_{11} is the $r \times r$ matrix appearing in the upper left-hand corner of (2.73), and so on. Since [see (A2.1.18t) of Appendix A2.1 and Problems 2.7(c) and (d)]

$$|E| = |E_{22}||E_{11} - E_{12}E_{22}^{-1}E_{21}|, \tag{2.74a}$$

it can be shown (see Problem 2.6) that the coefficient of τ_1^r on the left-hand side of (2.71b) is

$$\lambda^m(-1)^r c_1 \cdots c_r |\lambda I_{m-r} - \tau_2 G_{22}|. \tag{2.75}$$

But the right-hand side of (2.71b) has τ_1 appearing only in the first factor, which is the determinant of the matrix $\lambda I_m - \tau_1 C$, which is of the partitioned form

$$\left(\begin{array}{cccc|ccc} \lambda - \tau_1 c_1 & \cdots & 0 & & & & \\ \vdots & \ddots & \vdots & & & O & \\ & & 0 & & & & \\ 0 & \cdots & 0 & \lambda - \tau_1 c_r & & & \\ \hline & & & & \lambda & & \\ & O & & & & \ddots & \\ & & & & & & \lambda \end{array}\right) \tag{2.76}$$

so that the coefficient of τ_1^r on the right-hand side is

$$\lambda^{m-r} c_1 \cdots c_r (-1)^r |\lambda I_m - \tau_2 G|. \tag{2.76a}$$

Equating (2.75) and (2.76a),

$$\lambda^m(-1)^r c_1 \cdots c_r |\lambda I_{m-r} - \tau_2 G_{22}| = \lambda^{m-r} c_1 \cdots c_r (-1)^r |\lambda I_m - \tau_2 G| \tag{2.77}$$

or

$$\lambda^r |\lambda I_{m-r} - \tau_2 G_{22}| = |\lambda I_m - \tau_2 G|. \tag{2.77a}$$

This equation states that G_{22} and G have the same nonzero roots, so that, in particular, *the sum of squares of the elements of* G_{22} equals *the sum of squares of the elements of* G, since both equal the sum of squares of these roots. (Problem 2.7 gives the needed steps for a proof.) Hence

$$G_{11} = O, \qquad G_{12} = O, \qquad G_{21} = O. \tag{2.78}$$

Now recall that $C_{12} = O$, $C_{21} = O$, and $C_{22} = O$. This means that

$$CG = \left(\begin{array}{c|c} C_{11} & O \\ \hline O & O \end{array}\right)\left(\begin{array}{c|c} O & O \\ \hline O & G_{22} \end{array}\right) = \left(\begin{array}{c|c} O & O \\ \hline O & O \end{array}\right) = O. \tag{2.79}$$

From (2.71) and (2.71*a*), we now have

$$U'H_1UU'H_2U = CG = O \tag{2.80}$$

or, since U is orthogonal,

$$U'P'A_1PP'A_2PU = O, \tag{2.80a}$$

that is

$$U'P'A_1\Sigma A_2PU = O. \tag{2.80b}$$

But U, P, and P' are nonsingular, so that

$$A_1\Sigma A_2 = O, \tag{2.80c}$$

and the necessity is proven. This means Craig's theorem is now proved.

We have the following useful corollary to the above theorem.

Corollary 2.2.3.1. *Suppose* $\mathbf{y}$ *is* $m \times 1$, $\mathbf{y} = N(\boldsymbol{\mu}, \Sigma)$. *Let* $Q = (\mathbf{y} - \boldsymbol{\mu})'A(\mathbf{y} - \boldsymbol{\mu})$, *$A$ symmetric, and define the (scalar) random variable z by*

$$z = \mathbf{b}'(\mathbf{y} - \boldsymbol{\mu}) \tag{2.81}$$

where $\mathbf{b}' = (b_1, \ldots, b_m)$ *is an arbitrary* $1 \times m$ *vector of constants.*

Then, a necessary and sufficient condition that Q and z be statistically independent is that $\mathbf{b}'\Sigma A = \mathbf{0}' = (0, \ldots, 0)$.

The proof of Corollary 2.2.3.1 is left to the reader (see Problem 2.8).

Theorem 2.2.3 and its corollary (Corollary 2.2.3.1) enable us to look into the following question. Suppose $Q_1 = \mathbf{y}'A_1\mathbf{y}$ and $Q_2 = (\mathbf{y} - \boldsymbol{\mu})'A_2(\mathbf{y} - \boldsymbol{\mu})$, where $\mathbf{y} = N(\boldsymbol{\mu}, \Sigma)$. Then Q_1 and Q_2 are independent if and only if $A_1 \Sigma A_2 = \mathrm{O}$. (We are assuming A_1 and A_2 to be symmetric and positive definite.) We demonstrate this result as follows.

It is algebraically very easy to see that we may write

$$Q_1 = Q_1^* + l + c, \tag{2.82}$$

where

$$Q_1^* = (\mathbf{y} - \boldsymbol{\mu})' A_1 (\mathbf{y} - \boldsymbol{\mu}), \qquad l = 2\boldsymbol{\mu}' A_1 (\mathbf{y} - \boldsymbol{\mu}), \qquad c = \boldsymbol{\mu}' A_1 \boldsymbol{\mu} \tag{2.82a}$$

Furthermore, from Theorem 2.2.3 and Corollary 2.2.3.1, we have that Q_1^* and l, two random variables with a certain joint distribution, are such that both are independent of Q_2. Hence, any function of Q_1^* and l are independent of Q_2, *and in particular*, $Q_1 = Q_1^* + l + c$ *is independent of* Q_2. This simple fact will be very useful when discussing certain noncentral (e.g., Q_1) and central (e.g., Q_2) quadratic forms and their independence.

EXAMPLE 2.2.3

Suppose $y_1, \ldots, y_i, \ldots, y_N$ are N independent observations from $N(\eta, \sigma^2)$, that is, the $N \times 1$ vector $\mathbf{y} = (y_1, \ldots, y_N)'$ is such that $\mathbf{y} = N(\eta \mathbf{1}_N, \sigma^2 I_N)$.

Then $\Sigma_{i=1}^N (y_i - \bar{y})^2$ and $\bar{y}$ are statistically independent. To see this we apply Corollary 2.2.3.1. We have that

$$\bar{y} - \eta = N^{-1} \sum_{i=1}^{N} (y_i - \eta) = N^{-1} \mathbf{1}_N' (\mathbf{y} - \boldsymbol{\eta}), \qquad \boldsymbol{\eta} = \eta \mathbf{1}_N,$$

so that, in the language of the corollary, $\mathbf{b}' = N^{-1} \mathbf{1}_N'$. Furthermore, we have seen that

$$Q = \sum_{i=1}^{N} (y_i - \bar{y})^2 = (\mathbf{y} - \boldsymbol{\eta})' [I_N - N^{-1} \mathbf{1}_N \mathbf{1}_N'] (\mathbf{y} - \boldsymbol{\eta}), \qquad \boldsymbol{\eta} = \eta \mathbf{1}_N$$

and we have that $\Sigma = V(\mathbf{y}) = \sigma^2 I_N$. Hence

$$\mathbf{b}' \Sigma A = N^{-1} \mathbf{1}_N' (\sigma^2 I_N)(I_N - N^{-1} \mathbf{1}_N \mathbf{1}_N') = N^{-1} \sigma^2 (\mathbf{1}_N' - \mathbf{1}_N') = \mathbf{0}'$$

so that $\bar{y} - \eta$ and Q are independent; that is, $\bar{y}$ and $(N-1) S_y^2$ are independent, or, $\bar{y}$ and S_y^2 are independent.

Since $\bar{y} - \eta = N(0, \sigma^2/N)$ is independent of $S_y^2 = \sigma^2\chi^2_{N-1}/(N-1)$, then applying the definition of a Student's-t variable, we have that

$$N^{1/2}(\bar{y} - \eta)/\left(S_y^2\right)^{1/2} = t_{N-1}.$$

Note that the results of this example could have been found directly from application of Theorem 2.2.3; see Problem 2.9.

We now discuss a very important theorem concerning the distribution of quadratic forms in normal variables. Part of the theorem is due to Cochran (1934). Before we state it we need the following lemma, due to Loynes (1966).

Lemma 2.2.1. *Suppose M is symmetric and idempotent and P is symmetric and positive semidefinite. If $I_m - M - P$ is also positive semidefinite, then $MP = PM = \bigcirc$ (all matrices mentioned are of the same order $m \times m$).*

Proof. Suppose $\mathbf{x}$ is any $m \times 1$ vector, and let $\mathbf{y} = M\mathbf{x}$. Then

$$\mathbf{y}'M\mathbf{y} = \mathbf{y}'MM\mathbf{x} = \mathbf{y}'M\mathbf{x} = \mathbf{y}'\mathbf{y}, \tag{2.83}$$

so that

$$\mathbf{y}'(I_m - M - P)\mathbf{y} = -\mathbf{y}'P\mathbf{y}. \tag{2.83a}$$

But $I_m - M - P$ and P are positive semidefinite, so that we must have

$$\mathbf{y}'P\mathbf{y} = 0. \tag{2.83b}$$

Now since P is symmetric and positive semidefinite, $P = L'L$ for some L. Hence $\mathbf{y}'P\mathbf{y} = \mathbf{y}'L'L\mathbf{y} = 0$ implies $L\mathbf{y} = \mathbf{0}$, or $L'L\mathbf{y} = \mathbf{0}$; this in turn says that $P\mathbf{y} = PM\mathbf{x} = \mathbf{0}$. But this is true for any $\mathbf{x}$ so that

$$PM = \bigcirc. \tag{2.83c}$$

Hence we have

$$(PM)' = M'P' = MP = \bigcirc, \quad Q.E.D. \tag{2.83d}$$

Using the lemma, we may state and prove a theorem due to Graybill and Marsaglia (1957), which is the theorem needed to prove Cochran's theorem, discussed below.

Theorem 2.2.4. *(Graybill and Marsaglia). If $D_1, D_2, \ldots, D_q$ are symmetric $m \times m$ matrices, and if*

$$\text{(2.84)} \qquad \begin{aligned} &\text{(i)}\quad D_1, \ldots, D_q \textit{ are each idempotent} \\ &\text{(ii)}\quad D = D_1 + \cdots + D_q \textit{ idempotent,} \end{aligned}$$

then

$$\text{(2.84}a\text{)} \qquad \text{(iii)}\quad D_i D_j = \mathrm{O}, \textit{ for all } i \neq j.$$

Furthermore, any two of the above conditions imply the third.

Proof. If condition (ii) holds, then it is easy to see that $I - D$ is idempotent and hence positive semidefinite. Also, $D - D_i - D_j = \Sigma_{t \neq i, j} D_t$ is positive semidefinite if condition (i) holds. That is, when conditions (i) and (ii) hold, $I - D + D - D_i - D_j = I - D_i - D_j$ is positive semidefinite. Applying Loynes' lemma 2.2.1 to $I - D_i - D_j$, we have that $D_i D_j = \mathrm{O}$, that is, conditions (i) and (ii) imply (iii).

Now if conditions (i) and (iii) hold, we have that

$$DD = \sum_{i=1}^{m} D_i D_i + \sum_{i \neq j}\sum D_i D_j = \sum_{i=1}^{m} D_i D_i = \sum_{i=1}^{m} D_i,$$

that is, $DD = D$, so that (i) and (iii) imply condition (ii).

Suppose now that conditions (ii) and (iii) hold. Let $\mathbf{d}$ be a characteristic vector and α the corresponding root of D_j, for example, so that

$$\text{(2.85)} \qquad D_j \mathbf{d} = \alpha \mathbf{d}$$

and for $\alpha \neq 0$, $\mathbf{d} = D_j \mathbf{d}/\alpha$. Using (iii),

$$\text{(2.85}a\text{)} \qquad D_i \mathbf{d} = D_i D_j \mathbf{d}/\alpha = \mathbf{0}$$

for $i \neq j$. Hence

$$\text{(2.85}b\text{)} \qquad D\mathbf{d} = \sum_{i \neq j} D_i \mathbf{d} + D_j \mathbf{d} = D_j \mathbf{d} = \alpha \mathbf{d}$$

so that α is a characteristic root of D. But condition (ii) holds, that is, D is idempotent (and symmetric), so that $\alpha = 0$ or 1. Hence D_j, which is assumed symmetric, is idempotent, that is, conditions (ii) and (iii) imply (i).

We can now state and prove an historic and important theorem due to Cochran (1934). [Actually, we state and prove a somewhat stronger version —Cochran originally proved that conditions (i) and (ii) of (2.87) below are together equivalent to condition (iii). Further discussion of this important theorem may be found in Appendix VI of the book by Scheffe (1959).]

***Theorem* 2.2.5.** *(Cochran's Theorem). Suppose $y = N(0, 1)$ and that $(y_1,\ldots,y_N)$ are N independent observations on y, so that $\mathbf{y} = N(\mathbf{0}, I_N)$. Suppose further that*

(2.86) $$Q = \mathbf{y}'\mathbf{y} = Q_1 + \cdots + Q_k,$$

where

(2.86*a*) $$Q_t = \mathbf{y}'A_t\mathbf{y}$$

is a quadratic form of rank n_t, that is, $r(A_t) = n_t$, and A_t is $N \times N$ and symmetric, $t = 1,\ldots,k$.

Then, any one of the following three conditions implies the other two:

(i) $Q_1,\ldots,Q_k$ *are statistically independent;*

(2.87) (ii) $Q_1,\ldots,Q_k$ *are individually distributed as χ^2 variables;*

(iii) $n_1 + \cdots + n_k = N.$

Proof. We prove the theorem by showing that condition (i) implies condition (ii), that condition (ii) implies condition (iii), and that condition (iii) implies condition (i).

Part (a). Suppose condition (i) of (2.87) holds. We then have that

(2.88) $$Q_1, \qquad Q_2 + \cdots + Q_k$$

are statistically independent. Now

(2.89) $$\mathbf{y}'\mathbf{y} = \mathbf{y}'(A_1 + \cdots + A_k)\mathbf{y},$$

that is, $I = A_1 + \cdots + A_k$. Let $B = A_2 + \cdots + A_k$. We have that $\mathbf{y}'B\mathbf{y}$ is independent of $Q_1 = \mathbf{y}'A_1\mathbf{y}$. But $A_1 + B = I$, and using Craig's theorem, $A_1B = \mathrm{O} = A_1(I - A_1) = \mathrm{O}$, that is,

(2.89*a*) $$A_1 = A_1A_1$$

and A_1 is idempotent. But if A_1 is idempotent, Q_1 is a chi-square variable, with degrees of freedom equal to $r(A_1) = \text{tr}(A_1) = n_1$. Similarly, Q_j is $\chi^2_{n_j}$, and so on; that is, condition (i) implies condition (ii).

Part (*b*). Suppose condition (ii) of (2.87) holds, that is, $Q_j = \mathbf{y}'A_j\mathbf{y}$ is a chi-square variable. Then A_j is idempotent and $\text{tr}(A_j) = r(A_j) = n_j$. But $I = A_1 + \cdots + A_k$, so that

$$\text{tr}(I) = \text{tr}(A_1 + \cdots + A_k) = \text{tr}(A_1) + \cdots + \text{tr}(A_k),$$

that is,

$$N = n_1 + \cdots + n_k,$$

and hence, condition (ii) of (2.87) implies condition (iii).

Part (*c*). Suppose $n_1 + \cdots + n_k = N$, and we write

$$A_1 + B = I, \qquad B = A_2 + \cdots + A_k \tag{2.90}$$

so that

$$r(B) = r(A_2 + \cdots + A_k) \leq r(A_2) + \cdots + r(A_k) \tag{2.90a}$$

or

$$r(B) \leq n_2 + \cdots + n_k = N - n_1. \tag{2.90b}$$

Since A_1 is symmetric, there exists an orthogonal matrix P such that

$$P'A_1P = \begin{bmatrix} \alpha_1 & & & & & \\ & \ddots & & & & \\ & & \alpha_{n_1} & & & \\ & & & 0 & & \\ & & & & \ddots & \\ & & & & & 0 \end{bmatrix}, \tag{2.91}$$

since $r(A_1) = n_1$. Thus, from (2.90),

$$P'A_1P + P'BP = P'IP = I, \tag{2.92}$$

or

(2.92*a*)

$$\begin{bmatrix} \alpha_1 & & & & & \\ & \ddots & & & & \\ & & \alpha_{n_1} & & & \\ & & & 0 & & \\ & & & & \ddots & \\ & & & & & 0 \end{bmatrix} + P'BP = \begin{bmatrix} 1 & & & & & \\ & \ddots & & & & \\ & & 1 & & & \\ & & & 1 & & \\ & & & & \ddots & \\ & & & & & 1 \end{bmatrix}.$$

This says that $P'BP$ is diagonal and we may write

(2.92b)

$$\begin{bmatrix} \alpha_1 & & & & & \\ & \ddots & & & & \\ & & \alpha_{n_1} & & & \\ & & & 0 & & \\ & & & & \ddots & \\ & & & & & 0 \end{bmatrix} + \begin{bmatrix} \beta_1 & & & & & \\ & \ddots & & & & \\ & & \beta_{n_1} & & & \\ & & & \beta_{n_1+1} & & \\ & & & & \ddots & \\ & & & & & \beta_N \end{bmatrix} = \begin{bmatrix} 1 & & & & & \\ & \ddots & & & & \\ & & 1 & & & \\ & & & 1 & & \\ & & & & \ddots & \\ & & & & & 1 \end{bmatrix}.$$

Thus, $\beta_{n_1+1} = \cdots = \beta_N = 1$. Now $r(B) \leq N - n_1$, which in turn implies that

$$\beta_1 = \cdots = \beta_{n_1} = 0 \tag{2.92c}$$

so that

$$\alpha_1 = \cdots = \alpha_{n_1} = 1. \tag{2.92d}$$

But A_1 is symmetric, and the above states that each characteristic root of A_1 is either 0 or 1, so that A_1 is idempotent. Similarly, A_j are symmetric and idempotent, for all j. Now $A_1 + \cdots + A_k = I$, a symmetric idempotent matrix. Applying Graybill's theorem 2.2.4, we have $A_i A_j = O$ for all $i \neq j$, so that by Craig's theorem, Q_i, Q_j are independent for $i \neq j$; that is, condition (iii) implies condition (i) and this theorem is proven.

Before turning to the theory and applications of the least-squares method, we state and prove a theorem which is useful in many situations, relating the distribution of quadratic forms in normal variables when the variables are known to obey certain linear conditions.

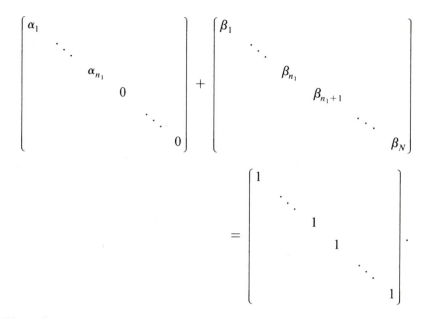

Theorem 2.2.6. *Suppose the* $m \times 1$ *vector* $\mathbf{y} = N(\boldsymbol{\eta}, \Sigma)$, *with* Σ *positive definite. We consider the conditional distribution of* $\mathbf{y}$, *given that*

$$A(\mathbf{y} - \boldsymbol{\eta}) = \mathbf{0}, \tag{2.93}$$

where A *is* $k \times m$, $k < m$ *with* $r(A) = k$.

Then, the conditional distribution of $\mathbf{y}$, *given* (2.93), *is such that*

$$Q = (\mathbf{y} - \boldsymbol{\eta})'\Sigma^{-1}(\mathbf{y} - \boldsymbol{\eta}) = \chi^2_{m-k}. \tag{2.93a}$$

Proof. Since A is of full rank, there exists a matrix B, of order $(m-k) \times m$, such that the *augmented* matrix

$$H = \left(\frac{A}{B}\right), \qquad \text{of order } m \times m \tag{2.94}$$

has rank $r(H) = m$. Consider the transformation

$$\mathbf{z} = H(\mathbf{y} - \boldsymbol{\eta}) \quad \text{or} \quad \left(\frac{\mathbf{z}_1}{\mathbf{z}_2}\right) = \left(\frac{A}{B}\right)(\mathbf{y} - \boldsymbol{\eta}), \tag{2.95}$$

where $\mathbf{z}_1$ is $k \times 1$ and $\mathbf{z}_2$ is $(m-k) \times 1$. We have that

$$\mathbf{z} = N(\mathbf{0}, H\Sigma H'). \tag{2.95a}$$

Let $V = H\Sigma H'$ be partitioned in the obvious fashion as

$$V = H\Sigma H' = \left(\begin{array}{c|c} V_{11} & V_{12} \\ \hline V_{21} & V_{22} \end{array}\right), \qquad V_{11} \text{ is } k \times k, \text{ and so on.} \tag{2.95b}$$

Let V^{-1} denote the inverse of V, and after similar partitioning, write

$$V^{-1} = \left(\begin{array}{c|c} C_{11} & C_{12} \\ \hline C_{21} & C_{22} \end{array}\right), \qquad C_{11} \text{ is } k \times k, \text{ and so on.} \tag{2.95c}$$

We note that

$$V^{-1} = (H')^{-1}\Sigma^{-1}H^{-1} \quad \text{or} \quad \Sigma^{-1} = H'V^{-1}H. \tag{2.95d}$$

Then

$$Q = (\mathbf{y} - \boldsymbol{\eta})'\Sigma^{-1}(\mathbf{y} - \boldsymbol{\eta}) = (\mathbf{y} - \boldsymbol{\eta})'H'V^{-1}H(\mathbf{y} - \boldsymbol{\eta}), \tag{2.96}$$

and from (2.95), we obtain

$$(2.96a)\qquad Q = \mathbf{z}'V^{-1}\mathbf{z} = \left(\mathbf{z}_1' \,\middle|\, \mathbf{z}_2'\right)\begin{pmatrix} C_{11} & C_{12} \\ C_{21} & C_{22} \end{pmatrix}\begin{pmatrix} \mathbf{z}_1 \\ \mathbf{z}_2 \end{pmatrix}$$

which may be written

$$(2.96b)\qquad Q = \mathbf{z}_1'V_{11}^{-1}\mathbf{z}_1 + \left(\mathbf{z}_2 + C_{22}^{-1}C_{21}\mathbf{z}_1\right)'C_{22}\left(\mathbf{z}_2 + C_{22}^{-1}C_{21}\mathbf{z}_1\right).$$

Now the conditional distribution of $\mathbf{z}_2$, given $\mathbf{z}_1$, is

$$(2.97)\qquad N\left(-C_{22}^{-1}C_{21}\mathbf{z}_1, C_{22}^{-1}\right)$$

so that the conditional distribution of $\mathbf{z}_2$, given that $\mathbf{z}_1 = \mathbf{0}$, is

$$(2.97a)\qquad N\left(\mathbf{0}, C_{22}^{-1}\right).$$

But this implies that, given $\mathbf{z}_1 = \mathbf{0}$,

$$(2.97b)\qquad Q = \mathbf{z}_2'C_{22}\mathbf{z}_2,$$

which is the quadratic form in the exponent of the distribution in (2.97*a*), that is, $Q = \chi^2_{m-k}$, given that $A(\mathbf{y} - \boldsymbol{\eta}) = \mathbf{0}$. *Q.E.D.*

Problems

2.1. If a $N \times N$ positive definite matrix Σ has characteristic roots $\lambda_1, \ldots, \lambda_N$, then show that Σ^{-1} has characteristic roots $1/\lambda_1, \ldots, 1/\lambda_N$, and the same latent vectors as Σ.

2.2. (a) By differentiating $\phi_{\mathbf{y}}(\mathbf{t})$, given by (2.15), with respect to t_j, then multiplying by $-i$ and setting $\mathbf{t} = \mathbf{0}$ in the result, show that $E(y_j) = \mu_j$ $(i = \sqrt{-1})$.

(b) By differentiating $\phi_{\mathbf{y}}(\mathbf{t})$, given by (2.15), twice with respect to t_j, then multiplying by $(-i)^2$ and setting $\mathbf{t} = \mathbf{0}$ in the result, show that $E(y_j^2) = \mu_j^2 + \sigma_{jj}$, so that $V(y_j) = \sigma_{jj}$, the j–j element of Σ.

(c) By differentiating $\phi_{\mathbf{y}}(\mathbf{t})$, given by (2.15), once with respect to t_j and then with respect to t_k, $k \neq j$, then multiplying the result by $(-i)^2$ and setting $\mathbf{t} = \mathbf{0}$ in the result, show that $E(y_j y_k) = \sigma_{jk} + \mu_j\mu_k$, so that $\operatorname{cov}(y_j, y_k) = \sigma_{jk}$, the j–k element of Σ. *Hint*: The derivative of $\mathbf{t}'\Sigma\mathbf{t}$ with respect to t_j is $2\Sigma_{s=1}^{m}\sigma_{js}t_s = 2\boldsymbol{\sigma}_{j\cdot}'\mathbf{t}$, where $\boldsymbol{\sigma}_{j\cdot}'$ is the jth row of Σ.

2.3. (a) Prove (2.24). *Hint*: $\Sigma\Sigma^{-1} = I$. [The proof of (2.24a) follows similarly, using $\Sigma^{-1}\Sigma = I$.]

(b) Show also that $-C_{22}^{-1}C_{21} = \Sigma_{21}\Sigma_{11}^{-1}$, by using the fact that $\Sigma^{-1}\Sigma = I$.

2.4. Derive (2.48c).

2.5. Prove (2.66b).

2.6. Show that the coefficient of τ_1^r on the left-hand side of (2.71b) is given by (2.75).

2.7. (a) Show that if A is symmetric, for example, $m \times m$, the sum of squares of the elements of A equals the sum of squares of the eigenroots of A. *Hints*: (i) Since A is symmetric, the ith diagonal element of A^2 is of the form $\Sigma_{j=1}^{m} a_{ij}^2$, so that $\text{tr}(A^2) = \Sigma\Sigma_{ij} a_{ij}^2$, the sum of squares of the elements of A. (ii) Since A is symmetric, there exists P such that $P'AP = \Lambda$, and Λ is the diagonal matrix of roots of A, with P orthogonal.

(b) More generally, show that for A symmetric, $\text{tr}(A^s) = \Sigma_{i=1}^{m} \lambda_i^s$, and $\text{tr}(A^{-1}) = \Sigma_{i=1}^{m} \lambda_i^{-1}$ when A is nonsingular.

(c) Prove (2.74a) for symmetric E. *Hints*:

(i) Pre- and post-multiply E by B and B', respectively, where

$$B = \left(\begin{array}{c|c} I & -E_{12}E_{22}^{-1} \\ \hline O & I \end{array}\right),$$

take determinants, and use parts (ii) and (iii) below. Show too that $|B| = 1$.

(ii) The determinant of any matrix of the form

$$\left(\begin{array}{c|c} C_{11} & O \\ \hline O & C_{22} \end{array}\right)$$

is $|C_{11}| \cdot |C_{22}|$, and this may be seen by taking determinants on both sides of the matrix equation

$$\left(\begin{array}{c|c} C_{11} & O \\ \hline O & I \end{array}\right)\left(\begin{array}{c|c} I & O \\ \hline O & C_{22} \end{array}\right) = \left(\begin{array}{c|c} C_{11} & O \\ \hline O & C_{22} \end{array}\right),$$

and using part (iii) below.

(iii) If E and F are square matrices of the same order, then $\det(EF) = (\det E) \times (\det F)$, and so on.

(d) Prove (A2.1.18*t*), given that the partitioned matrix of (A2.1.18*t*) is not symmetric. *Hint*: Part (c) of this problem proves (A2.1.18*t*) for the case of a symmetric matrix.

2.8. (a) Prove Corollary 2.2.3.1. *Hint*: Prove that $z^2 = (\mathbf{y} - \boldsymbol{\mu})'\mathbf{b}\mathbf{b}'(\mathbf{y} - \boldsymbol{\mu})$ and Q are independent if and only if $\mathbf{b}\mathbf{b}'\Sigma A = \bigcirc$.

(b) The random variable $\mathbf{y} = (y_1, y_2)'$ is distributed as $N(\mathbf{0}, I_2)$. Show that $y_1^2 + 2ay_1y_2 + y_2^2$ cannot be independent of $y_1^2 + 2by_1y_2 + y_2^2$ unless a and b are both of absolute value 1 and of opposite sign.

2.9. Show that $Q_1 = N(\bar{y} - \mu)^2$ is independent of $Q_2 = (N-1)S_y^2$, where $\mathbf{y} = N(\mu\mathbf{1}, \sigma^2 I_N)$. Show that $Q_1/[Q_2/(N-1)] = F_{1, N-1}$. (*Hint*: Apply Theorem 2.2.3.)

2.10. Let $f(x_1, x_2) = (\sqrt{3}/2\pi)\exp[-\frac{1}{2}(2x_1^2 + 2x_2^2 + 2x_1x_2 - 8x_1 - 10x_2 + 14)]$, for $-\infty < x_1, x_2 < \infty$.

(a) Determine the mean of x_1 and x_2. *Hint*: Transform to new variables $w_1 = x_1 - a$, $w_2 = x_2 - b$, choosing a and b so that $Q = 2x_1^2 + 2x_2^2 + 2x_1x_2 - 8x_1 - 10x_2 + 14$ is a quadratic form in w_1 and w_2 that does not contain linear terms in w_1 and w_2. It is easy to verify that if the linear terms are to vanish then

$$\left.\frac{\partial Q}{\partial x_1}\right|_{a,b} = \left.\frac{\partial Q}{\partial x_2}\right|_{a,b} = 0.$$

(b) Find the variance–covariance matrix of $\mathbf{w} = (w_1, w_2)'$.

(c) Find the characteristic roots of the matrix A of the quadratic form appearing in the distribution of $\mathbf{w}$.

(d) Obtain an orthogonal matrix U such that $U'AU = D$, where D is diagonal.

(e) Draw contours of the distribution $f(x_1, x_2)$, determined by $Q = c$ for $c = 1$, $c = 4$. [A *contour* is a set of points $\{x_1, x_2\}$ for which f is of constant value, i.e., $f(x_1, x_2) = k$, for all $(x_1, x_2) \in \{x_1, x_2\}$. *Hint*: Use part (d) to rotate axes about (a, b) into a new coordinate system, for example, (v_1, v_2), for which Q takes the form $Q = ev_1^2 + gv_2^2$, and so on.]

2.11. Suppose the random vector $\mathbf{y} = (y_1, y_2, y_3)'$ is $N(\boldsymbol{\mu}, \Sigma)$, where $\boldsymbol{\mu} = (2, 0, 1)'$, and

$$\Sigma = \begin{bmatrix} 2 & 1 & 3 \\ 1 & 1 & 0 \\ 3 & 0 & 10 \end{bmatrix}.$$

Partition $\mathbf{y} = (y_1, y_2 | y_3)' = (\mathbf{y}_1' | y_3)$ and, correspondingly, $\boldsymbol{\mu} = (2, 0 | 1)'$, and

$$\Sigma = \left(\begin{array}{c|c} \Sigma_{11} & \Sigma_{12} \\ \hline \Sigma_{21} & \Sigma_{22} \end{array}\right) = \left(\begin{array}{cc|c} 2 & 1 & 3 \\ 1 & 1 & 0 \\ \hline 3 & 0 & 10 \end{array}\right).$$

(a) Find Σ_{11}^{-1}.

(b) Find $\boldsymbol{\mu}_{2\cdot1} = \mu_{2\cdot1}$, a scalar.

(c) Find $\Sigma_{2\cdot1} = \Sigma_{22} - \Sigma_{21}\Sigma_{11}^{-1}\Sigma_{12}$, a scalar.

(d) Using the above, write down the conditional density of y_3, given (y_1, y_2).

(e) Obtain the marginal distribution of (y_1, y_2).

(f) Obtain the conditional distribution of y_2, given $y_1 = 4$.

(g) Obtain the joint distribution of $x_1 = y_1 - y_2$ and $x_2 = y_1 + y_2$, using any method you wish.

2.12. Let $\mathbf{y} = N(\eta\mathbf{1}, \Sigma)$, where Σ is the diagonal matrix

$$\sigma^2 \begin{pmatrix} 1/n_1 & & \\ & \ddots & \\ & & 1/n_k \end{pmatrix}.$$

If $Q = \Sigma_{i=1}^{k} n_i(y_i - \bar{y})^2$, where $\bar{y} = N^{-1}\Sigma_1^k n_i y_i$ and $N = \Sigma n_i$, find the distribution of Q.

2.13. For the necessity part of the proof of Craig's theorem, discuss in detail the case $r = m$.

2.14. Prove: If $V = E[(\mathbf{y} - \boldsymbol{\mu})(\mathbf{y} - \boldsymbol{\mu})']$, where $\mathbf{y}$ has expectation $\boldsymbol{\mu}$, then V is positive semidefinite at least. Use this result to show that if (y_1, y_2, y_3) have correlations given by ρ_{ij}, then

$$1 + 2\rho_{12}\rho_{13}\rho_{23} \geq \rho_{12}^2 + \rho_{13}^2 + \rho_{23}^2.$$

2.15. If $(y_1, y_2) = N(\mathbf{0}, \Sigma)$,

$$\Sigma = \begin{pmatrix} 1 & \rho \\ \rho & 1 \end{pmatrix},$$

obtain the density of $w = y_1/y_2$.

2.16. Let $\mathbf{y} = N(\eta \mathbf{1}_N, \Sigma)$, with the $N \times N$ matrix Σ such that $\Sigma = [(1-\rho)I_N + \rho \mathbf{1}_N \mathbf{1}'_N]$. Show that if Σ is positive definite for all N, then

$$-\frac{1}{N-1} < \rho < 1.$$

Determine the distribution of $N(\bar{y} - \eta)^2 / \Sigma (y_i - \bar{y})^2$.

2.17. Let $\mathbf{y} = N(\mathbf{0}, \sigma^2 I)$, where $\mathbf{y}$ is 3×1.

(a) What is the distribution of $Q = \mathbf{y}' A \mathbf{y}$, where

$$A = \frac{1}{3}\begin{bmatrix} 2 & -1 & -1 \\ -1 & 2 & -1 \\ -1 & -1 & 2 \end{bmatrix}?$$

(b) What is the distribution of Q if $\mathbf{y} = N(\mathbf{0}, \sigma^2 \Sigma)$ where

$$\Sigma = \begin{bmatrix} 1 & \rho & \rho \\ \rho & 1 & \rho \\ \rho & \rho & 1 \end{bmatrix}?$$

(c) If

$$B = \begin{bmatrix} 1 & 1 & 1 \\ 1 & 1 & 1 \\ 1 & 1 & 2 \end{bmatrix},$$

and $\mathbf{y}$ has the distribution specified in (b), are the two quadratic forms $\mathbf{y}' A \mathbf{y}$ and $\mathbf{y}' B \mathbf{y}$ independent?

2.18. If $\mathbf{x} = N(\mathbf{0}, \Sigma)$, where $\mathbf{x}$ is $N \times 1$ and Σ is positive definite of rank N, show that a necessary and sufficient condition that $\mathbf{x}' A \mathbf{x}$ is distributed as χ^2_s is that $(\Sigma^{-1} - A)$ has rank $N - s$, with $A\Sigma$, or ΣA, idempotent.

2.19. Verify (2.96b).

2.20. Prove the following extension of Graybill's theorem 2.2.4. (Assume Theorem 2.2.4 proved.) Suppose $D_1, \ldots, D_q$ are symmetric, for example, $m \times m$, with D_i of rank k_i, and that $D = \Sigma_{i=1}^q D_i$ is of rank k. Consider the following four conditions: (i) $D_1, \ldots, D_q$ are idempotent. (ii) D idempotent. (iii) $D_i D_j = \bigcirc$ for $i \neq j$. (iv) $k = \Sigma_1^q k_i$. Then, any two of (i), (ii), and (iii) implies (i), (ii), (iii), and (iv).

2.21. A characterization of the multivariate normal distribution is that $\mathbf{y} = N(\boldsymbol{\mu}, \Sigma)$, if and only if $\mathbf{a}'\mathbf{y}$ has a univariate normal distribution (any real $\mathbf{a}'$, and so on). Using this as a *definition* of the multivariate normal, derive the characteristic function of $\mathbf{y}$ from that of the univariate normal. [*Hint*: $\phi_{\mathbf{y}}(\mathbf{t}) = \phi_{\mathbf{t}'\mathbf{y}}(1)$.]

2.22. If $(x_1, \ldots, x_r)$ is a random variable having the r-dimensional spherical normal $N(\boldsymbol{\mu}, \sigma^2 I)$, and if $W = [(x_1 - \mu_1)^2 + \cdots + (x_r - \mu_r)^2]^{1/2}$, find $E(W)$ and $V(W)$.

2.23. Let

$$\Sigma = \begin{pmatrix} \sigma_1^2 & \rho\sigma_1\sigma_2 \\ \rho\sigma_1\sigma_2 & \sigma_2^2 \end{pmatrix}$$

be a positive definite variance–covariance matrix.

(a) Find the general expression for its characteristic roots.

(b) If $\sigma_1 = \sigma_2 = \sigma$, find the characteristic roots.

(c) If $\sigma_1 = \sigma_2 = \sigma$, find a matrix P such that $P'\Sigma P = D$, a diagonal matrix. (*Hint*: for P, take columns that are normalized latent vectors of Σ.)

2.24. Prove: $E(\mathbf{y}'A\mathbf{y}) = \boldsymbol{\eta}'A\boldsymbol{\eta} + \operatorname{tr}(AV)$, where $E(\mathbf{y}) = \boldsymbol{\eta}$, $V(\mathbf{y}) = V$.

2.25. Prove (A2.1.18v) using Lemma A2.1.1 of Appendix 2.

2.26. Suppose $\mathbf{y}_1$ and $\mathbf{y}_2$ are independent $m \times 1$ random vectors such that $\mathbf{y}_i = N(\boldsymbol{\eta}_i, \Sigma_i)$, where the $m \times m$ matrices Σ_i are positive definite. Show that $\mathbf{y}_s = \mathbf{y}_1 + \mathbf{y}_2$ is distributed as $N(\boldsymbol{\eta}_s, \Sigma_s)$, where $\boldsymbol{\eta}_s = \boldsymbol{\eta}_1 + \boldsymbol{\eta}_2$ and $\Sigma_s = \Sigma_1 + \Sigma_2$. *Hint*: Use the fact that $\phi_{\mathbf{y}_s}(\mathbf{t}) = \prod_1^2 \phi_{\mathbf{y}_i}(t)$, and so on.

2.27. Consider the partitioned $k \times k$ matrix

$$A = \left[\begin{array}{c|c} A_{11} & A_{12} \\ \hline A_{21} & A_{22} \end{array}\right]$$

where A is nonsingular, and let

$$A^{-1} = B = \left[\begin{array}{c|c} B_{11} & B_{12} \\ \hline B_{21} & B_{22} \end{array}\right].$$

Define $A_{1\cdot2}$ and $A_{2\cdot1}$ by

$$A_{1\cdot2} = A_{11} - A_{12}A_{22}^{-1}A_{21}, \qquad A_{2\cdot1} = A_{22} - A_{21}A_{11}^{-1}A_{12}.$$

(a) Verify that $M_1A = I$, where

$$M_1 = \left[\begin{array}{c|c} A_{1\cdot2}^{-1} & -A_{1\cdot2}^{-1}A_{12}A_{22}^{-1} \\ \hline -A_{2\cdot1}^{-1}A_{21}A_{11}^{-1} & A_{2\cdot1}^{-1} \end{array}\right],$$

so that M_1 is A^{-1} [see (2.24a) and Problem 2.3 for the case where $A = \Sigma$ is symmetric positive definite, and $B = \Sigma^{-1}$].

(b) Verify that $AM_2 = I$ where

$$M_2 = \left[\begin{array}{c|c} A_{1\cdot2}^{-1} & -A_{11}^{-1}A_{12}A_{2\cdot1}^{-1} \\ \hline -A_{22}^{-1}A_{21}A_{1\cdot2}^{-1} & A_{2\cdot1}^{-1} \end{array}\right],$$

so that $M_2 = A^{-1}$.

(c) Verify that $AM_3 = I$, where

$$M_3 = \left[\begin{array}{c|c} A_{11}^{-1} + A_{11}^{-1}A_{12}A_{2\cdot1}^{-1}A_{21}A_{11}^{-1} & -A_{11}^{-1}A_{12}A_{2\cdot1}^{-1} \\ \hline -A_{2\cdot1}^{-1}A_{21}A_{11}^{-1} & A_{2\cdot1}^{-1} \end{array}\right],$$

so that $M_3 = A^{-1}$.

(d) Verify that $AM_4 = I$, where

$$M_4 = \left[\begin{array}{c|c} A_{1\cdot2}^{-1} & -A_{1\cdot2}^{-1}A_{12}A_{22}^{-1} \\ \hline -A_{22}^{-1}A_{21}A_{1\cdot2}^{-1} & A_{22}^{-1} + A_{22}^{-1}A_{21}A_{1\cdot2}^{-1}A_{12}A_{22}^{-1} \end{array}\right],$$

so that $M_4 = A^{-1}$. Note: Since A is assumed nonsingular, A^{-1} is unique so that, for example, we have the following:

(e) $A_{1\cdot2}^{-1} = (A_{11} - A_{12}A_{22}^{-1}A_{21})^{-1} = A_{11}^{-1} + A_{11}^{-1}A_{12}A_{2\cdot1}^{-1}A_{21}A_{11}^{-1}$, and so on. Note too that M_3 does not require A_{22}^{-1} to exist, while M_4 does not require A_{11}^{-1} to exist. Also, part (a) is given in Problem 2.3 for the case $A = \Sigma$, a symmetric positive definite matrix, as mentioned above. Finally, if we let

$$A = \left[\begin{array}{c|c} A_{11} & A_{12} \\ \hline A_{21} & A_{22} \end{array}\right] = \left[\begin{array}{c|c} I & M \\ \hline N & I \end{array}\right],$$

we have on applying (e) that the following relationship holds:

(f) $(I - MN)^{-1} = I + M(I - NM)^{-1}N$, which is a quick proof of Lemma A2.1.1 of Appendix A2.1 at the end of this chapter.

2.28. Suppose the 3×3 matrix A, written in partitioned form, is as follows

$$A = \left(\begin{array}{cc|c} 1 & 1 & 0 \\ 1 & 1 & -1 \\ \hline 2 & 3 & 1 \end{array}\right).$$

Noting that A_{11} is singular, use part (d) of Problem 2.27 to verify that the inverse of A is

$$A^{-1} = \begin{bmatrix} 4 & -1 & -1 \\ -3 & 1 & 1 \\ 1 & -1 & 0 \end{bmatrix}.$$

APPENDIX A2.1. REVIEW OF MATRICES

A rectangular array of quantities, for example, a_{ij}, with $i = 1,\ldots,m$ and $j = 1,\ldots,n$, of m rows and n columns, A, where

$$\text{(A2.1.1)} \quad A = \begin{bmatrix} a_{11} & a_{12} & \cdots & a_{1j} & \cdots & a_{1n} \\ a_{21} & a_{22} & \cdots & a_{2j} & \cdots & a_{2n} \\ \vdots & \vdots & & \vdots & & \vdots \\ a_{i1} & a_{i2} & \cdots & a_{ij} & \cdots & a_{in} \\ \vdots & \vdots & & \vdots & & \vdots \\ a_{m1} & a_{m2} & \cdots & a_{mj} & \cdots & a_{mn} \end{bmatrix} = (a_{ij})$$

is said to constitute a $m \times n$ matrix with a_{ij} as the element for the ith row and jth column. We say that the order of the matrix A is $m \times n$. A matrix of order $1 \times n$ is called a row vector and denoted by

$$\mathbf{b}' = (b_1, \ldots, b_n) \tag{A2.1.1a}$$

while a column vector of order $m \times 1$ is denoted by

$$\mathbf{c} = \begin{pmatrix} c_1 \\ \vdots \\ c_m \end{pmatrix} \tag{A2.1.1b}$$

and is a matrix of order $m \times 1$. Referring to the above, we now list the properties and a calculus of matrices.

1. *Equality.* Two matrices $A = (a_{ij})$ and $B = (b_{ij})$, of the same order, $m \times n$, are said to be equal if $a_{ij} = b_{ij}$, for all (i, j). We then write $A = B$.

2. *Transpose.* The transpose of the $m \times n$ matrix $A = (a_{ij})$ is denoted by A' and is a $n \times m$ matrix given by

$$A' = \begin{pmatrix} a_{11} & a_{21} & \cdots & a_{m1} \\ & & \vdots & \\ a_{1n} & a_{2n} & \cdots & a_{mn} \end{pmatrix}, \tag{A2.1.2}$$

that is, $B = (b_{rs}) = A'$ if $b_{rs} = a_{sr}$, $r = 1, \ldots, n$, $s = 1, \ldots, m$. Note that $(b_1, \ldots, b_n)$ of (A2.1.1a) has as its transpose the $n \times 1$ column vector

$$\begin{pmatrix} b_1 \\ \vdots \\ b_n \end{pmatrix},$$

while the transpose of the $m \times 1$ column vector $\mathbf{c}$ of (A2.1.1b) is the $1 \times m$ row vector $(c_1, \ldots, c_m)$.

3. *Addition.* The sum of the two $m \times n$ matrices $A = (a_{ij})$ and $B = (b_{ij})$ is the $m \times n$ matrix $C = (c_{ij})$, where

$$c_{ij} = a_{ij} + b_{ij}. \tag{A2.1.3}$$

4. *Multiplication.* The product AB of the two matrices $A = (a_{ij})$, of order $m \times n$, and $B = (b_{ij})$, of order $n \times p$, is the $m \times p$ matrix $C = (c_{ij})$,

where

(A2.1.4) $$c_{ij} = \sum_{k=1}^{n} a_{ik} b_{kj},$$

where the subscripts (i, j) for c_{ij} are such that $i = 1,\ldots,m$, $j = 1,\ldots,p$. Note that the definition of matrix multiplication is valid only when the number of columns of A is the same as the number of rows of B, and indeed the product BA, for example, does not exist when $p \neq m$. For $C = AB$, we sometimes say that A is *post-multiplied* by B or B is *pre-multiplied* by A.

In the special case where A is a row vector, $A = \mathbf{a}' = (a_1,\ldots,a_n)$, and B is a column vector, $B = \mathbf{b} = (b_1,\ldots,b_n)'$, we say that the scalar quantity (i.e., a 1×1 matrix)

(A2.1.4a) $$c = \mathbf{a}'\mathbf{b} = (a_1,\ldots,a_n)\begin{pmatrix} b_1 \\ \vdots \\ b_n \end{pmatrix} = \sum_{i=1}^{n} a_i b_i$$

is the inner product of $\mathbf{a}'$ and $\mathbf{b}$, and we will say that c is the inner product of $\mathbf{a}$ and $\mathbf{b}$. Thus, we may refer to the composition of $C = AB$ above as being made up of inner products, and in particular the $(i\text{-}j)$th element of C is formed by taking the inner product of the ith row of A (thought of as a row vector of order $1 \times n$) with the jth column of B (thought of as a column vector of order $n \times 1$, and so on). Referring to the transpose of a matrix, note that if $C = AB$, then, as the reader should verify,

(A2.1.4b) $$C' = B'A'.$$

The following should be noted:

14(i). If A is $m \times n$ and B is $n \times p$, with $m \neq p$, then AB is defined, but BA is not defined.

14(ii). If A is $m \times n$ and B is $n \times m$, $n \neq m$, then AB is $m \times m$ and BA is $n \times n$.

14(iii). If both A and B are of order $n \times n$, then even though AB and BA are both defined, it is not necessarily true that $AB = BA$. For example, if

$$A = \begin{bmatrix} 1 & 1 \\ 0 & 1 \end{bmatrix}, \quad B = \begin{bmatrix} 1 & 0 \\ 0 & 2 \end{bmatrix}$$

then $C_1 = AB$ and $C_2 = BA$ are such that

$$\begin{bmatrix} 1 & 2 \\ 0 & 2 \end{bmatrix} = C_1 \neq C_2 = \begin{bmatrix} 1 & 1 \\ 0 & 2 \end{bmatrix}$$

5. *Scalar multiplication.* If α is a real number and $C = (c_{ij})$ is a $m \times n$ matrix, then the $m \times n$ matrix D, where $D = (d_{ij}) = \alpha C$, is such that

(A2.1.5) $d_{ij} = \alpha c_{ij}$ for all $(i, j), i = 1,\ldots,m,\ j = 1,\ldots,n.$

6. *Linear independence.* Let $\mathbf{x}_1,\ldots,\mathbf{x}_n$ be n (column) vectors, each of order $m \times 1$, and suppose, for a set of n real numbers $(a_1,\ldots,a_k)$

(A2.1.6) $$a_1\mathbf{x}_1 + a_2\mathbf{x}_2 + \cdots + a_n\mathbf{x}_n = \mathbf{z}$$

where $\mathbf{z}$ is a $n \times 1$ vector. Then we say that $\mathbf{z}$ is a *linear combination* of the n vectors $\mathbf{x}_1,\ldots,\mathbf{x}_n$.

We say that $\mathbf{x}_1,\ldots,\mathbf{x}_n$ are *linearly independent* if

(A2.1.7) $$a_1\mathbf{x}_1 + \cdots + a_n\mathbf{x}_n = \mathbf{0} = \begin{pmatrix} 0 \\ \vdots \\ 0 \end{pmatrix}$$

implies that

(A2.1.7a) $$a_1 = a_2 = \cdots = a_n = 0,$$

that is, the vectors $\mathbf{x}_j$ are linearly independent if none of the vectors $\mathbf{x}_j$ is a linear combination of the others.

We will say that $\mathbf{x}_1,\ldots,\mathbf{x}_n$ are *linearly dependent* if they are not linearly independent.

7. *Rank of A.* The rank $r(A)$ of the $m \times n$ matrix A is defined to be

$r(A)$ = maximum number of linearly independent rows (columns) in A

(A2.1.8) = order of the first nonvanishing determinant in A (see paragraph 13 below).

Using the first definition in (A2.1.8), we are able to prove

$$
\text{(A2.1.8}a) \qquad
\begin{aligned}
r(A) &\le \min(m, n) \\
r(A + B) &\le r(A) + r(B) \\
r(AB) &\le \min[r(A), r(B)] \\
r(A) &= r(A').
\end{aligned}
$$

8. *Square matrix.* When a matrix A has the same number of rows as columns, for example, n, then we say that (the $n \times n$ matrix) A is a *square* matrix. If $A = (a_{ij})$ is a square matrix and such that

$$\text{(A2.1.9)} \qquad a_{ij} = a_{ji}, \qquad i \neq j = 1,\ldots,n,$$

then we say that the matrix A is a *symmetric* matrix.

9. *Trace.* If A is of order $n \times n$, then the trace of A, $\text{tr}(A)$, is defined to be the sum of the *diagonal elements* a_{ii}, for $i = 1,\ldots,n$ of A, that is,

$$\text{(A2.1.10)} \qquad \text{tr}(A) = \sum_{i=1}^{n} a_{ii}.$$

It is easy to show that

$$\text{(A2.1.10}a) \qquad \text{tr}(A + B) = \text{tr}(A) + \text{tr}(B)$$

and if C is of order $n \times m$ and D is $m \times n$ then

$$\text{(A2.1.10}b) \qquad \text{tr}(CD) = \text{tr}(DC).$$

10. *Identity matrix.* The identity matrix of order n, denoted by I_n or I, is the $n \times n$ matrix with diagonal elements each equal to 1, and all off-diagonal elements equal to zero, that is,

$$\text{(A2.1.11)} \qquad I_n = \begin{bmatrix} 1 & 0 & 0 & \cdots & 0 \\ 0 & 1 & 0 & \cdots & 0 \\ \vdots & \vdots & \vdots & & \vdots \\ 0 & 0 & 0 & \cdots & 1 \end{bmatrix}.$$

We note that for any matrix A, of order $m \times n$,

$$\text{(A2.1.11}a) \qquad AI_n = A, \qquad I_m A = A.$$

11. *Inverse of a matrix.* If A is of order $n \times n$ and rank $r(A) = n$ (we sometimes say that A is then of full rank), there exists a $n \times n$ matrix B such that

$$BA = I_n, \tag{A2.1.12}$$

and B is called the inverse of A; we write

$$B = A^{-1}. \tag{A2.1.12a}$$

The following should be noted:

11 (i). We have from (A2.1.12) and (A2.1.12a) trivially that

$$A^{-1}A = I_n. \tag{A2.1.12b}$$

11 (ii). Suppose now that the $n \times n$ matrix T is such that

$$AT = I. \tag{A2.1.12c}$$

Then,

$$A^{-1}AT = A^{-1} \tag{A2.1.12d}$$

and from (A.2.1.12b)

$$T = A^{-1}. \tag{A2.1.12e}$$

Substituting (A2.1.12e) in (A2.1.12c), we have

$$AA^{-1} = I.$$

In summary, then

$$AA^{-1} = I = A^{-1}A. \tag{A2.1.12f}$$

11(iii). We note from the results above that

$$A = (A^{-1})^{-1}. \tag{A2.1.12g}$$

11(iv). If F and G are both $n \times n$ matrices of full rank, then

$$(FG)^{-1} = G^{-1}F^{-1}. \tag{A2.1.12h}$$

Because if $P = FG$, then $r(P) = n$ and P^{-1} exists, so that $P^{-1}P = I$. That is,

$$(FG)^{-1}FG = I \quad \text{or} \quad (FG)^{-1}F = G^{-1}, \tag{A2.1.12i}$$

or

$$(FG)^{-1} = G^{-1}F^{-1}. \tag{A2.1.12j}$$

Note too that if A is of order $n \times n$ with $r(A) = n$, then

$$(A')^{-1} = (A^{-1})', \tag{A2.1.12k}$$

because from $AA^{-1} = I$ we have $(A^{-1})'A' = I' = I$, and (A2.1.12k) then follows.

12. *Rank of a matrix product.* If the $n \times n$ matrix A is full, and if B is a $m \times n$ matrix with rank $r(B) = k$, then

$$r(BA) = k = r(B). \tag{A2.1.13}$$

To see this we note that (min denotes minimum)

$$r(BA) \le \min(r(B), r(A)) \le r(B). \tag{A2.1.13a}$$

(Of course, the last part of the above follows from the definition of the *min*imum of two quantities.) Now if $C = BA$, C is of order $m \times n$, and we have, since A is of full rank, that $B = CA^{-1}$. Thus,

$$r(BA) \le r(B) = r(CA^{-1}) \le r(C) \tag{A2.1.13b}$$

or

$$r(BA) \le r(B) \le r(BA), \tag{A2.1.13c}$$

which is to say

$$r(BA) = r(B). \tag{A2.1.13d}$$

We note that if *both* A and B are $n \times n$ matrices, with $r(A) = n$ and

$r(B) = k \leq n$, then, as is easily verified in the manner used above,

(A2.1.13e) $$r(BA) = r(AB) = r(B) = k.$$

It is also to be noted that if, for a $m \times n$ matrix Q,

(A2.1.13f) $$Q\mathbf{b} = Q\mathbf{c}$$

($\mathbf{b}$ and $\mathbf{c}$ are both $n \times 1$ vectors), then

(A2.1.13g) $$Q(\mathbf{b} - \mathbf{c}) = \mathbf{0}$$

and the above *does not* in general imply that

(A2.1.13h) $$\mathbf{b} - \mathbf{c} = \mathbf{0} \quad \text{or} \quad \mathbf{b} = \mathbf{c}.$$

However, if Q is of order $n \times n$ and of *full rank*, that is, $r(Q) = n$, then if (A2.1.13g) holds, we have

(A2.1.13i) $$Q^{-1}Q(\mathbf{b} - \mathbf{c}) = \mathbf{b} - \mathbf{c} = Q^{-1}\mathbf{0} = \mathbf{0}$$

or

(A2.1.13j) $$\mathbf{b} = \mathbf{c}.$$

Again, however, if Q is of order $n \times n$ with $r(Q) < n$, then it does not necessarily follow from (A2.1.13f) that $\mathbf{b} = \mathbf{c}$.

Finally, we note for this section that it may be proven that if A is $n_1 \times n_2$, with $r(A) = k \leq \min(n_1, n_2)$, then

(A2.1.13k) $$r(A) = r(A') = r(A'A) = r(AA') = k.$$

[For the proof of this statement, the interested reader is referred to Seber (1977, pp. 384–385), and/or Srivastava and Khatri (1979, pp. 9–10).]

13. *Determinants.* If A is a matrix of order $n \times n$, the determinant of A, denoted by $|A|$ or det A is

(A2.1.14) $$|A| = \det A = \Sigma'(-1)^{\sigma} a_{1i_1} a_{2i_2} \cdots a_{ni_n},$$

where $(i_1, \ldots, i_n)$ is a permutation of the n integers $(1, 2, \ldots, n)$, σ is the least number of inversions of $(1, \ldots, n)$ needed to arrive at a particular permutation $(i_1, \ldots, i_n)$ of $(1, \ldots, n)$ (see below), and Σ' denotes summation over all $n!$ permutations of $(1, \ldots, n)$. As an example of "inversions," if $n = 4$, the permutation $(1, 2, 3, 4)$ is in natural order, so that $\sigma = 0$ since no inversions are needed to bring $(1, 2, 3, 4)$ into $(1, 2, 3, 4)$. However, six inversions are needed to bring $(1, 2, 3, 4)$ into the permutation $(4, 3, 2, 1)$, and, as the reader may verify, six is the minimum number required. In fact, from $(1, 2, 3, 4)$ we have that these inversions are:

$$(1,2,4,3);\quad (1,4,2,3);\quad (4,1,2,3);\quad (4,1,3,2);\quad (4,3,1,2);\quad (4,3,2,1).$$

To illustrate (A2.1.14), suppose $n = 3$; then

$$\det A = \det\begin{pmatrix} a_{11} & a_{12} & a_{13} \\ a_{21} & a_{22} & a_{23} \\ a_{31} & a_{32} & a_{33} \end{pmatrix}$$

$$\text{(A2.1.14}a\text{)} \qquad = (-1)^0 a_{11}a_{22}a_{33} + (-1)^1 a_{11}a_{23}a_{32} + (-1)^1 a_{12}a_{21}a_{33}$$

$$= (-1)^2 a_{12}a_{23}a_{31} + (-1)^2 a_{13}a_{21}a_{32} + (-1)^3 a_{13}a_{22}a_{31}.$$

Before listing some properties of determinants, we need some additional terminology. If A is of order $n \times n$, $A = (a_{ij})$, and if we delete the ith row and jth column, we are left with a $(n-1) \times (n-1)$ matrix, M_{ij}. Then the *determinant* of M_{ij} is called the *minor* of a_{ij}. Furthermore, the *signed minor* of a_{ij} is called the *cofactor* of a_{ij} and denoted by A_{ij}, that is

$$\text{(A2.1.14}b\text{)} \qquad (-1)^{i+j}|M_{ij}| = A_{ij} = \text{cofactor of } a_{ij}.$$

Consider now the $n \times n$ matrix of cofactors, (A_{ij}), and consider the *transpose* of this matrix. Its (i-j)th element is A_{ji}, and the matrix $(A_{ij})' = (A_{ji})$ is called the adjoint of A and is denoted by adj A, that is

$$\text{(A2.1.14}c\text{)} \qquad \operatorname{adj} A = (A_{ij})' = (A_{ji}).$$

We now list properties of determinants. A denotes a $n \times n$ matrix, and so on.

13 (i). $|A| = |A'|$.

13 (ii). If two rows or columns are interchanged, the determinant changes sign.

13 (iii). If elements of a row or column are all zero, then $|A| = 0$.

13 (iv). If elements of a row or column are all multiplied by a scalar α, then the determinant is multiplied by α.

13 (v). If $r(A) = n$, then $|A| \neq 0$, but if $r(A) < n$, $|A| = 0$. If $|A| \neq 0$, we say that A is *nonsingular*, and if $|A| = 0$, we say that A is *singular*.

13 (vi). It may be proved that the following row expansions hold:

$$\sum_{j=1}^{n} a_{ij}A_{kj} = \begin{cases} |A|, & \text{if } k = i \\ 0, & \text{if } k \neq i, \end{cases} \tag{A2.1.14d}$$

$i = 1,\ldots,n$, and similarly, we have the column expansions:

$$\sum_{i=1}^{n} a_{ij}A_{ik} = \begin{cases} |A|, & \text{if } k = j \\ 0, & \text{otherwise}, \end{cases} \tag{A2.1.14e}$$

$j = 1,\ldots,n$.

13 (vii). It may be verified that, assuming $r(A) = n$ so that A^{-1} exists,

$$A^{-1} = \frac{1}{|A|} \text{adj } A. \tag{A2.1.14f}$$

13(viii). If A and B are both (square) of order n, then

$$|AB| = |A||B| = |BA| \tag{A2.1.14g}$$

and if A is nonsingular

$$|A^{-1}| = \frac{1}{|A|} \quad \text{or} \quad |A^{-1}| = (|A|)^{-1}. \tag{A2.1.14h}$$

13 (ix). If to the elements of any row (or column) we add the corresponding elements of another row (or column), each multiplied by the same number k, the determinant is unchanged. As an example, suppose $n = 3$, then

$$\det\begin{pmatrix} a_{11} & a_{12} & a_{13} \\ a_{21} & a_{22} & a_{23} \\ a_{31} & a_{32} & a_{33} \end{pmatrix} = \det\begin{pmatrix} a_{11} + ka_{12} & a_{12} & a_{13} \\ a_{21} + ka_{22} & a_{22} & a_{23} \\ a_{31} + ka_{32} & a_{32} & a_{33} \end{pmatrix}.$$

13 (x). If Q is a $n \times n$ matrix, and if $\mathbf{d}$ is a $n \times 1$ nontrivial vector [that is, $\mathbf{d}' \neq (0,\ldots,0)$], and if

$$Q\mathbf{d} = \mathbf{0}, \tag{A2.1.14i}$$

then it may be shown that $r(Q) < n$, so that $|Q| = 0$.

13 (xi). If A is of order $m \times n$ and B is $n \times m$, then

$$|I_m - AB| = |I_n - BA|. \tag{A2.1.14j}$$

We defer a proof of (A2.1.14j) for now, but see (A2.1.18n)–(A2.1.18s) of Section 15 of this appendix.

14. *Characteristic roots and vectors.* Let A be a $n \times n$ matrix. If there exists a nontrivial $\mathbf{x}$ [i.e., $\mathbf{x}' \neq (0, 0, \ldots, 0)$] and a scalar α such that $A\mathbf{x} = \alpha\mathbf{x}$, then α is called a *characteristic root* and $\mathbf{x}$ is called a *characteristic vector* of A corresponding to (the characteristic root) α. Sometimes a characteristic root and characteristic vector are referred to as *eigenroot* and *eigenvector*, or *latent* root and *latent* vector.

We note that $(A - \alpha I)\mathbf{x} = \mathbf{0}$, and since $\mathbf{x}$ is nontrivial, we have [see (A2.1.14i)]

$$|A - \alpha I| = 0 \tag{A2.1.15}$$

or

$$\det\begin{pmatrix} a_{11} - \alpha & a_{12} & \cdots & a_{1n} \\ a_{21} & a_{22} - \alpha & \cdots & a_{2n} \\ \vdots & \vdots & \ddots & \vdots \\ a_{n1} & a_{n2} & \cdots & a_{nn} - \alpha \end{pmatrix} = 0$$

or

$$a_0 + a_1\alpha + a_2\alpha^2 + \cdots + a_n\alpha^n = 0. \tag{A2.1.15a}$$

The polynomial on the left-hand side of (A2.1.15a) is often referred to as the characteristic polynomial and the equation (A2.1.15a) is called the characteristic equation. The n roots of the characteristic equation are, of course, the n characteristic roots. If we denote the characteristic roots by $\alpha_1, \ldots, \alpha_n$, we have

$$(\alpha - \alpha_1)(\alpha - \alpha_2) \cdots (\alpha - \alpha_n) = 0 \tag{A2.1.15b}$$

and from this equation we quickly see that the coefficient of $-\alpha^{n-1}$ on the right-hand side of (A2.1.15b) is $\Sigma_{i=1}^n \alpha_i$. Now (A2.1.15) is equivalent to $\det(\alpha I - A)$ having value zero, and, if we expand using (A2.1.14), it is easily seen that the coefficient of $-\alpha^{n-1}$ is $\Sigma_{i=1}^{n} a_{ii}$, that is

$$\sum_{i=1}^{n} \alpha_i = \text{tr}(A). \tag{A2.1.15c}$$

We need some terminology at this point. Two $n \times 1$ vectors $\mathbf{x}$ and $\mathbf{y}$ are *orthogonal* if their inner product is zero, that is, if $\mathbf{x}'\mathbf{y} = 0$. We will also say that the two vectors $\mathbf{x}$ and $\mathbf{y}$ are *orthonormal* if they are orthogonal and are unit vectors, that is, if $\mathbf{x}'\mathbf{y} = 0$ with $\mathbf{x}'\mathbf{x} = 1 = \mathbf{y}'\mathbf{y}$.

Additionally, we will say that the $n \times n$ matrix P is *orthogonal* if $PP' = P'P = I_n$, that is, if all rows or columns are orthonormal. Note that if P is orthogonal, then

$$P' = P^{-1} \quad \text{or} \quad P = (P^{-1})' \quad \text{and} \quad P = (P')^{-1}. \tag{A2.1.16}$$

Furthermore, we note that if P is orthogonal, then $|P||P'| = 1$; that is, $|P|^2 = 1$ or, if P is orthogonal,

$$|P| = \pm 1. \tag{A2.1.16a}$$

A class of orthogonal matrices, for example, of order $n \times n$, that are very useful are the so-called *permutation* matrices. A permutation matrix of order n may be constructed from an identity matrix I_n of order n by writing down the rows (or columns) of I_n in different sequence. For example, the 3×3 matrix P,

$$P = \begin{bmatrix} 0 & 0 & 1 \\ 1 & 0 & 0 \\ 0 & 1 & 0 \end{bmatrix}, \tag{A2.1.16b}$$

is obtained by listing the third, first, and second rows of I_3 as the first, second, and third rows of P. The designation P (for permutation) derives from the fact that, for $A = (a_{ij})$, PA is such that

$$PA = \begin{bmatrix} a_{31} & a_{32} & a_{33} \\ a_{11} & a_{12} & a_{13} \\ a_{21} & a_{22} & a_{23} \end{bmatrix}, \tag{A2.1.16c}$$

that is, PA is the matrix found by *permuting* rows 1, 2, and 3 of A so that a new matrix (PA) is obtained whose first, second, and third rows are the third, first, and second rows, respectively, of A. Note that $P'P = PP' = I$. The reader should also verify that AP' produces a matrix derived from A

with columns of A similarily permuted; namely, the first, second, and third columns of AP' are the third, first, and second columns of A.

We now list some properties of characteristic roots and vectors.

14 (i). If A is a $n \times n$ nonsingular matrix, and if $\alpha_1, \ldots, \alpha_n$ are the characteristic roots of A, then $1/\alpha_1, \ldots, 1/\alpha_n$ are the characteristic roots of A^{-1}.

14 (ii). If A is a symmetric matrix, the roots are all real.

14(iii). $\operatorname{tr} A = \sum_{i=1}^{n} \alpha_i$. [See (A2.1.15$c$).]

14(iv). $|A| = \alpha_1 \alpha_2 \cdots \alpha_n$.

14 (v). If A is a *symmetric* matrix of order $n \times n$, and such that $r(A) = n$, with no repeated characteristic roots, then the characteristic vectors corresponding to different roots are orthogonal. We have, if $\mathbf{x}_j$ is the characteristic vector of A corresponding to the characteristic root α_j, that, for example

(A2.1.17) $$A\mathbf{x}_1 - \alpha_1 \mathbf{x}_1 = \mathbf{0}$$

or

(A2.1.17a) $$\mathbf{x}_2' A \mathbf{x}_1 - \alpha_1 \mathbf{x}_2' \mathbf{x}_1 = \mathbf{x}_2' \mathbf{0} = 0.$$

Furthermore, we also have

(A2.1.17b) $$\mathbf{x}_1' A \mathbf{x}_2 - \alpha_2 \mathbf{x}_1' \mathbf{x}_2 = 0$$

and transposing on the left-hand side and using the fact that A is symmetric so that $A' = A$,

(A2.1.17c) $$\mathbf{x}_2' A \mathbf{x}_1 - \alpha_2 \mathbf{x}_2' \mathbf{x}_1 = 0.$$

Thus, from (A2.1.17a) and (A2.1.17c), we obtain

(A2.1.17d) $$(\alpha_2 - \alpha_1)\mathbf{x}_2' \mathbf{x}_1 = 0,$$

and assuming that α_1 and α_2 are different roots, we have

(A2.1.17e) $$\mathbf{x}_2' \mathbf{x}_1 = 0.$$

That is, $\mathbf{x}_1$ and $\mathbf{x}_2$ are orthogonal.

14(vi). If A is a $n \times n$ symmetric matrix, there exists an orthogonal matrix P such that

(A2.1.17f) $$P'AP = D = \begin{pmatrix} \alpha_1 & & \bigcirc \\ & \ddots & \\ \bigcirc & & \alpha_n \end{pmatrix},$$

where D is a (diagonal) matrix whose ith diagonal entry is α_i—the ith root of A (and D has zeros in the off-diagonal positions). Indeed, the columns of P are the characteristic vectors corresponding to the α_j terms, where these characteristic vectors are normalized so that they are of unit length. Note, then, that the use of (A2.1.17f) leads to proofs of 14(iii) and 14(iv) for symmetric matrices A.

15. *Matrix partitions.* If A is of order $m \times n$, then the $m \times n_1$ matrix A_1 and the $m \times n_2$ matrix A_2—where A_1 consists of n_1 columns which are the first n_1 columns of A, and A_2 consists of $n_2 = n - n_1$ columns which are the last n_2 columns of A—are called partitions of A, and we write

$$A = \left[\begin{array}{c|c} \overset{m \times n_1}{A_1} & \overset{m \times n_2}{A_2} \end{array}\right], \qquad n_1 + n_2 = n. \tag{A2.1.18}$$

Similarly, and in obvious notation, a matrix B, of order $r \times s$, may be written

$$B = \left[\begin{array}{c} \overset{r_1 \times s}{B_1} \\ \hline \overset{r_2 \times s}{B_2} \end{array}\right], \qquad r_1 + r_2 = r. \tag{A2.1.18a}$$

We note, referring to (A2.1.18), that the transpose of A may be written

$$A' = \left[\begin{array}{c} \overset{n_1 \times m}{A_1'} \\ \hline \overset{n_2 \times m}{A_2'} \end{array}\right], \qquad \text{and so on.} \tag{A2.1.18b}$$

Also, it can be verified that if C is of order $m \times n$, and D is $n \times p$, with

$$C = \left[\begin{array}{c} \overset{m_1 \times n}{C_1} \\ \hline \overset{m_2 \times n}{C_2} \end{array}\right], \qquad D = \left[\begin{array}{c|c} \overset{n \times p_1}{D_1} & \overset{n \times p_2}{D_2} \end{array}\right] \tag{A2.1.18c}$$

where $m = m_1 + m_2$ and $p = p_1 + p_2$, then the $m \times p$ matrix CD is such that

$$CD = \left[\begin{array}{c|c} C_1 D_1 & C_1 D_2 \\ \hline C_2 D_1 & C_2 D_2 \end{array}\right], \tag{A2.1.18d}$$

where C_1D_1 is of order $m_1 \times p_1$, and so on. Furthermore, if E is of order $m \times n$ and F is $n \times p$, with

(A2.1.18e)
$$E = \left[\begin{array}{c|c} m \times n_1 & m \times n_2 \\ E_1 & E_2 \end{array}\right], \qquad F = \left[\begin{array}{c} n_1 \times p \\ F_1 \\ \hline n_2 \times p \\ F_2 \end{array}\right]$$

then the $m \times p$ matrix EF is given by

(A2.1.18f)
$$EF = E_1F_1 + E_2F_2.$$

We also have that

(A2.1.18g)
$$\left\{ \begin{array}{c} \\ m_1 \\ m_2 \end{array} \overset{\begin{array}{cc} n_1 & n_2 \end{array}}{\left[\begin{array}{c|c} E_{11} & E_{12} \\ \hline E_{21} & E_{22} \end{array}\right]} \right\}_{m \times n} \left\{ \begin{array}{c} \\ n_1 \\ n_2 \end{array} \overset{\begin{array}{cc} p_1 & p_2 \end{array}}{\left[\begin{array}{c|c} F_{11} & F_{12} \\ \hline F_{21} & F_{22} \end{array}\right]} \right\}_{n \times p}$$

$$= \begin{array}{c} \\ m_1 \\ m_2 \end{array} \underset{m \times p}{\overset{\begin{array}{cc} p_1 & p_2 \end{array}}{\left[\begin{array}{c|c} E_{11}F_{11} + E_{12}F_{21} & E_{11}F_{12} + E_{12}F_{22} \\ \hline E_{21}F_{11} + E_{22}F_{21} & E_{21}F_{12} + E_{22}F_{22} \end{array}\right]}}.$$

Suppose now that A is a (square) matrix of order $n \times n$ and that $r(A)$ is n. Suppose further that A is partitioned as follows:

(A2.1.18h)
$$A = \begin{array}{c} \\ n_1 \\ n_2 \end{array} \overset{\begin{array}{cc} n_1 & n_2 \end{array}}{\left[\begin{array}{c|c} A_{11} & A_{12} \\ \hline A_{21} & A_{22} \end{array}\right]}, \qquad \text{with } A^{-1} = B = \begin{array}{c} \\ n_1 \\ n_2 \end{array} \overset{\begin{array}{cc} n_1 & n_2 \end{array}}{\left[\begin{array}{c|c} B_{11} & B_{12} \\ \hline B_{21} & B_{22} \end{array}\right]},$$

where A_{11}, A_{22}, B_{11}, and B_{22} are assumed nonsingular. Then it may be proved that

(A2.1.18i)
$$\begin{aligned} A_{11} &= \left(B_{11} - B_{12}B_{22}^{-1}B_{21}\right)^{-1} \\ A_{12} &= \left(B_{12}B_{22}^{-1}B_{21} - B_{11}\right)^{-1}B_{12}B_{22}^{-1} \\ A_{21} &= \left(B_{21}B_{11}^{-1}B_{12} - B_{22}\right)^{-1}B_{21}B_{11}^{-1} \\ A_{22} &= \left(B_{22} - B_{21}B_{11}^{-1}B_{12}\right)^{-1}. \end{aligned}$$

To see this, we have from $AB = I = AA^{-1}$, or

$$\text{(A2.1.18}j\text{)} \qquad \left[\begin{array}{c|c} A_{11} & A_{12} \\ \hline A_{21} & A_{22} \end{array}\right]\left[\begin{array}{c|c} B_{11} & B_{12} \\ \hline B_{21} & B_{22} \end{array}\right] = \left[\begin{array}{c|c} I_{n_1} & \bigcirc \\ \hline \bigcirc & I_{n_2} \end{array}\right],$$

that

$$\text{(A2.1.18}k\text{)} \qquad \begin{aligned} &\text{(i)} \quad A_{11}B_{11} + A_{12}B_{21} = I_{n_1} \\ &\text{(ii)} \quad A_{21}B_{11} + A_{22}B_{21} = \bigcirc \\ &\text{(iii)} \quad A_{11}B_{12} + A_{12}B_{22} = \bigcirc \\ &\text{(iv)} \quad A_{21}B_{12} + A_{22}B_{22} = I_{n_2}. \end{aligned}$$

From (iii) above we have that $A_{11}B_{12} = -A_{12}B_{22}$ or

$$\text{(v)} \quad A_{12} = -A_{11}B_{12}B_{22}^{-1}.$$

Substituting (v) into (i) of (A2.1.18k), we find

$$\text{(vi)} \quad A_{11}B_{11} - A_{11}B_{12}B_{22}^{-1}B_{21} = I_{n_1}$$

so that

$$\text{(vii)} \quad A_{11}(B_{11} - B_{12}B_{22}^{-1}B_{21}) = I_{n_1}$$

so that the first result of (A2.1.18i) follows. Using this result in (v) gives the second result of (A2.1.18i). The fourth and third results of (A2.1.18i) may be derived in similar fashion. (We are, of course, assuming that A_{11}, A_{22}, B_{11}, and B_{22} are all nonsingular; but see Problems 2.27 and 2.28.)

Using $BA = I$, we obtain in particular that

$$\text{(A2.1.18}l\text{)} \quad B_{21}A_{11} + B_{22}A_{21} = \bigcirc \quad \text{or} \quad B_{22}^{-1}B_{21}A_{11} + A_{21} = \bigcirc,$$

so that

$$\text{(A2.1.18}m\text{)} \qquad B_{22}^{-1}B_{21} = -A_{21}A_{11}^{-1},$$

an often used result.

We are now in a position to prove (A2.1.14j). Suppose we let the $(m+n)\times(m+n)$ matrix M be such that

$$M=\left[\begin{array}{c|c} I_m & A \\ \hline B & I_n \end{array}\right]. \tag{A2.1.18n}$$

Then it is easy to see that

$$M\cdot\left[\begin{array}{c|c} I_m & -A \\ \hline \begin{matrix} n\times m \\ \bigcirc \end{matrix} & I_n \end{array}\right]=\left[\begin{array}{c|c} I_m & \begin{matrix} m\times n \\ \bigcirc \end{matrix} \\ \hline B & I_n-BA \end{array}\right]. \tag{A2.1.18o}$$

Taking determinants on both sides of the equation and using previous properties and row expansions for determinants, we find

$$(\det M)\times(\det I_m)=\det(I_n-BA). \tag{A2.1.18p}$$

Similarly, we have

$$M\cdot\left[\begin{array}{c|c} I_m & \begin{matrix} m\times n \\ \bigcirc \end{matrix} \\ \hline -B & I_n \end{array}\right]=\left[\begin{array}{c|c} I_m-AB & A \\ \hline \begin{matrix} n\times m \\ \bigcirc \end{matrix} & I_n \end{array}\right] \tag{A2.1.18q}$$

so that

$$(\det M)\times(\det I_n)=\det(I_m-AB) \tag{A2.1.18r}$$

which together with (A2.1.18p) yields

$$\det(I_m-AB)=\det(I_n-BA)=\det M. \tag{A2.1.18s}$$

Another useful result for determinants of partitioned matrices [see Problems 2.7(c) and 2.7(d)] is

$$\det\left[\begin{array}{c|c} A & B \\ \hline C & D \end{array}\right]=(\det D)\times\left[\det(A-BD^{-1}C)\right] \tag{A2.1.18t}$$

We conclude this section by stating a very useful lemma quoted in Tocher (1952).

Lemma A2.1.1. *If A and B are of order $m \times n$ and $n \times m$, respectively, then*

$$(I_m - AB)^{-1} = I_m + A(I_n - BA)^{-1}B. \tag{A2.1.18u}$$

Proof. From the identity $I - BA = I - BA$ we quickly find that

$$I = (I - BA)^{-1} - (I - BA)^{-1}BA,$$

and upon pre- and post-multiplying both sides by A and B, respectively, and rearranging we have

$$\bigcirc = -AB + A(I - BA)^{-1}B - A(I - BA)^{-1}BAB.$$

Adding an identity of order m to both sides, we find, after some algebra, that

$$I = I - AB + A(I - BA)^{-1}B(I - AB),$$

so that post-multiplication by $(I - AB)^{-1}$ on both sides of the above yields the desired result. (For another interesting proof of this result, see Problem 2.27.)

We remark that if $n < m$, the result (A2.1.18u) states that in order to find the inverse of a certain $m \times m$ matrix, we need only find the inverse of a $n \times n$ matrix. The result also has many useful theoretical applications.

Corollary A2.1.1.1. *Assuming the products and inverses mentioned exist, then*

$$(C - EF)^{-1} = C^{-1} + C^{-1}E(I - FC^{-1}E)^{-1}FC^{-1}. \tag{A2.1.18v}$$

The proof is left to the reader.

16. *Positive definite quadratic forms and matrices.* Q is said to be a *quadratic form* in the n variables $(x_1, \ldots, x_n)$ if

$$Q = \sum_{i=1}^{n} \sum_{j=1}^{n} a_{ij}x_ix_j = \mathbf{x}'A\mathbf{x}. \tag{A2.1.19}$$

We refer to $A = (a_{ij})$ as the matrix of the quadratic form and we will assume that A is symmetric.

A matrix A is said to be *positive definite* if the quadratic form $Q = \mathbf{x}'A\mathbf{x} > 0$ for all $\mathbf{x} \neq \mathbf{0}$ (Q is said to be positive definite if $Q > 0$ for all $\mathbf{x} \neq \mathbf{0}$). A matrix A is *positive semidefinite* if $Q = \mathbf{x}'A\mathbf{x} \geq 0$ for all $\mathbf{x} \neq \mathbf{0}$ (and we say that Q is a positive semidefinite quadratic form).

Some examples, by way of illustration, are:

(a) $$n = 2: \quad Q_1 = \mathbf{x}'\begin{pmatrix} 3 & -1 \\ -1 & 3 \end{pmatrix}\mathbf{x} = 3x_1^2 + 3x_2^2 - 2x_1x_2$$

is positive definite, that is,

$$A_1 = \begin{pmatrix} 3 & -1 \\ -1 & 3 \end{pmatrix}$$

is positive definite, since we may write, after some algebra,

$$Q_1 = 3\left[\left(x_1 - \frac{x_2}{3}\right)^2 + \frac{8x_2^2}{9}\right]$$

which is clearly greater than zero for all $\mathbf{x} = (x_1, x_2)' \neq (0, 0)$.

(b) $$n = 2: \quad Q_2 = \mathbf{x}'\begin{pmatrix} 1 & -1 \\ -1 & 1 \end{pmatrix}\mathbf{x}$$

is positive semidefinite, that is,

$$A_2 = \begin{pmatrix} 1 & -1 \\ -1 & 1 \end{pmatrix}$$

is positive semidefinite, since we may write

$$Q_2 = x_1^2 - 2x_1x_2 + x_2^2 = (x_1 - x_2)^2 \geq 0$$

for all $(x_1, x_2) \neq (0, 0)$, and so on.

Suppose now that A is positive definite. Then we note the following:

16 (i). All diagonal elements are positive. To see this, suppose, for example, we put $\mathbf{x} = (1, 0, \ldots, 0)'$. Then $\mathbf{x}'A\mathbf{x} > 0$ implies $a_{11} > 0$, and so on.

16 (ii). All principal submatrices are positive definite.

16(iii). A^{-1} is positive definite.

16(iv). $|A| > 0$ and all characteristic roots α_j are positive so that

$$|A| = \alpha_1 \times \cdots \times \alpha_n.$$

17. *Differentiation of matrices, and linear and quadratic forms.* If $f = \mathbf{a}'\mathbf{x} = \mathbf{x}'\mathbf{a} = x_1 a_1 + \cdots + x_n a_n$, then f is sometimes said to be a linear form in the n variables x_j, with coefficients a_j. It is easy to see that

$$\frac{\partial f}{\partial x_i} = a_i \quad \text{or} \quad \frac{\partial f}{\partial x_i} = \mathbf{a}'\frac{\partial \mathbf{x}}{\partial x_i} = \frac{\partial \mathbf{x}'}{\partial x_i}\mathbf{a} \tag{A2.1.20}$$

where $\partial \mathbf{x}/\partial x_i$ is a $n \times 1$ vector of zeros, except for the ith component which has the value 1.

Now, similarly, if X is of order $m \times n$, for example, $X = (x_{ij})$, then $\partial X/\partial x_{ij}$ is a $m \times n$ matrix of zeros, except for the $(i\text{-}j)$th element which has value 1.

Now suppose we consider $Q = \mathbf{x}'A\mathbf{x}$, where A is of order $n \times n$, and so on. Then

$$\frac{\partial Q}{\partial x_i} = \frac{\partial \mathbf{x}'}{\partial x_i}A\mathbf{x} + \mathbf{x}'A\frac{\partial \mathbf{x}}{\partial x_i}. \tag{A2.1.20a}$$

But

(A2.1.20b)

$$\frac{\partial \mathbf{x}'}{\partial x_i}A = (0,0,\ldots,0,1,0,\ldots,0)\begin{bmatrix} a_{11} & \cdots & a_{1i} & \cdots & a_{1n} \\ & \vdots & & \vdots & \\ a_{i1} & \cdots & a_{ii} & \cdots & a_{in} \\ & \vdots & & \vdots & \\ a_{n1} & \cdots & a_{ni} & \cdots & a_{nn} \end{bmatrix},$$

and it is easy to see that

$$\frac{\partial \mathbf{x}'}{\partial x_i}A = (a_{i1},\ldots,a_{in}) = \mathbf{a}'_{i\cdot} \tag{A2.1.20c}$$

and $\mathbf{a}'_{i\cdot}$ is the ith row of A. Similarly,

$$A\frac{\partial \mathbf{x}}{\partial x_i} = \begin{bmatrix} a_{11} & \cdots & a_{1i} & \cdots & a_{1n} \\ & \vdots & & \vdots & \\ a_{i1} & \cdots & a_{ii} & \cdots & a_{in} \\ & \vdots & & \vdots & \\ a_{n1} & \cdots & a_{ni} & \cdots & a_{nn} \end{bmatrix}\begin{pmatrix} 0 \\ \vdots \\ 0 \\ 1 \\ 0 \\ \vdots \\ 0 \end{pmatrix} = \mathbf{a}_{\cdot i} \tag{A2.1.20d}$$

and $\mathbf{a}_{\cdot i}$ is the ith column of A, so that

(A2.1.20e)
$$\frac{\partial Q}{\partial x_i} = \mathbf{a}'_{i\cdot}\mathbf{x} + \mathbf{x}'\mathbf{a}_{\cdot i}.$$

Now if A is symmetric, $\mathbf{a}'_{\cdot i} = \mathbf{a}'_{i\cdot}$ so that

(A2.1.20f)
$$\frac{\partial Q}{\partial x_i} = 2\mathbf{a}'_{i\cdot}\mathbf{x} = 2\mathbf{x}'\mathbf{a}_{\cdot i}.$$

APPENDIX A2.2. IDEMPOTENT MATRICES

1. *A Definition and some properties.* We will begin with the following definition.

Definition A2.2.1. *A matrix A, of order $m \times m$, is said to be idempotent if*

(A2.2.1)
$$AA = A.$$

We note that if A is *idempotent*, each of its eigenroots is either 0 or 1. To see this, let α denote the eigenroot of A corresponding to the eigenvector $\mathbf{x}$, that is,

(A2.2.2)
$$A\mathbf{x} = \alpha\mathbf{x}.$$

Then, since A is idempotent, we have that

(A2.2.3)
$$\alpha\mathbf{x} = A\mathbf{x} = AA\mathbf{x} = A\alpha\mathbf{x} = \alpha A\mathbf{x} = \alpha^2\mathbf{x},$$

that is,

(A2.2.3a)
$$(\alpha - \alpha^2)\mathbf{x} = \mathbf{0} \quad \text{or} \quad (\alpha - \alpha^2) = 0$$

so that each of the roots of A is either 0 or 1. The converse need not be true. A simple example is

(A2.2.3b)
$$A = \begin{pmatrix} 1 & 1 \\ -1 & -1 \end{pmatrix}$$

which has roots both equal to zero, but $AA \neq A$. However, we have the following lemma.

Lemma A2.2.1. *If A is a symmetric $m \times m$ matrix, then a necessary and sufficient condition for A to be idempotent is that each of the eigenroots of A is* 0 *or* 1.

Proof. (i) *Necessity.* This has been proved above—see (A2.2.2)–(A2.2.3a).

(ii) *Sufficiency.* Here we assume that the roots of the symmetric matrix A are 0's or 1's. Now since A is symmetric, there exists an orthogonal $m \times m$ matrix, U, such that

$$U'AU = \Lambda = \begin{bmatrix} 1 & & & & & & \\ & 1 & & & & & \\ & & \ddots & & & & \\ & & & 1 & & & \\ & & & & 0 & & \\ & & & & & \ddots & \\ & & & & & & 0 \end{bmatrix} \tag{A2.2.4}$$

where Λ is the diagonal matrix of the roots of A, assumed to be 0's and 1's. Then

$$AA = U\Lambda U'U\Lambda U' = U\Lambda\Lambda U' = U\Lambda U' = A, \tag{A2.2.5}$$

so that A is idempotent, and the sufficiency part is proved.

Note that

$$r(A) = \operatorname{tr}(A). \tag{A2.2.6}$$

This may be seen easily from the fact that the rank of any square matrix A is the number of nonzero roots, and thus for A idempotent (so that the roots are 1's or 0's), the rank is clearly the sum of the roots. But for any square matrix $A = (a_{ij})$ of order m, with roots α_i,

$$\sum_{i=1}^{m} \alpha_i = \sum_{i=1}^{m} a_{ii} = \operatorname{tr}(A), \tag{A2.2.6a}$$

since $\sum_{i=1}^{m} a_{ii}$ is the coefficient of $-\alpha^{m-1}$ in the expansion of $\det(\alpha I - A) = 0$. Hence, for A idempotent,

$$r(A) = \sum_{i=1}^{m} \alpha_i = \operatorname{tr}(A), \tag{A2.2.6b}$$

proving (A2.2.6).

Also, note that the only *nonsingular idempotent* $m \times m$ matrix is I_m—the identity matrix of order m. For after all, we have from (A2.3.1) that if A is nonsingular and idempotent,

$$A^{-1}AA = A^{-1}A; \tag{A2.2.7}$$

that is,

$$A = I_m. \tag{A2.2.7a}$$

It is of interest to note also that if A is idempotent, then the matrix $I_m - A$ is also idempotent, since we have

$$(I - A)(I - A) = I - A - A + AA = I - A \tag{A2.2.8}$$

since $AA = A$. That is, $I - A$ is idempotent if A is idempotent.

The reader may verify that the matrices A_j below are idempotent, and that if $I - A_j$ is formed that these too are idempotent.

$$\text{(i)}\quad A_1 = \frac{1}{3}\begin{bmatrix} 2 & -1 & -1 \\ -1 & 2 & -1 \\ -1 & -1 & 2 \end{bmatrix}, \qquad \text{(ii)}\quad A_2 = \frac{1}{3}\begin{bmatrix} 1 & 1 & 1 \\ 1 & 1 & 1 \\ 1 & 1 & 1 \end{bmatrix},$$

$$\text{(iii)}\quad A_3 = \frac{1}{3}\begin{bmatrix} 1 & 0 & -\sqrt{2} \\ 0 & 0 & 0 \\ -\sqrt{2} & 0 & 2 \end{bmatrix}, \qquad \text{(iv)}\quad A_4 = \frac{1}{3}\begin{bmatrix} 2 & 0 & \sqrt{2} \\ 0 & 3 & 0 \\ \sqrt{2} & 0 & 1 \end{bmatrix},$$

$$\text{(v)}\quad A_5 = \begin{bmatrix} 2 & 4 & 6 \\ 4 & 8 & 12 \\ -3 & -6 & -9 \end{bmatrix}.$$

We will return to idempotent matrices in Section 3 below. We turn now to the following topics.

2. *Spans, image spaces, and projections.* The reader will recall for x_j and y_j real, $j = 1,\ldots,n$, that $\mathbf{x} = (x_1,\ldots,x_n)'$ and $\mathbf{y} = (y_1,\ldots,y_n)'$ denote (column) vectors in R^n. R^n can be made into an Euclidean space by defining an inner product that gives meaning to "length" and "angle." Recall that the inner product $\mathbf{x}'\mathbf{y}$ of $\mathbf{x}$ and $\mathbf{y}$ is defined to be

$$\mathbf{x}'\mathbf{y} = \sum_{i=1}^{n} x_i y_i = \mathbf{y}'\mathbf{x}. \tag{A2.2.9}$$

We may further define the length $|\mathbf{x}|$ of the vector $\mathbf{x}$ as

$$|\mathbf{x}| = (\mathbf{x}'\mathbf{x})^{1/2} = \left(\sum_{1}^{n} x_i^2 \right)^{1/2}. \tag{A2.2.9a}$$

Furthermore, we define the angle θ between the vectors $\mathbf{x}$ and $\mathbf{y}$ by

$$\theta = \cos^{-1}\left(\frac{\mathbf{x}'\mathbf{y}}{|\mathbf{x}|\,|\mathbf{y}|} \right); \tag{A2.2.10}$$

that is, $\cos\theta$ is the inner product of the unit vectors $(1/|\mathbf{x}|)\mathbf{x}$ and $(1/|\mathbf{y}|)\mathbf{y}$. We state, it may be recalled, that $\mathbf{x}$ and $\mathbf{y}$ are orthogonal if $\cos\theta = 0$, that is, if $\mathbf{x}'\mathbf{y} = 0$. (These definitions correspond to reality in R^2 and R^3.)

Now in R^n, we can find n linearly independent vectors, $\mathbf{v}_1, \ldots, \mathbf{v}_n$, and write any other vector $\mathbf{x}$ as a linear combination of these; that is,

$$\mathbf{x} = a_1\mathbf{v}_1 + \cdots + a_n\mathbf{v}_n. \tag{A2.2.11}$$

The vectors $\mathbf{v}_1, \ldots, \mathbf{v}_n$ are then said to be a *basis* for R^n. The standard choice for such a set of basis vectors is

$$\mathbf{v}_1 = \begin{pmatrix} 1 \\ 0 \\ 0 \\ \vdots \\ 0 \end{pmatrix}, \quad \mathbf{v}_2 = \begin{pmatrix} 0 \\ 1 \\ 0 \\ \vdots \\ 0 \end{pmatrix}, \quad \ldots, \quad \mathbf{v}_n = \begin{pmatrix} 0 \\ \vdots \\ 0 \\ 1 \end{pmatrix}, \tag{A2.2.11a}$$

but of course there is an infinite number of basis sets available. We note that for the set (A2.2.11*a*), the $\mathbf{v}_j$ are linearly independent and

$$\mathbf{v}_i'\mathbf{v}_j = \begin{cases} 1, & \text{if } i = j \\ 0, & \text{if } i \neq j. \end{cases} \tag{A2.2.11b}$$

Now any k linearly independent vectors $\mathbf{x}_1, \ldots, \mathbf{x}_k$, where $\mathbf{x}_1, \ldots, \mathbf{x}_k$ are vectors in R^n and $k \le n$, are said to *span* a k-dimensional subspace S_k, $S_k \subset R^n$.

Suppose we now consider the $n \times n$ matrix A. We have the following terminology.

2(i). $\mathfrak{M}$ is said to be the *column* or *image space* of A, $\mathfrak{M} \subset R^n$, if a $n \times 1$ vector $\mathbf{z}$ belonging to $\mathfrak{M}$ is such that $\mathbf{z} = A\mathbf{x}$, where $\mathbf{x}$ is a $n \times 1$ vector in R^n. (We note that $\mathbf{z}$ is a linear combination of the columns of A, that is, $\mathbf{z} = \mathbf{a}_{\cdot 1}x_1 + \cdots + \mathbf{a}_{\cdot n}x_n$.) In symbols, we have that

$$\mathfrak{M} = \{\mathbf{z} | \mathbf{z} = A\mathbf{x}, \mathbf{x} \in R^n\}.$$

2(ii). The vector $\mathbf{x}$ is said to be *mapped* or *transformed* into the vector $\mathbf{z}$ by the matrix A if $\mathbf{z} = A\mathbf{x}$. [We sometimes speak of the point P in R^n, whose coordinates are $(x_1, \ldots, x_n) = \mathbf{x}'$, being mapped into the point Q whose coordinates are $\mathbf{z}'$, where $\mathbf{z} = A\mathbf{x}$.] This mapping of $\mathbf{x}'$ into $\mathbf{z}'$, that is, of $\mathbf{x}$ into $\mathbf{z}$, is sometimes referred to as the *projection* of $\mathbf{x}'$ onto $\mathbf{z}'$.

2(iii). Consider the matrix A and its column or image space $\mathfrak{M}$. We say that A projects R^n *orthogonally* onto $\mathfrak{M}$ if $\mathbf{x} - A\mathbf{x}$ is orthogonal to each vector in $\mathfrak{M}$, for all $\mathbf{x}$, and we then say that A defines an *orthogonal projection*.

In Figure A2.2.1, we present a geometrical interpretation of an orthogonal projection. The matrix A maps the point P of R^n whose coordinates are $\mathbf{x}'$ to the point Q in $\mathfrak{M}$, where Q has coordinates $\mathbf{z}' = (A\mathbf{x})'$, in such a way that the so-called residual vector from P to Q, that is, $\mathbf{x} - \mathbf{z} = \mathbf{x} - A\mathbf{x} = (I - A)\mathbf{x}$, is orthogonal to each vector in $\mathfrak{M}$, and in particular to the vector $\mathbf{z} = A\mathbf{x}$. We then say that A defines an orthogonal projection.

3. *Idempotent matrices and orthogonal projections.* An interesting fact about orthogonal projections is their intimate connection with idempotent matrices, as given in the following important theorem.

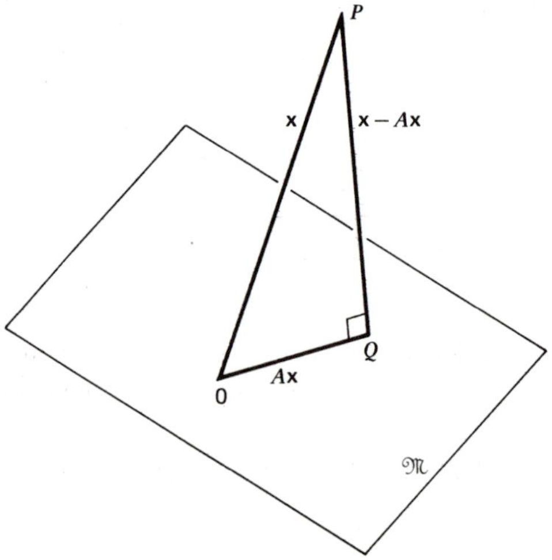

FIGURE A2.2.1. Geometrical interpretation of an orthogonal projection of the point P into $\mathfrak{M}$, the column on image space of A.

Theorem A 2.2.1. *The $n \times n$ matrix A defines an orthogonal projection if and only if A is symmetric and idempotent.*

Proof. (i) *Necessity.* Suppose A projects R^n orthogonally onto $\mathfrak{M}$. Then for any $\mathbf{x} \in R^n$, the residual vector $\mathbf{x} - A\mathbf{x}$ is orthogonal to $\mathfrak{M}$ and, in particular, orthogonal to the vector $A\mathbf{y}$, where $\mathbf{y}$ is any vector belonging to R^n. That is,

$$(A\mathbf{y})'(\mathbf{x} - A\mathbf{x}) = \mathbf{0} \tag{A2.2.12}$$

or

$$\mathbf{y}'A'\mathbf{x} = \mathbf{y}'A'A\mathbf{x} \tag{A2.2.12a}$$

for all $\mathbf{x}, \mathbf{y} \in R^n$. Thus,

$$A' = A'A \quad \text{or} \quad A = A'A; \tag{A2.2.12b}$$

that is,

$$A' = A = A'A \tag{A2.2.12c}$$

so that A is *symmetric*. Since $A' = A$,

$$A = A'A = AA, \tag{A2.2.12d}$$

so that A, which is symmetric, is also idempotent.

(ii) *Sufficiency.* Here we assume A is such that $A = A'$ and $A = AA$. Hence $A' = A'A$ so that

$$\mathbf{y}'A'\mathbf{x} = \mathbf{y}'A'A\mathbf{x} \tag{A2.2.12e}$$

for any vectors $\mathbf{x}, \mathbf{y} \in R^n$, so that

$$(A\mathbf{x})'(\mathbf{x} - A\mathbf{x}) = 0. \tag{A2.2.12f}$$

That is, $A\mathbf{x}$ is orthogonal to $\mathbf{x} - A\mathbf{x}$, so that A defines an orthogonal projection.

3 The Linear Regression Model: Full-Rank Case

Much experimentation deals with the problem of obtaining information about the functional relationship between variables. For example, it may be of interest to have information about how, in a certain process, the temperature, time, pressure, and amount of catalyst used affect the yield of a certain by-product. The experimenter may postulate a relationship of the following form:

$$\eta = f(\mathbf{z}, \boldsymbol{\theta}), \tag{3.1}$$

where η is the yield (in kilograms, e.g.) of the by-product when the process is run at a temperature of z_1 (degrees centigrade) for a time z_2 (hours) at a pressure of z_3 [in pascals (newtons divided by meters squared)] and in the presence of z_4 liters of the catalyst. We usually term η the *response variable* and think of it as the response to the variables z_j, which are referred to as the *factors* of the experiment. The components of $\boldsymbol{\theta}$ are unknown constants linking the effects of the z_j to η and are called parameters.

In general, there may be s factors that affect the response variable so that $\mathbf{z}$ is a $s \times 1$ vector. Now in many situations, we can write (3.1) as a linear form in parameters which we will denote as $\boldsymbol{\beta}$ [see (1.1) of Chapter 1], that is,

$$\eta = f(\mathbf{z}, \boldsymbol{\theta}) = \sum_{i=1}^{k} \beta_i x_i, \tag{3.2}$$

where the x_i are functions of the z_j. For example, in a certain investigation the postulated model may be

$$\eta = \theta_0 + \theta_1(z_1 - z_2) + \theta_2 z_3^2, \tag{3.3}$$

and this may be rewritten as

$$\eta = \beta_1 x_1 + \beta_2 x_2 + \beta_3 x_3 \tag{3.3a}$$

with

$$\begin{aligned} &x_1 = 1, \quad x_2 = z_1 - z_2, \quad x_3 = z_3^2 \\ &\beta_1 = \theta_0, \quad \beta_2 = \theta_1, \quad \beta_3 = \theta_2. \end{aligned} \tag{3.3b}$$

Note that in the (linear) formulation of (3.2), k may be large or small. This linear formulation of the response functional relationship of η to various factors z_j under the experimenter's control is very important for several reasons. *Firstly*, it is important because very often the relationship may be written exactly as in (3.2). *Secondly*, if f is not linear in the parameters, then it often turns out that in a sufficiently small range of the factors $z_1,\ldots,z_s$, and hence of $x_1,\ldots,x_k$, where the ranges cover the usual set of values employed in practice or experimentation (called the region of experimentation in factor space), then f is, to good approximation, suitably described by a linear form of the type (3.2). *Lastly*, the precise form of f is often unknown but can be represented by a graduating function, for example, by a polynomial in the z_j's, and polynomials are trivially of the linear form (3.2).

At this point, the experimenter may be interested in either of the following situations (compare Sections 1.1 and 1.7 of Chapter 1).

(i) Given that (3.2) holds, or is, to good approximation, an adequate representation over the region of experimentation in factor space, *estimate* the parameter $\boldsymbol{\beta}$ by "running experiments." The running of experiments implies that for each of N given sets of values $z_{u1},\ldots,z_{us}$, and hence $x_{u1},\ldots,x_{uk}$, $u = 1,\ldots,N$, an observation y_u is made on the response variable η. We note here that the observation y_u is subject to experimental error, and, in fact, that

$$y_u - \sum_{1}^{k} \beta_j x_{uj} = y_u - \eta_u = \varepsilon_u, \tag{3.4}$$

where we assume that ε_u are (continuous) random experimental errors, identically and independently distributed, with mean 0 and variance σ^2.

That is,

$$(3.4a) \qquad y_u = \eta_u + \varepsilon_u = \sum_1^k \beta_j x_{uj} + \varepsilon_u, \qquad u = 1,\ldots,N$$

so that

$$(3.4b) \qquad E(y_u) = \eta_u, \qquad V(y_u) = \sigma^2$$

where, to repeat, the y_u's (or ε_u's) are independent. The description (3.2), (3.4a), and (3.4b) is often termed the *linear regression model.* For brevity we will refer to $x_1,\ldots,x_k$ as the *factors*. (As mentioned in Chapter 1, we could initially assume the y_u's are mutually uncorrelated. If we also assume normality, the y_u's are then mutually independent.)

There are many reasons for the presence of experimental error. One situation that occurs often is due to the fact that an experimenter cannot control all the factors that may affect the response, so that indeed, we may have observations y on the response such that

$$(3.5) \qquad y = \sum_{j=1}^{k} \beta_j x_j + \sum_{k+1}^{t} \beta_j x_j.$$

By controlling only $(x_1,\ldots,x_k)$, it is apparent that an observation y on the response is made up of

$$(3.6) \qquad y = \sum_{i=1}^{k} \beta_j x_j + \varepsilon$$

where ε denotes experimental error, induced because the terms in $(x_{k+1},\ldots,x_t)$ cannot be taken into account. These factors may change values each time we run an experiment, which is done by choosing *factor levels* for $(x_1,\ldots,x_k)$, such as $(x_{u1},\ldots,x_{uk})$ for the uth experiment. Indeed, the effect of the factors $(x_{k+1},\ldots,x_t)$ may vary at random and, if $t-k$ is large, a central limit effect would be built up: This is a justification for the normality assumption which at certain points later on will be imposed. In any event, we make the assumption that ε is a random variable, with mean 0 and variance σ^2 and indeed that (3.4a) and (3.4b) hold.

(ii) There may be another interest of the experimenter or at least a collateral one to that of estimating the β_j's. The question may be asked: If for the moment the assumption

$$E(y_u) = \eta_u = \sum_{j=1}^{k} \beta_j x_{uj} \tag{3.6a}$$

is made, and we estimate the β_j's, thereby fitting the model (3.6), say

$$\hat{\eta} = \sum_{j=1}^{k} \hat{\beta}_j x_j, \tag{3.6b}$$

then can we examine the assumption (3.6)? Put in a more concise way, can we examine the question: *Is the model right*? We sometimes refer to (3.6*b*) as the fitted regression plane, where $\hat{\beta}_j$ is an estimate of β_j. We will look at both these problems (i) and (ii), and the reader is invited to reread Sections 1.1 and 1.7 of Chapter 1. We assume, unless otherwise stated, that

$$N > k. \tag{3.6c}$$

(iii) A third interest of the experimenter might be to *predict*, on the basis of data y_u observed at $(x_{u1},\ldots,x_{uk})$, $u = 1,\ldots,N$, what a future response may be when $(x_1,\ldots,x_k)$ is set at $(x_{01},\ldots,x_{0k}) = \mathbf{x}'_0$. We have dealt with this problem for the simple model of Chapter 1, Section 1.8. We refer the reader to Problem 3.53 for the case of prediction in the more general linear model situation, after completing the reading of this chapter.

3.1. LEAST SQUARES: SAMPLE, ESTIMATION, AND ERROR SPACES

As we have mentioned, when the experimenter is satisfied that the model (3.4*a*) and (3.4*b*) is the one that should be used, then the taking of data implies the setting of the factors $(x_1,\ldots,x_k)$ at $(x_{u1},\ldots,x_{uk})$ and observing the corresponding response—unhappily in the presence of experimental error. We suppose this will be done N times, that is, $u = 1,\ldots,N$, in such a way as to guarantee that the y_u's are uncorrelated. We may record the selection of the sets of $(x_1,\ldots,x_k)$, that is, $(x_{u1},\ldots,x_{uk})$, $u = 1,\ldots,N$, in a matrix, X. That is, X is of order $N \times k$ and such that

$$X = \begin{bmatrix} x_{11} & x_{12} & \cdots & x_{1k} \\ \vdots & \vdots & & \vdots \\ x_{u1} & x_{u2} & \cdots & x_{uk} \\ \vdots & \vdots & & \vdots \\ x_{N1} & x_{N2} & \cdots & x_{Nk} \end{bmatrix}. \tag{3.7}$$

The uth row of the matrix X is the uth selection of the k values of $x_1,\ldots,x_k$ used to generate the uth observation y_u. Now clearly the N observations $(y_1,\ldots,y_N) = \mathbf{y}'$ may be thought of as a point in the N-dimensional sample space

$$(3.8) \qquad R^N = \{\mathbf{y} \mid -\infty < y_u < \infty, u = 1,\ldots,N\},$$

and in matrix notation, our model (3.4a) may be rewritten

$$(3.9) \qquad \mathbf{y} = X\boldsymbol{\beta} + \boldsymbol{\varepsilon},$$

where $\boldsymbol{\varepsilon} = (\varepsilon_1,\ldots,\varepsilon_u,\ldots,\varepsilon_N)'$ are the N experimental errors which are assumed to be such that

$$(3.9a) \qquad E(\boldsymbol{\varepsilon}) = \mathbf{0}, \qquad V(\boldsymbol{\varepsilon}) = \sigma^2 I_N,$$

so that

$$(3.9b) \qquad E(\mathbf{y} \mid X) = X\boldsymbol{\beta} = \boldsymbol{\eta} \quad \text{and} \quad V(\mathbf{y}) = \sigma^2 I_N.$$

Often, the vector $\boldsymbol{\beta}$ is called the vector of *regression parameters*. Note also that (3.7) may be thought of as

$$(3.10) \qquad X = (\mathbf{x}_1,\ldots,\mathbf{x}_j,\ldots,\mathbf{x}_k)$$

where the $N \times 1$ column vector $\mathbf{x}_j$ has as its components the N settings of the jth factor x_j. In this chapter we are assuming that the k vectors $\mathbf{x}_j$ are linearly independent, or, in more familiar terminology, the rank of the $N \times k$ matrix X, denoted by $r(X)$, is such that

$$(3.11) \qquad r(X) = k < N.$$

When (3.11) holds we say that we are dealing with the *full rank model*. We note that the sample space of the N-variate random variable $\mathbf{y}$ is R^N. Now when the full rank model is appropriate, we say that the vectors $\mathbf{x}_j$, $j = 1,\ldots,k$, *span* a k-dimensional subspace S_k, where $S_k \subset R^N$, is of the form

$$(3.12) \qquad S_k = \{\theta_1\mathbf{x}_1 + \cdots + \theta_k\mathbf{x}_k \mid \theta_1,\ldots,\theta_k \in R\}.$$

Any point $\boldsymbol{\tau}' = (\tau_1,\ldots,\tau_N)$ belongs to the subspace S_k if it is a linear combination of $\mathbf{x}_1,\ldots,\mathbf{x}_k$, that is, if we may write $\boldsymbol{\tau}$ in the form

$$\boldsymbol{\tau} = \sum_1^k \theta_j \mathbf{x}_j, \tag{3.12a}$$

for some real numbers $\theta_1,\ldots,\theta_k$. The subspace S_k so generated is often called the *regression* or *parameter plane* or *solution surface*, and we will most often refer to it as the *estimation space*. What interests us at the moment in the estimation space is that the point $\boldsymbol{\eta}$, where

$$\boldsymbol{\eta} = E(\mathbf{y} \mid X) = X\boldsymbol{\beta} = \beta_1 \mathbf{x}_1 + \cdots + \beta_k \mathbf{x}_k, \tag{3.13}$$

is obviously a point in this *estimation space* (see Figure 3.1.1). We illustrate with a simple example.

Example 3.1.1.

Suppose $E(y) = \beta_0 + \beta_1 x$, and that N experiments are run at $x = x_1,\ldots,x_N$. Then

$$X = \begin{bmatrix} 1 & x_1 \\ \vdots & \vdots \\ 1 & x_u \\ \vdots & \vdots \\ 1 & x_N \end{bmatrix} = [\mathbf{1}, \mathbf{x}]. \tag{3.13a}$$

If $N = 3$, then $\mathbf{x}_1 = (1, 1, 1)'$ and $\mathbf{x}_2 = \mathbf{x} = (x_1, x_2, x_3)'$, and

$$S_2 = \{\boldsymbol{\tau} = \theta_1 \mathbf{x}_1 + \theta_2 \mathbf{x}_2 = \theta_1 \mathbf{1}_3 + \theta_2 \mathbf{x} \mid \theta_1, \theta_2 \in R\}, \tag{3.14}$$

which is a plane (two dimensions) in R^3 and contains the point $\boldsymbol{\eta} = \beta_0 \mathbf{1} + \beta_1 \mathbf{x}$.

In general, if there is no experimental error, then we would have $y_u = \eta_u = \sum_{j=1}^k \beta_j x_{uj}$, that is, the point

$$(y_1,\ldots,y_u,\ldots,y_N) = (\eta_1,\ldots,\eta_u,\ldots,\eta_N) \tag{3.15}$$

would be in the subspace S_k. But of course the y_u's are subject to experimental error, and we have

$$\mathbf{y} = \sum_{j=1}^k \beta_j \mathbf{x}_j + \boldsymbol{\varepsilon} \tag{3.16}$$

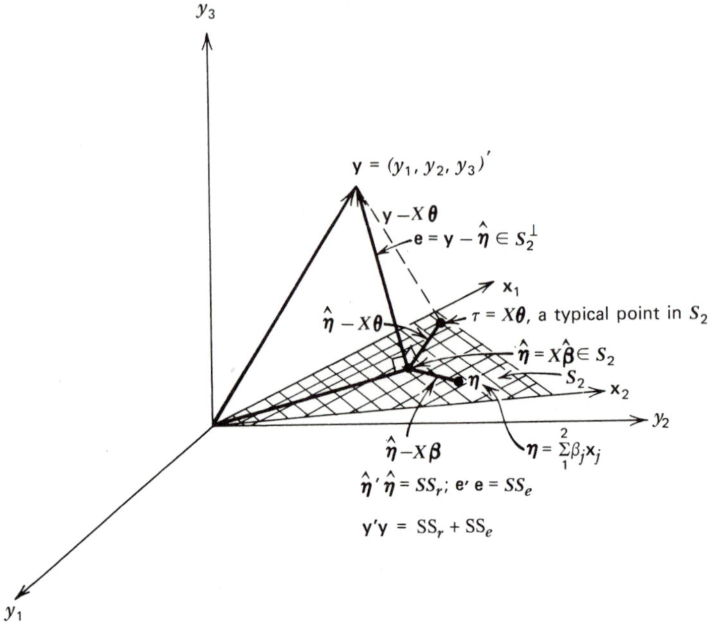

FIGURE 3.1.1. Sample space R^3 containing the subspace $S_2 = \{\boldsymbol{\tau} = X\boldsymbol{\theta} = \theta_1\mathbf{x}_1 + \theta_2\mathbf{x}_2 \mid \theta_1, \theta_2 \in R\}$. In the diagram we are supposing that $\mathbf{y} = (y_1, y_2, y_3)'$ has been observed and plotted as shown.

or

$$(3.16a) \qquad y_u = \eta_u + \varepsilon_u = \sum_{j=1}^{k} \beta_j x_{uj} + \varepsilon_u$$

so that, with probability 1 (ε_u are continuous random variables), the point $\mathbf{y}$ does not lie in S_k. This being the case, we may wish to choose a point in the *estimation space* S_k which is in some sense the *best representation*, that is, it serves as *best estimator* of $\boldsymbol{\eta} = X\boldsymbol{\beta}$.

A natural choice, whose properties turn out to be highly desirable, is the orthogonal projection of $\mathbf{y}$ into the estimation space; we denote this projection by $\hat{\boldsymbol{\eta}}$. (We do this since we will use this as an estimator of $\boldsymbol{\eta}$.) Since this projection lies in the estimation space, we may write it as

$$(3.17) \qquad \hat{\boldsymbol{\eta}} = \sum_{j=1}^{k} \hat{\beta}_j \mathbf{x}_j = X\hat{\boldsymbol{\beta}}$$

where $\hat{\boldsymbol{\beta}} = (\hat{\beta}_1, \ldots, \hat{\beta}_k)'$ (see Figure 3.1.1, for the case $k = 2$). We note that because of the above construction, $\mathbf{y} - \hat{\boldsymbol{\eta}}$ is orthogonal to S_k, which means that $\mathbf{y} - \hat{\boldsymbol{\eta}}$ is orthogonal to all vectors lying in S_k. This means, in particular, that $\mathbf{y} - \hat{\boldsymbol{\eta}}$ is orthogonal to the $\mathbf{x}_j$, for all j, so that we may write

$$X'(\mathbf{y} - \hat{\boldsymbol{\eta}}) = \mathbf{0}_k, \tag{3.18}$$

where $\mathbf{0}_k$ is a $k \times 1$ vector of zeros. Using (3.17),

$$X'\mathbf{y} - X'\hat{\boldsymbol{\eta}} = X'\mathbf{y} - X'X\hat{\boldsymbol{\beta}} = \mathbf{0} \tag{3.18a}$$

or

$$X'X\hat{\boldsymbol{\beta}} = X'\mathbf{y}. \tag{3.18b}$$

These are the so-called *normal equations*. [We note in passing that they may be derived whether or not X (and hence $X'X$) is of full rank, a point that we return to in Chapter 4, when we discuss the non-full-rank case, i.e., when $r(X) = p < k < N$.]

We have assumed the $N \times k$ matrix X is of full rank, so that the $k \times k$ matrix $X'X$ is of full rank, and we immediately have that the solution to equations (3.18*b*) is given by

$$\hat{\boldsymbol{\beta}} = (X'X)^{-1}X'\mathbf{y}. \tag{3.19}$$

This means that the foot of the projection of $\mathbf{y}$ into the estimation space is given by (see Figure 3.1.1)

$$\hat{\boldsymbol{\eta}} = X\hat{\boldsymbol{\beta}} = X(X'X)^{-1}X'\mathbf{y} = \mathcal{R}\,\mathbf{y} \tag{3.19a}$$

where $\mathcal{R} = X(X'X)^{-1}X'$ is easily seen to be *idempotent and symmetric*; that is,

$$\mathcal{R}'\mathcal{R} = \mathcal{R}, \qquad \mathcal{R}\mathcal{R} = \mathcal{R}. \tag{3.19b}$$

This is no accident for, as we have discussed in Appendix A2.2, a matrix that is symmetric and idempotent defines an *orthogonal projection*, that is, the *matrix* $\mathcal{R}$ *is a projection matrix*. Note that the transformation $\mathcal{R}$ projects $\mathbf{y}$ into S_k, since $\hat{\boldsymbol{\eta}} = \mathcal{R}\,\mathbf{y} = X\hat{\boldsymbol{\beta}}$, a linear combination of $\mathbf{x}_j$, and hence $\hat{\boldsymbol{\eta}} \in S_k$. This is very general, and it may be phrased as follows. Given a vector $\mathbf{y}$ and a space S_k spanned by the columns of a full-rank matrix X, then the projection of $\mathbf{y}$ into X, that is, into S_k, or the regression of $\mathbf{y}$ on X, is the vector $\hat{\boldsymbol{\eta}}$, where

$$\hat{\boldsymbol{\eta}} = \mathcal{R}(X)\mathbf{y}, \qquad \mathcal{R}(X) = X(X'X)^{-1}X'. \tag{3.19c}$$

Note that $\mathcal{R}(X)$ is symmetric and idempotent. This immediately implies (see Appendix A2.2) that $I_N - \mathcal{R}$ is symmetric and idempotent. Thus, $(I_N - \mathcal{R})$ is the matrix of an orthogonal projection matrix, and since

$$X'(I_N - \mathcal{R}(X)) = \bigcirc, \tag{3.19d}$$

$(I_N - \mathcal{R}(X))$ orthogonally projects R^N onto the space orthogonal to S_k. [The left-hand side of (3.19d) is $X' - X'[X(X'X)^{-1}X'] = X' - X' = \bigcirc$.] We call the set

$$S_k^{\perp} = \{\mathbf{a} \mid X'\mathbf{a} = \mathbf{0}\}, \tag{3.20}$$

the *error space*; that is, $S_k^{\perp}$ is orthogonal to the estimation space S_k. Note that the vector $(\mathbf{y} - \hat{\boldsymbol{\eta}})$ belongs to the error space, as indicated in (3.18). Furthermore, we have

$$(\mathbf{y} - \hat{\boldsymbol{\eta}}) = (I_N - \mathcal{R}(X))\mathbf{y}, \tag{3.20a}$$

so it also follows from (3.19d) that $(\mathbf{y} - \hat{\boldsymbol{\eta}}) \in S_k^{\perp}$.

The solution $\hat{\boldsymbol{\beta}}$ [given by (3.19)] to the normal equations (3.18b) is called the *least-squares solution*, or the *least-squares estimator*, and has various properties which we will discuss below. Before doing that, however, we need to record some terminology, algebraic facts, and assumptions which we will wish to invoke at certain points.

Terminology

(a) We shall say that

$$\hat{\boldsymbol{\beta}} = (X'X)^{-1}X'\mathbf{y} \tag{3.21}$$

is the *least-squares solution* [to the normal equations (3.18b), or equivalently, to the problem (see Property 1 below) of finding the point in S_k, and hence the value of $\boldsymbol{\theta}$, that minimizes $(\mathbf{y} - X\boldsymbol{\theta})'(\mathbf{y} - X\boldsymbol{\theta})$, the squared distance between $\mathbf{y}$ and points in S_k]. It is also often called the least-squares *estimator* of $\boldsymbol{\beta}$; see Properties 1 and 2 below.

(b) The foot of the orthogonal projection, $\hat{\boldsymbol{\eta}}$, of $\mathbf{y}$ into S_k, where

$$(3.21a)\quad \hat{\boldsymbol{\eta}} = X\hat{\boldsymbol{\beta}} = X(X'X)^{-1}X'\mathbf{y} = \mathcal{R}\mathbf{y}, \qquad \mathcal{R} = X(X'X)^{-1}X',$$

is called the vector of fitted values, or *fitted values* for short.

(c) The vector $(\mathbf{y} - \hat{\boldsymbol{\eta}}) = \mathbf{e}$ which is such that

$$(3.21b)\quad \begin{aligned} \mathbf{e} &= (\mathbf{y} - \hat{\boldsymbol{\eta}}) = \mathbf{y} - \mathcal{R}\mathbf{y} = (I_N - \mathcal{R})\mathbf{y} = \mathcal{M}\mathbf{y}, \\ \mathcal{M} &= (I_N - \mathcal{R}), \end{aligned}$$

is called the *residual* vector. The squared length of $\mathbf{e}$—the sum of squares

$$(3.21c)\quad \mathbf{e}'\mathbf{e} = (\mathbf{y} - \hat{\boldsymbol{\eta}})'(\mathbf{y} - \hat{\boldsymbol{\eta}}) = \mathrm{SS}_e$$

is called the *residual sum of squares*, or sometimes, the *error sum of squares*. The squared length of the vector $\hat{\boldsymbol{\eta}}$,

$$(3.21d)\quad \hat{\boldsymbol{\eta}}'\hat{\boldsymbol{\eta}} = \hat{\boldsymbol{\beta}}'X'X\hat{\boldsymbol{\beta}} = \mathbf{y}'X(X'X)^{-1}X'\mathbf{y} = \mathbf{y}'\mathcal{R}\mathbf{y} = \mathrm{SS}_r,$$

is called the *regression sum of squares*. The term derives from the fact that $\hat{\boldsymbol{\eta}}$ lies in the *regression plane*, that is, in the *estimation space* S_k. Note that SS_r may be thought of as the inner product of $\mathbf{y}$ with $\mathcal{R}\mathbf{y}$, and $\mathcal{R}\mathbf{y}$ is of course the projection of $\mathbf{y}$ into the estimation space. Similarly, using (3.21b), we have that

$$(3.21e)\quad \mathrm{SS}_e = \mathbf{e}'\mathbf{e} = \mathbf{y}'(I - \mathcal{R})'(I - \mathcal{R})\mathbf{y},$$

and since $\mathcal{M} = I - \mathcal{R}$ is symmetric and idempotent (Problem 3.2), we have that

$$(3.21f)\quad \mathrm{SS}_e = \mathbf{y}'(I - \mathcal{R})\mathbf{y}.$$

Hence SS_e may be looked at as the inner product of $\mathbf{y}$ with $(I_N - \mathcal{R})\mathbf{y}$, and as we have seen [(3.19d)], $(I_N - \mathcal{R})\mathbf{y}$ is a projection of $\mathbf{y}$ into the error space.

Algebraic Facts

(a) $\mathcal{R}$ and $\mathcal{M} = I_N - \mathcal{R}$ are *symmetric* and *idempotent*, that is,

$$(3.22)\quad \mathcal{R}' = \mathcal{R}, \quad \mathcal{R}\mathcal{R} = \mathcal{R}, \qquad \mathcal{M}' = \mathcal{M}, \quad \mathcal{M}\mathcal{M} = \mathcal{M}$$

(see Problem 3.2). Then, using the identity $\mathrm{tr}(AB) = \mathrm{tr}(BA)$, we find, since

$\mathcal{R}$ and $\mathcal{M}$ are idempotent, that

$$r(\mathcal{R}) = \operatorname{tr}(\mathcal{R}) = \operatorname{tr}\left[X'X(X'X)^{-1}\right] = \operatorname{tr}(I_k) = k \tag{3.23}$$

and

$$r(\mathcal{M}) = \operatorname{tr}(\mathcal{M}) = \operatorname{tr}(I_N - \mathcal{R}) = \operatorname{tr}(I_N) - \operatorname{tr}(\mathcal{R}) = N - k. \tag{3.23a}$$

(b) The matrix $\mathcal{R} = X(X'X)^{-1}X'$ is such that

$$\mathcal{R}X = X, \qquad X'\mathcal{R} = X' \tag{3.24}$$

and we then have that the matrix $\mathcal{M}$ is such that

$$\begin{gathered} X'\mathcal{M} = \bigcirc, \qquad \mathcal{M}'X = \mathcal{M}X = \bigcirc \\ X'(I_N - \mathcal{R}) = \bigcirc \quad \text{or} \quad (I_N - \mathcal{R})'X = (I_N - \mathcal{R})X = \bigcirc \end{gathered} \tag{3.24a}$$

where $\bigcirc$ denotes a matrix of appropriate order, whose elements are all zeros. The property (3.24a) should be expected because, after all, from (3.18) we have that

$$X'\mathbf{e} = X'(I - \mathcal{R})\mathbf{y} = \mathbf{0}_k \tag{3.25}$$

for all $\mathbf{y}$. We note this separately below.

(c) We have that

$$X'\mathbf{e} = \mathbf{0}_k \quad \text{or} \quad \mathbf{e}'X = \mathbf{0}'_k. \tag{3.25a}$$

($\mathbf{0}_k$ is a $k \times 1$ vector of zeros.) Of course, the *geometric* interpretation of (3.25) and (3.25a) is immediate: $\mathbf{e} \in S_k^{\perp}$, the error space, that is, $\mathbf{e}$ is orthogonal to the estimation space S_k. Similarly, the geometric interpretation of (3.24) is that the projection of any vector belonging to S_k, for example, $X\boldsymbol{\gamma}$, is $X\boldsymbol{\gamma}$ itself, since $\mathcal{R}X\boldsymbol{\gamma} = X\boldsymbol{\gamma}$.

Assumptions

These will be labeled as on the left.

$AE(\mathbf{y})$	1.	$E(\mathbf{y}) = E(\mathbf{y} \mid X) = X\boldsymbol{\beta} = \boldsymbol{\eta}$ or $E(\boldsymbol{\varepsilon}) = \mathbf{0}$, $\boldsymbol{\varepsilon} = \mathbf{y} - X\boldsymbol{\beta}$.
$AV(\mathbf{y})$	2.	$V(\mathbf{y}) = V(\mathbf{y} \mid X) = \sigma^2 I_N$.
$AD\mathbf{y}$	3.	$\mathbf{y}$ is distributed normally, that is, $\mathbf{y} = N(\boldsymbol{\eta}, \sigma^2 I_N)$ where $\boldsymbol{\eta} = X\boldsymbol{\beta}$. Equivalently, $\boldsymbol{\varepsilon} = N(\mathbf{0}, \sigma^2 I_N)$.

These assumptions will now be invoked one at a time in order to determine statistical properties of the least-squares solution. We now list the various properties of least squares.

PROPERTY 1. $\hat{\boldsymbol{\eta}} = X\hat{\boldsymbol{\beta}}$ is the *closest point* of S_k to $\mathbf{y}$, where $\hat{\boldsymbol{\beta}}$ and $\hat{\boldsymbol{\eta}}$ are given by (3.19) and (3.19*a*), respectively.

Geometrically, the above statement is obviously true. After all, the orthogonal projection of a vector (in this case $\mathbf{y}$) into a linear subspace (in this case the estimation space) does minimize the distance between the endpoint of the vector and the subspace. Thus, for $\boldsymbol{\tau} \in S_k$, where the estimation space S_k is defined in (3.12), we have, geometrically, that

$$(3.26) \qquad \min_{\boldsymbol{\tau} \in S_k} (\mathbf{y} - \boldsymbol{\tau})'(\mathbf{y} - \boldsymbol{\tau}) = (\mathbf{y} - \hat{\boldsymbol{\eta}})'(\mathbf{y} - \hat{\boldsymbol{\eta}}), \qquad \hat{\boldsymbol{\eta}} = \mathcal{R}(X)\mathbf{y},$$

with $\hat{\boldsymbol{\eta}} = (\hat{\eta}_1, \ldots, \hat{\eta}_N)' \in S_k$. Now because we have minimized the sum of squares $(\mathbf{y} - \boldsymbol{\tau})'(\mathbf{y} - \boldsymbol{\tau})$, $\hat{\boldsymbol{\eta}}$ is called the *least-squares solution*. Since $\hat{\boldsymbol{\eta}} \in S_k$, $\hat{\boldsymbol{\eta}} = X\hat{\boldsymbol{\beta}}$, where $\hat{\boldsymbol{\beta}}$ is a solution of the normal equations (3.18*b*). Thus, in the full-rank case, $\hat{\boldsymbol{\beta}}$ is given by (3.19), so that there is a one-to-one correspondence between $\hat{\boldsymbol{\beta}}$ and the least-squares solution $\hat{\boldsymbol{\eta}}$; $\hat{\boldsymbol{\beta}}$ itself is often called the least-squares solution as mentioned before. Of course, if we write the squared distance between $\mathbf{y}$ and any point $\boldsymbol{\tau} \in S_k$ as

$$(3.27) \qquad (\mathbf{y} - X\boldsymbol{\theta})'(\mathbf{y} - X\boldsymbol{\theta}) = (\mathbf{y} - \boldsymbol{\tau})'(\mathbf{y} - \boldsymbol{\tau}),$$

then

$$(3.27a) \qquad \min_{\boldsymbol{\theta}} (\mathbf{y} - X\boldsymbol{\theta})'(\mathbf{y} - X\boldsymbol{\theta}) = (\mathbf{y} - X\hat{\boldsymbol{\beta}})'(\mathbf{y} - X\hat{\boldsymbol{\beta}}),$$

where $\hat{\boldsymbol{\beta}}$ is given by (3.19).

Now some people do not like geometric reasoning, and so a proof by calculus follows. Let

$$(3.28) \qquad Q = (\mathbf{y} - X\boldsymbol{\theta})'(\mathbf{y} - X\boldsymbol{\theta}) = \mathbf{y}'\mathbf{y} - 2\boldsymbol{\theta}'X'\mathbf{y} + \boldsymbol{\theta}'X'X\boldsymbol{\theta}.$$

Differentiating with respect to $\boldsymbol{\theta}$ yields ($X'X$ is symmetric)

$$(3.28a) \qquad \frac{\partial Q}{\partial \boldsymbol{\theta}} = -2X'y + 2X'X\boldsymbol{\theta}.$$

Setting $\partial Q/\partial\boldsymbol{\theta} = \mathbf{0}_k$ and denoting the root by $\hat{\boldsymbol{\beta}}$, we arrive at the normal equations (3.18*b*), which for the case discussed in this chapter (X and hence $X'X$ of full rank k) leads to the least-squares solution (3.19).

An alternative proof, which is illustrative and interesting—half algebraic and half *geometric*—is as follows. We have that the distance squared between $\mathbf{y}$ and any point $\boldsymbol{\tau} = X\boldsymbol{\theta}$ in the estimation space is

(3.29)

$$Q = (\mathbf{y} - X\boldsymbol{\theta})'(\mathbf{y} - X\boldsymbol{\theta}) = [(\mathbf{y} - \hat{\boldsymbol{\eta}}) + (\hat{\boldsymbol{\eta}} - X\boldsymbol{\theta})]'[(\mathbf{y} - \hat{\boldsymbol{\eta}}) + (\hat{\boldsymbol{\eta}} - X\boldsymbol{\theta})],$$

where $\hat{\boldsymbol{\eta}}$ is the orthogonal projection of $\mathbf{y}$ into S_k. Of course, this means that $\hat{\boldsymbol{\eta}} - X\boldsymbol{\theta}$ lies in S_k (see Figure 3.1.1), so that $\mathbf{y} - \hat{\boldsymbol{\eta}}$ is orthogonal to $\hat{\boldsymbol{\eta}} - X\boldsymbol{\theta}$. Using this fact, and expanding (3.29), we obtain

(3.29*a*)

$$Q = (\mathbf{y} - \hat{\boldsymbol{\eta}})'(\mathbf{y} - \hat{\boldsymbol{\eta}}) + 2(\hat{\boldsymbol{\eta}} - X\boldsymbol{\theta})'(\mathbf{y} - \hat{\boldsymbol{\eta}}) + (\hat{\boldsymbol{\eta}} - X\boldsymbol{\theta})'(\hat{\boldsymbol{\eta}} - X\boldsymbol{\theta}),$$

that is,

(3.29*b*) $$Q = (\mathbf{y} - \hat{\boldsymbol{\eta}})'(\mathbf{y} - \hat{\boldsymbol{\eta}}) + (\hat{\boldsymbol{\eta}} - X\boldsymbol{\theta})'(\hat{\boldsymbol{\eta}} - X\boldsymbol{\theta}).$$

See Problem 3.1 for a strictly algebraic proof of the fact that the cross-term vanishes. [The reader will no doubt recognize (3.29*b*) as a version of the Pythagoras theorem; see Figure 3.1.1.] The last term of (3.29*b*) vanishes at $\boldsymbol{\theta} = \hat{\boldsymbol{\beta}}$, where $\hat{\boldsymbol{\beta}}$ is chosen so that $X\hat{\boldsymbol{\beta}} = \hat{\boldsymbol{\eta}}$ as in (3.17). We note again [see (3.27*a*), or put $\boldsymbol{\theta} = \hat{\boldsymbol{\beta}}$ in the second term of (3.29*b*)] that

(3.29*c*) $$\min Q = (\mathbf{y} - \hat{\boldsymbol{\eta}})'(\mathbf{y} - \hat{\boldsymbol{\eta}}).$$

PROPERTY 2. (a) If the assumption $AE(\mathbf{y})$ about expectation of $\mathbf{y}$ holds, then

(3.30)

(i) $\hat{\boldsymbol{\beta}}$ is an unbiased estimator for $\boldsymbol{\beta}$, that is, $E(\hat{\boldsymbol{\beta}}) = \boldsymbol{\beta}$.

(ii) $\hat{\boldsymbol{\eta}}$ is unbiased for $\hat{\boldsymbol{\eta}} = X\boldsymbol{\beta}$, that is, $E(\hat{\boldsymbol{\eta}}) = \boldsymbol{\eta}$.

(iii) $\mathbf{e}$ is unbiased for $\mathbf{0}$, that is, $E(\mathbf{e}) = \mathbf{0}$.

The proofs of (i)–(iii) above are straightforward and left as an exercise for the reader; see Problem 3.3.

(b) If $AV(\mathbf{y})$ holds, then

(3.31)

(i) $V(\hat{\boldsymbol{\beta}}) = \sigma^2(X'X)^{-1}$,

(ii) $V(\hat{\boldsymbol{\eta}}) = \sigma^2\mathfrak{R}$,

(iii) $V(\mathbf{e}) = \sigma^2(I_N - \mathfrak{R})$.

For proofs, see Problem 3.4. It is interesting to note that the $N \times N$ matrices $V(\hat{\boldsymbol{\eta}})$ and $V(\mathbf{e})$ have ranks k and $N-k$, respectively, and since $0 < k < N$, they are *singular* [see (3.23) and (3.27*a*)].

The next statistical property justifies the use of least squares.

PROPERTY 3.

Theorem 3.1.1. (*Gauss Theorem*). *If $AE(\mathbf{y})$ and $AV(\mathbf{y})$ hold, and if interest lies in a (scalar) linear function of $\boldsymbol{\beta}$, for example, $\gamma = \mathbf{a}'\boldsymbol{\beta}$, where $\mathbf{a}' = (a_1, \ldots, a_k)$ is a set of k constants, then, among all linear (in the y_j's) unbiased estimators of γ, the estimator $\hat{\gamma} = \mathbf{a}'\hat{\boldsymbol{\beta}}$, where $\hat{\boldsymbol{\beta}}$ is the least-squares estimator of $\boldsymbol{\beta}$, has minimum variance.*

Proof. We have that $\hat{\gamma}$ is linear in $\mathbf{y}$, since, as is easily verified,

$$\hat{\gamma} = \mathbf{d}'\mathbf{y}, \qquad \mathbf{d}' = \mathbf{a}'(X'X)^{-1}X' \tag{3.32}$$

and using $AV(\mathbf{y})$, it is also easy to verify that the variance of $\hat{\gamma}$ is

$$V(\hat{\gamma}) = \sigma^2\mathbf{a}'(X'X)^{-1}\mathbf{a}. \tag{3.33}$$

Suppose we have a different estimator, t, linear in the y's, which is also unbiased for γ. Specifically, let

$$t = \mathbf{c}'\mathbf{y}, \qquad \mathbf{c}' \neq \mathbf{d}'. \tag{3.34}$$

Since t is unbiased for γ, and since $AE(\mathbf{y})$ holds,

$$\gamma = \mathbf{a}'\boldsymbol{\beta} = E(t) = \mathbf{c}'X\boldsymbol{\beta} \tag{3.35}$$

for any value that $\boldsymbol{\beta}$ may have. Thus we obtain from (3.35) that

$$\mathbf{a}' = \mathbf{c}'X. \tag{3.35a}$$

Now the covariance of t and $\hat{\gamma}$ is

$$\operatorname{cov}(\mathbf{c}'\mathbf{y}, \mathbf{d}'\mathbf{y}) = E[\mathbf{c}'(\mathbf{y}-\boldsymbol{\eta})(\mathbf{y}-\boldsymbol{\eta})'\mathbf{d}] \tag{3.36}$$

and since $AV(\mathbf{y})$ holds, we have, using (3.32),

$$\operatorname{cov}(t, \hat{\gamma}) = \sigma^2\mathbf{c}'\mathbf{d} = \sigma^2\mathbf{c}'X(X'X)^{-1}\mathbf{a}, \tag{3.36a}$$

so that from (3.35*a*) we obtain

$$\text{cov}(t, \hat{\gamma}) = \sigma^2 \mathbf{a}'(X'X)^{-1}\mathbf{a}. \tag{3.36b}$$

From (3.33) we have

$$\text{cov}(t, \hat{\gamma}) = V(\hat{\gamma}). \tag{3.36c}$$

Now

$$0 \le V(t - \hat{\gamma}) = V(t) + V(\hat{\gamma}) - 2\,\text{cov}(t, \hat{\gamma}) \tag{3.37}$$

or, in view of (3.36*c*),

$$0 \le V(t - \hat{\gamma}) = V(t) - V(\hat{\gamma}) \tag{3.37a}$$

or

$$V(\hat{\gamma}) \le V(t) \tag{3.37b}$$

with equality if and only if $t = \hat{\gamma}$, ruled out by (3.34). Hence

$$V(\hat{\gamma}) < V(t) \tag{3.37c}$$

and the theorem is proved.

Because of the above theorem and, in particular, because (3.37*c*) holds uniformly (i.e., for all possible values of $\boldsymbol{\beta}$), we refer to the estimator $\hat{\gamma} = \mathbf{a}'\hat{\boldsymbol{\beta}}$ of $\gamma = \mathbf{a}'\boldsymbol{\beta}$ as the UMV unbiased estimator, where UMV stands for uniformly minimum variance. We note that, as the proof of the theorem suggests, $\hat{\gamma}$ is *unique*. We may see this as follows.

If we start off with an unbiased estimator t of γ of the form $t = \mathbf{c}'\mathbf{y}$, then we have from (3.35*a*) that

$$\mathbf{a}' = \mathbf{c}'X, \tag{3.38}$$

and using $A\,V(\mathbf{y})$,

$$V(t) = \sigma^2 \mathbf{c}'\mathbf{c}. \tag{3.38a}$$

If we wish to minimize (3.38*a*) subject to (3.38), we find [Problem 3.5(a)] by using a Lagrange multiplier argument that

$$\min_{\mathbf{c}} V(\mathbf{c}'\mathbf{y}) = V(\mathbf{d}'\mathbf{y}), \qquad \mathbf{d}' = \mathbf{a}'(X'X)^{-1}X'; \tag{3.38b}$$

that is, in order for $t = \mathbf{c}'\mathbf{y}$ to be minimum variance unbiased, $\mathbf{c}' = \mathbf{d}'$.

One consequence of the Gauss theorem is the following: Let

$$\mathbf{a}' = (0,\ldots,0,1,0,\ldots,0) \tag{3.39}$$

where the "1" occurs in the ith place. Then *the Gauss theorem states that* $\hat{\beta}_i$, *the* ith *component of* $\hat{\boldsymbol{\beta}} = (X'X)^{-1}X'\mathbf{y}$, *is the uniformly minimum variance unbiased estimator of* β_i, *the* ith *component of* $\boldsymbol{\beta}$. Lastly, by setting $\mathbf{a}' = \mathbf{x}'$, where $\mathbf{x}'\boldsymbol{\beta} = \eta_{\mathbf{x}}$—the expected response to the setting of the k factors at the values $(x_1,\ldots,x_k)$—we also have that $\mathbf{x}'\hat{\boldsymbol{\beta}}$ *is the* UMV *unbiased estimator for* $\eta_{\mathbf{x}}$.

We have not yet used $AD\mathbf{y}$, which assumes that the y_j's are *normal*, independent, and such that

$$E(\mathbf{y}) = X\boldsymbol{\beta}, \qquad V(\mathbf{y}) = \sigma^2 I_N. \tag{3.40}$$

If $AD\mathbf{y}$ holds, then

$$\begin{aligned} &\text{(i)} \quad \hat{\boldsymbol{\beta}} = N\left(\boldsymbol{\beta}, \sigma^2(X'X)^{-1}\right) \\ &\text{(ii)} \quad \hat{\boldsymbol{\eta}} = N(\boldsymbol{\eta}, \sigma^2\mathcal{R}), \qquad \mathcal{R} = X(X'X)^{-1}X' \\ &\text{(iii)} \quad \mathbf{e} = (\mathbf{y} - \hat{\boldsymbol{\eta}}) = N(\mathbf{0}, \sigma^2\mathcal{M}), \qquad \mathcal{M} = I - \mathcal{R}. \end{aligned} \tag{3.41}$$

In fact, if $AD\mathbf{y}$ holds, then it turns out that the least-squares estimator $\hat{\boldsymbol{\beta}}$ is the *maximum likelihood estimator* of $\boldsymbol{\beta}$, since the log likelihood function of $(\boldsymbol{\beta}, \sigma^2)$, l, is

$$l\left(\boldsymbol{\beta}, \sigma^2 \mid \mathbf{y}; X\right) = K - \frac{N}{2}\ln\sigma^2 - \frac{1}{2\sigma^2}(\mathbf{y} - X\boldsymbol{\beta})'(\mathbf{y} - X\boldsymbol{\beta}). \tag{3.42}$$

It is apparent, even at this point, that maximizing l with respect to $\boldsymbol{\beta}$ involves the minimization of the sum of squares $(\mathbf{y} - X\boldsymbol{\beta})'(\mathbf{y} - X\boldsymbol{\beta})$. In fact, we have that [Problem 3.5(b)]

$$\begin{aligned} \frac{\partial l}{\partial \boldsymbol{\beta}} &= \frac{1}{2\sigma^2}(2X'\mathbf{y} - 2X'X\boldsymbol{\beta}) \\ \frac{\partial l}{\partial \sigma^2} &= -\frac{N}{2\sigma^2} + \frac{1}{2(\sigma^2)^2}(\mathbf{y} - X\boldsymbol{\beta})'(\mathbf{y} - X\boldsymbol{\beta}). \end{aligned} \tag{3.43}$$

Setting the derivatives in (3.43) equal to zero, and denoting the roots of the resulting equations by $\hat{\boldsymbol{\beta}}_{ml}$ and $\hat{\sigma}^2_{ml}$, we find

$$X'X\hat{\boldsymbol{\beta}}_{ml} = X'\mathbf{y} \tag{3.44}$$

and

$$\hat{\sigma}^2_{ml} = \frac{1}{N}(\mathbf{y} - X\hat{\boldsymbol{\beta}}_{ml})'(\mathbf{y} - X\hat{\boldsymbol{\beta}}_{ml}). \tag{3.44a}$$

That is, the maximum likelihood estimator $\hat{\boldsymbol{\beta}}_{ml}$ is the solution to the normal equations (3.18*b*); $\hat{\boldsymbol{\beta}}_{ml} = \hat{\boldsymbol{\beta}}$ and also the maximum likelihood estimator of σ^2 is

$$\hat{\sigma}^2_{ml} = \frac{1}{N}(\mathbf{y} - X\hat{\boldsymbol{\beta}})'(\mathbf{y} - X\hat{\boldsymbol{\beta}}). \tag{3.44b}$$

We will see presently that $\hat{\sigma}^2_{ml}$ is biased for the error variance σ^2.

3.2. DETAILED LOOK AT THE REGRESSION AND RESIDUAL SUM OF SQUARES

Once we have found $\hat{\boldsymbol{\beta}}$ by minimizing the squared distance between $\mathbf{y}$ and a point in the estimation space S_k, the minimum distance squared is

$$\mathbf{e}'\mathbf{e} = (\mathbf{y} - \hat{\boldsymbol{\eta}})'(\mathbf{y} - \hat{\boldsymbol{\eta}}) = (\mathbf{y} - X\hat{\boldsymbol{\beta}})'(\mathbf{y} - X\hat{\boldsymbol{\beta}}), \tag{3.45}$$

and it is easy to see (Problem 3.7) that

$$\mathbf{e}'\mathbf{e} = \mathbf{y}'\mathbf{y} - \hat{\boldsymbol{\beta}}'X'\mathbf{y} = \mathbf{y}'\mathbf{y} - \hat{\boldsymbol{\beta}}'X'X\hat{\boldsymbol{\beta}} = \mathbf{y}'\mathbf{y} - \hat{\boldsymbol{\eta}}'\hat{\boldsymbol{\eta}}. \tag{3.46}$$

That is, the residual sum of squares, after fitting $X\boldsymbol{\beta}$ to the data, is the sum of squares of the y's, minus the inner product of the least-squares solutions with the right-hand side of the normal equations (3.18*b*). An alternative form for (3.46) is, as is easily verified (Problem 3.7),

$$\mathbf{e}'\mathbf{e} = \mathbf{y}'\mathbf{y} - \mathbf{y}'\mathcal{R}\,\mathbf{y} \tag{3.46a}$$

or, if we wish,

$$\begin{aligned} \mathbf{y}'\mathbf{y} &= \mathbf{y}'\mathcal{R}\,\mathbf{y} + \mathbf{e}'\mathbf{e} \\ &= \hat{\boldsymbol{\eta}}'\hat{\boldsymbol{\eta}} + \mathbf{e}'\mathbf{e}. \end{aligned} \tag{3.46b}$$

(See Figure 3.2.1.)

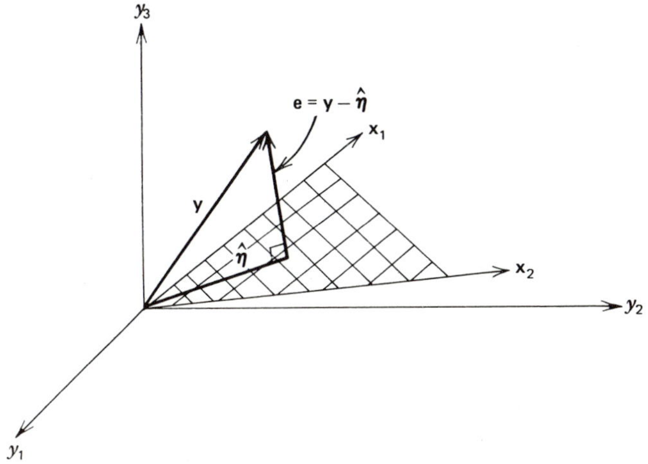

FIGURE 3.2.1. Projection of $\mathbf{y}$ into S_k at $\hat{\boldsymbol{\eta}}$ induces the decomposition $\mathbf{y}'\mathbf{y} = \hat{\boldsymbol{\eta}}'\hat{\boldsymbol{\eta}} + \mathbf{e}'\mathbf{e}$ or $\text{SS}_t = \text{SS}_r + \text{SS}_e$. $\text{SS}_r = \mathbf{y}'\mathfrak{R}\,\mathbf{y} = \hat{\boldsymbol{\eta}}'\hat{\boldsymbol{\eta}}$. $\text{SS}_e = \mathbf{y}'(I - \mathfrak{R})\mathbf{y}$. $\mathbf{y}'\mathbf{y} = (\mathbf{y} - \hat{\boldsymbol{\eta}})'(\mathbf{y} - \hat{\boldsymbol{\eta}}) + \hat{\boldsymbol{\eta}}'\hat{\boldsymbol{\eta}}$, by Pythagoras.

We recall that $\mathbf{e} = (\mathbf{y} - \hat{\boldsymbol{\eta}}) = (I - \mathfrak{R})\mathbf{y}$ so that we may write [cf. (1.47) of Chapter 1]

$$\mathbf{y}'\mathbf{y} = \mathbf{y}'\mathfrak{R}\,\mathbf{y} + \mathbf{y}'(I - \mathfrak{R})\mathbf{y} \tag{3.46c}$$

or

$$\text{SS}_t = \text{SS}_r + \text{SS}_e, \tag{3.46d}$$

where $\text{SS}_t = \mathbf{y}'\mathbf{y}$ is the total sum of squares of the observations y_i. Equation (3.46c) is often referred to as the *breakdown* of the total sum of squares—indeed, *the total sum of squares* $\mathbf{y}'\mathbf{y}$ *has been broken down into a part* $\mathbf{y}'\mathfrak{R}\,\mathbf{y}$ (*due to the fact that we have fitted the data* $X\boldsymbol{\beta}$ *to the data* $\mathbf{y}$, *a process called regressing* $\mathbf{y}$ *on* X) *and a part due to error*, $\mathbf{y}'(I - \mathfrak{R})\mathbf{y}$. To substantiate this statement further, recall that

$$\hat{\boldsymbol{\eta}} = X\hat{\boldsymbol{\beta}}, \quad E(\hat{\boldsymbol{\eta}}) = \boldsymbol{\eta}, \quad \text{SS}_r = \hat{\boldsymbol{\beta}}'X'X\hat{\boldsymbol{\beta}} = \hat{\boldsymbol{\eta}}'\hat{\boldsymbol{\eta}} \tag{3.47}$$

and

$$\mathbf{e} = \mathbf{y} - \hat{\boldsymbol{\eta}} = (I - \mathfrak{R})\mathbf{y}, \quad E(\mathbf{e}) = \mathbf{0}, \quad \text{SS}_e = \mathbf{e}'\mathbf{e}. \tag{3.48}$$

That is, $\hat{\boldsymbol{\eta}}$ contains information about the model $\boldsymbol{\eta} = X\boldsymbol{\beta}$; $\mathbf{e}$ reflects information about error only and hence $\mathbf{e}'\mathbf{e}$ should give us information about σ^2. Using the result of Problem 2.24, we have

$$E(\mathrm{SS}_r) = E(\mathbf{y}'\mathcal{R}\mathbf{y}) = \boldsymbol{\beta}'X'\mathcal{R}X\boldsymbol{\beta} + \mathrm{tr}\big[\mathcal{R}(\sigma^2 I)\big]. \tag{3.49}$$

Now from (3.28), (3.25a), and the fact that $\mathrm{tr}[X(X'X)^{-1}X'] = \mathrm{tr}[X'X(X'X)^{-1}] = \mathrm{tr}(I_k) = k$, we have that

$$\begin{aligned} E(\mathrm{SS}_r) &= \boldsymbol{\beta}'X'X\boldsymbol{\beta} + \sigma^2\mathrm{tr}\big[X(X'X)^{-1}X'\big] \\ &= \boldsymbol{\beta}'X'X\boldsymbol{\beta} + \sigma^2 k. \end{aligned} \tag{3.50}$$

Hence

$$E\left(\frac{\mathrm{SS}_r}{k}\right) = \frac{k\sigma^2 + \boldsymbol{\beta}'X'X\boldsymbol{\beta}}{k}. \tag{3.51}$$

If we denote the mean square for regression, SS_r/k, by MS_r, we have

$$E(\mathrm{MS}_r) = \sigma^2 + \frac{1}{k}\boldsymbol{\beta}'X'X\boldsymbol{\beta}. \tag{3.51a}$$

Furthermore, we have that

$$E(\mathrm{SS}_e) = E[\mathbf{y}'(I-\mathcal{R})\mathbf{y}] = \boldsymbol{\beta}'X'(I-\mathcal{R})X\boldsymbol{\beta} + \mathrm{tr}(I-\mathcal{R})(\sigma^2 I); \tag{3.52}$$

but we have seen in (3.24a) that $(I - \mathcal{R})X = \mathrm{O}$, so that

$$\begin{aligned} E(\mathrm{SS}_e) &= \sigma^2[\mathrm{tr}(I_N) - \mathrm{tr}(\mathcal{R})] \\ &= \sigma^2(N-k). \end{aligned} \tag{3.52a}$$

Denoting the mean square for error, $\mathrm{SS}_e/(N-k)$, by MS_e, we have

$$E(\mathrm{MS}_e) = \sigma^2 \tag{3.53}$$

so that MS_e is unbiased for σ^2. We usually refer to k as the degrees of freedom of the *regression sum of squares* SS_r. More properly, it is the dimension of the estimation space S_k; after all, any vector confined to the subspace S_k is free to move only in k dimensions, since it must belong to S_k. Similarly, we speak of $N-k$ as the degrees of freedom of the *error sum of squares* SS_e. More precisely, it is the dimension of the error space $S_k^\perp$, since

TABLE 3.2.1. Analysis of Variance for $y = X\beta + \varepsilon$ (Interest in $\beta = 0$)

Source	Degrees of Freedom	Sum of Squares	Mean Square	Expected Mean Square
Regression	k	$SS_r = \hat{\beta}'X'X\hat{\beta}$ $= \mathbf{y}'\mathcal{R}\mathbf{y}$	$MS_r = SS_r/k$	$\sigma^2 + \frac{1}{k}\beta'X'X\beta$
Error	$N-k$	$SS_e = \mathbf{y}'\mathbf{y} - \hat{\beta}'X'X\hat{\beta}$ $= \mathbf{y}'(I-\mathcal{R})\mathbf{y}$	$MS_e = SS_e/(N-k)$	σ^2
Total	N	$SS_t = \mathbf{y}'\mathbf{y}$		

a vector belonging to the error space is free to move in only $N-k$ dimensions, if it is to lie in the error space.

We usually tabulate the breakdown (3.46c) in an analysis of variance table; see Table 3.2.1.

The expected mean square (EMS) column is quite informative. Suppose we wish to examine the statement $\boldsymbol{\beta} = \mathbf{0}$. If this statement were true, both MS_r and MS_e would be unbiased estimators of σ^2, or put another way, both would only contain and/or be reflecting information about error alone, and a natural way to compare them would be by their ratio. As we shall see (Section 3.7), under normality, the likelihood ratio test criterion of the hypothesis $\boldsymbol{\beta} = \mathbf{0}$ uses the ratio MS_r/MS_e. If this ratio were unusually large, it would tend to cast doubt on the statement $\boldsymbol{\beta} = \mathbf{0}$ (see the expected mean square column of Table 3.2.1).

Of course, interest does not always focus on $\boldsymbol{\beta} = \mathbf{0}$; indeed, the point $\boldsymbol{\beta} = \boldsymbol{\beta}_0$ may be of central interest. However, *by simple transformation, we may put the problem of examining the statement* $\boldsymbol{\beta} = \boldsymbol{\beta}_0$ *into a problem of examining a statement involving the nullity of certain regression parameters by using a table of the form of Table* 3.2.1.

Geometrically, we have that the distance squared between $\mathbf{y}$ and $\boldsymbol{\eta}_0 = X\boldsymbol{\beta}_0 \in S_k$ is (see Figure 3.2.2)

(3.54)

$$(\mathbf{y} - X\boldsymbol{\beta}_0)'(\mathbf{y} - X\boldsymbol{\beta}_0) = (\mathbf{y} - X\hat{\boldsymbol{\beta}})'(\mathbf{y} - X\hat{\boldsymbol{\beta}}) + (\hat{\boldsymbol{\beta}} - \boldsymbol{\beta}_0)'X'X(\hat{\boldsymbol{\beta}} - \boldsymbol{\beta}_0).$$

Suppose we now let

(3.55) $$\mathbf{z} = \mathbf{y} - X\boldsymbol{\beta}_0,$$

that is, we are changing the origin of the data from $\mathbf{0}$ to $X\boldsymbol{\beta}_0$. Then

(3.55a) $$\mathbf{z} = X\boldsymbol{\beta} + \boldsymbol{\varepsilon} - X\boldsymbol{\beta}_0 = X(\boldsymbol{\beta} - \boldsymbol{\beta}_0) + \boldsymbol{\varepsilon}$$

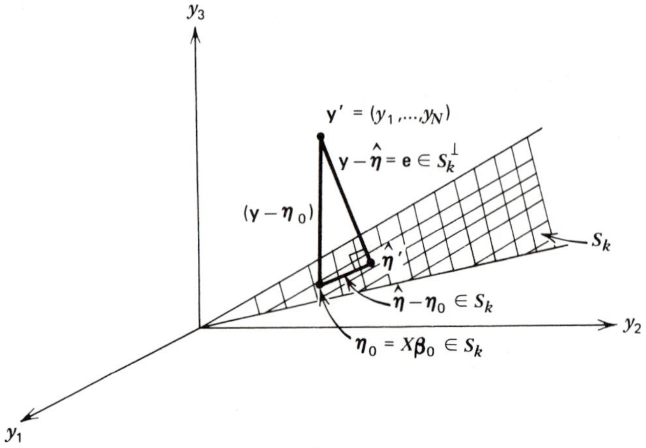

FIGURE 3.2.2. The squared distance between $\mathbf{y}$ and $\boldsymbol{\eta}_0 = X\boldsymbol{\beta}_0$ may be decomposed into SS_e and a regression sum of squares, since by the Pythagoras theorem

$$\begin{aligned}(\mathbf{y} - X\boldsymbol{\beta}_0)'(\mathbf{y} - X\boldsymbol{\beta}_0) &= \mathbf{e}'\mathbf{e} + (\hat{\boldsymbol{\eta}} - X\boldsymbol{\beta}_0)'(\hat{\boldsymbol{\eta}} - X\boldsymbol{\beta}_0) \\ &= (\mathbf{y} - \hat{\boldsymbol{\eta}})'(\mathbf{y} - \hat{\boldsymbol{\eta}}) + (\hat{\boldsymbol{\beta}} - \boldsymbol{\beta}_0)'X'X(\hat{\boldsymbol{\beta}} - \boldsymbol{\beta}_0).\end{aligned}$$

or

$$\mathbf{z} = X\boldsymbol{\phi} + \boldsymbol{\varepsilon}, \qquad \boldsymbol{\phi} = \boldsymbol{\beta} - \boldsymbol{\beta}_0. \tag{3.55b}$$

We note that the statement $\boldsymbol{\beta} = \boldsymbol{\beta}_0$ is equivalent to $\boldsymbol{\phi} = \mathbf{0}$. If we regress $\mathbf{z}$ on X using (3.55) and (3.55b), we have

$$\hat{\boldsymbol{\phi}} = (X'X)^{-1}X'\mathbf{z} = \hat{\boldsymbol{\beta}} - \boldsymbol{\beta}_0, \tag{3.55c}$$

and letting $X\boldsymbol{\phi} = \boldsymbol{\delta}$,

$$\hat{\boldsymbol{\delta}} = X\hat{\boldsymbol{\phi}} = X(\hat{\boldsymbol{\beta}} - \boldsymbol{\beta}_0) = \hat{\boldsymbol{\eta}} - \boldsymbol{\eta}_0. \tag{3.55d}$$

Thus,

$$(\mathbf{y} - X\boldsymbol{\beta}_0)'(\mathbf{y} - X\boldsymbol{\beta}_0) = \mathbf{z}'\mathbf{z} \tag{3.55e}$$

$$\begin{aligned}\mathrm{SS}_{e(\mathbf{y})} = (\mathbf{y} - \hat{\boldsymbol{\eta}})'(\mathbf{y} - \hat{\boldsymbol{\eta}}) &= [(\mathbf{y} - X\boldsymbol{\beta}_0) - (\hat{\boldsymbol{\eta}} - \boldsymbol{\eta}_0)]' \\ &\quad \times [(\mathbf{y} - X\boldsymbol{\beta}_0) - (\hat{\boldsymbol{\eta}} - \boldsymbol{\eta}_0)] \\ &= (\mathbf{z} - \hat{\boldsymbol{\delta}})'(\mathbf{z} - \hat{\boldsymbol{\delta}}) = (\mathbf{z} - X\hat{\boldsymbol{\phi}})'(\mathbf{z} - X\hat{\boldsymbol{\phi}}) = \mathrm{SS}_{e(\mathbf{z})}\end{aligned} \tag{3.55f}$$

$$(\hat{\boldsymbol{\beta}} - \boldsymbol{\beta}_0)'X'X(\hat{\boldsymbol{\beta}} - \boldsymbol{\beta}_0) = \hat{\boldsymbol{\phi}}'X'X\hat{\boldsymbol{\phi}} = \mathrm{SS}_{r(\mathbf{z})}. \tag{3.55g}$$

It is easy to see that (see Figure 3.2.2)

$$\mathbf{z}'\mathbf{z} = \mathrm{SS}_{e(\mathbf{z})} + \mathrm{SS}_{r(\mathbf{z})}. \tag{3.55h}$$

After the translation (3.55), the analysis of variance table derived from (3.54) is replaced by a table derived from (3.55e)–(3.55g) which is of the same form as Table 3.2.1. Note that the hypothesis $\boldsymbol{\beta} = \boldsymbol{\beta}_0$ is *equivalent to the hypothesis* $\boldsymbol{\phi} = \mathbf{0}$. This means that we compare $\mathrm{SS}_{r(\mathbf{z})}$ with $\mathrm{SS}_{e(\mathbf{z})} = \mathrm{SS}_{e(\mathbf{y})}$ in the manner discussed below and in Section 3.7.

From this point on, we assume that interest lies in the statement that a suggested value for the regression parameters is zero: If it isn't at the outset we merely transform simply as in (3.55) and work with the data $\mathbf{z}$, and so on.

Now, how exactly can we proceed to examine the statement $\boldsymbol{\beta} = \mathbf{0}$? We first impose the assumption $AD\mathbf{y}$, that is, the y_u are *independent* normals, with means $\mathbf{x}'_u\boldsymbol{\beta}$, where $\mathbf{x}'_u = (x_{u1}, \ldots, x_{uk})$, and common variance σ^2.

Now if the statement $\boldsymbol{\beta} = \mathbf{0}$ is true, then

$$E(\mathbf{y}) = \mathbf{0}, \qquad V(\mathbf{y}) = \sigma^2 I_N. \tag{3.56}$$

Also SS_r is independent of SS_e. To see the latter, we note that SS_r and SS_e are quadratic forms in normal central variables [$E(\mathbf{y}) = \mathbf{0}$ when, as we have assumed, $\boldsymbol{\beta} = \mathbf{0}$] with matrices

$$\mathcal{R} = X(X'X)^{-1}X' \quad \text{and} \quad \mathcal{M} = I - \mathcal{R}, \tag{3.57}$$

respectively. We know from Craig's theorem that these two quadratic forms are independent if and only if

$$\mathcal{R}V(\mathbf{y})\mathcal{M} = \bigcirc. \tag{3.57a}$$

But the left-hand side of (3.57a) is

$$\sigma^2\mathcal{R}\mathcal{M} = \sigma^2(\mathcal{R} - \mathcal{R}\mathcal{R}) = \sigma^2(\mathcal{R} - \mathcal{R}) = \bigcirc, \tag{3.57b}$$

since $\mathcal{R}$ is idempotent. Hence SS_r and SS_e are independent.

Furthermore, $\mathcal{R}$ and $\mathcal{M}$ are idempotent of ranks k and $N-k$, respectively, so that, from Theorem 2.2.1 of Chapter 2, we have

$$\mathrm{SS}_r = \sigma^2\chi^2_k, \qquad \mathrm{SS}_e = \sigma^2\chi^2_{N-k}, \tag{3.57c}$$

where the χ^2 variables are independent. Hence the ratio

$$F = \frac{\mathrm{MS}_r}{\mathrm{MS}_e} = F_{k,N-k} \tag{3.57d}$$

when $\boldsymbol{\beta} = \mathbf{0}$. From the entries in the EMS column of Table 3.2.1, if $\boldsymbol{\beta} \neq \mathbf{0}$, we expect the observed value of F to be large, at least larger than 1, roughly speaking, and so a procedure for examining the statement $\boldsymbol{\beta} = \mathbf{0}$ is

(3.58)
Reject $\boldsymbol{\beta} = \mathbf{0}$ at level α if the observed value of $F > F_{k, N-k; \alpha}$;
Accept otherwise.

(We note here that the use of the test statistic F is obtained naturally when investigating the likelihood ratio criterion for a test of $\boldsymbol{\beta} = \mathbf{0}$; see Section 3.7.)

Alternatively one may proceed as follows. We have

(3.59)
$$(\mathbf{y} - X\boldsymbol{\beta})'(\mathbf{y} - X\boldsymbol{\beta}) = (\mathbf{y} - X\hat{\boldsymbol{\beta}})'(\mathbf{y} - X\hat{\boldsymbol{\beta}}) + (\hat{\boldsymbol{\beta}} - \boldsymbol{\beta})'X'X(\hat{\boldsymbol{\beta}} - \boldsymbol{\beta}).$$

Now the first term on the right-hand side is

$$\mathbf{e}'\mathbf{e} = \mathbf{y}'(I - \mathcal{R})\mathbf{y}, \tag{3.60}$$

and it is easy to verify (see Problem 3.9) that

$$\mathbf{e}'\mathbf{e} = (\mathbf{y} - X\boldsymbol{\beta})'(I - \mathcal{R})(\mathbf{y} - X\boldsymbol{\beta}). \tag{3.60a}$$

Also (see Problem 3.9), the second term on the right-hand side of (3.59) is

$$(\hat{\boldsymbol{\beta}} - \boldsymbol{\beta})'X'X(\hat{\boldsymbol{\beta}} - \boldsymbol{\beta}) = (\mathbf{y} - X\boldsymbol{\beta})'\mathcal{R}(\mathbf{y} - X\boldsymbol{\beta}). \tag{3.61}$$

Thus, (3.59) becomes

$$\begin{aligned} \frac{1}{\sigma^2}(\mathbf{y} - X\boldsymbol{\beta})'(\mathbf{y} - X\boldsymbol{\beta}) &= \frac{1}{\sigma^2}(\mathbf{y} - X\boldsymbol{\beta})'(I - \mathcal{R})(\mathbf{y} - X\boldsymbol{\beta}) \\ &\quad + \frac{1}{\sigma^2}(\mathbf{y} - X\boldsymbol{\beta})'\mathcal{R}(\mathbf{y} - X\boldsymbol{\beta}), \end{aligned} \tag{3.62}$$

where $r(I - \mathcal{R}) + r(\mathcal{R}) = r(I)$. Applying Theorem 2.2.5, we have that the two quadratic forms on the right-hand side of (3.62) are independent; furthermore, σ^2 times their matrices are idempotent and, as we have seen, of rank $N - k$ and k. Hence (Problem 3.10 gives details),

$$\frac{(\mathbf{y} - X\boldsymbol{\beta})'\mathcal{R}(\mathbf{y} - X\boldsymbol{\beta})/k}{\mathbf{e}'\mathbf{e}/(N-k)} = \frac{\sigma^2\chi_k^2/k}{\sigma^2\chi_{N-k}^2/(N-k)} = F_{k, N-k}, \tag{3.63}$$

that is, using (3.61),

$$(3.63a)\qquad (\boldsymbol{\beta} - \hat{\boldsymbol{\beta}})'X'X(\boldsymbol{\beta} - \hat{\boldsymbol{\beta}}) = k\,\mathrm{MS}_e\,F_{k,\,N-k}.$$

Thus, a confidence region $C_{\boldsymbol{\beta}}$ of confidence level $1 - \alpha$ for $\boldsymbol{\beta}$ is

$$(3.64)\qquad C_{\boldsymbol{\beta}} = \{\boldsymbol{\beta} \mid (\boldsymbol{\beta} - \hat{\boldsymbol{\beta}})'X'X(\boldsymbol{\beta} - \hat{\boldsymbol{\beta}}) \leq k\,\mathrm{MS}_e\,F_{k,\,N-k;\,\alpha}\}$$

which is an ellipsoid in R^k. If the set (3.64) contains the point $\boldsymbol{\beta}_0$ we would accept the statement $\boldsymbol{\beta} = \boldsymbol{\beta}_0$.

We now state and prove a useful theorem.

Theorem 3.2.1. *Consider the linear model*

$$(3.65)\qquad \mathbf{y} = X\boldsymbol{\beta} + \boldsymbol{\varepsilon} = \boldsymbol{\eta} + \boldsymbol{\varepsilon},$$

where X is of order $N \times k$ and of full rank $k < N$, with $E(\boldsymbol{\varepsilon}) = \mathbf{0}$ and $V(\boldsymbol{\varepsilon}) = \sigma^2 I_N$. If H is of order $k \times k$ and nonsingular, and if $\mathbf{y}$ is regressed on $W = XH$, then the regression and error sum of squares are invariant.

Proof. Since H is nonsingular,

$$(3.66)\qquad \mathbf{y} = XHH^{-1}\boldsymbol{\beta} + \boldsymbol{\varepsilon} = W\boldsymbol{\phi} + \boldsymbol{\varepsilon}, \quad \boldsymbol{\phi} = H^{-1}\boldsymbol{\beta}, \quad W = XH.$$

If we regress $\mathbf{y}$ on W (see Problem 3.11), we find

$$(3.67)\qquad \hat{\boldsymbol{\phi}} = (W'W)^{-1}W'\mathbf{y} = H^{-1}\hat{\boldsymbol{\beta}}$$

and from the Gauss theorem, we have that $\hat{\boldsymbol{\phi}}$ is UMV unbiased for $\boldsymbol{\phi}$ among all unbiased estimators for $\boldsymbol{\phi}$ which are linear in the y_j's. When regressing $\mathbf{y}$ on W, the projection matrix is

$$(3.68)\qquad \mathcal{R}(W) = W(W'W)^{-1}W'$$

and (see Problem 3.11)

$$(3.68a)\qquad \mathcal{R}(W) = X(X'X)^{-1}X' = \mathcal{R}(X).$$

Now regressing $\mathbf{y}$ on W gives rise to

$$(3.69)\qquad \begin{aligned} &\text{(i)}\quad \mathrm{SS}_r(W) = \mathbf{y}'\mathcal{R}(W)\mathbf{y} = \mathbf{y}'\mathcal{R}(X)\mathbf{y} = \mathrm{SS}_r(X) \\ &\text{(ii)}\quad \mathrm{SS}_e(W) = \mathbf{y}'[I_N - \mathcal{R}(W)]\mathbf{y} = \mathbf{y}'[I_N - \mathcal{R}(X)]\mathbf{y} = \mathrm{SS}_e(X), \end{aligned}$$

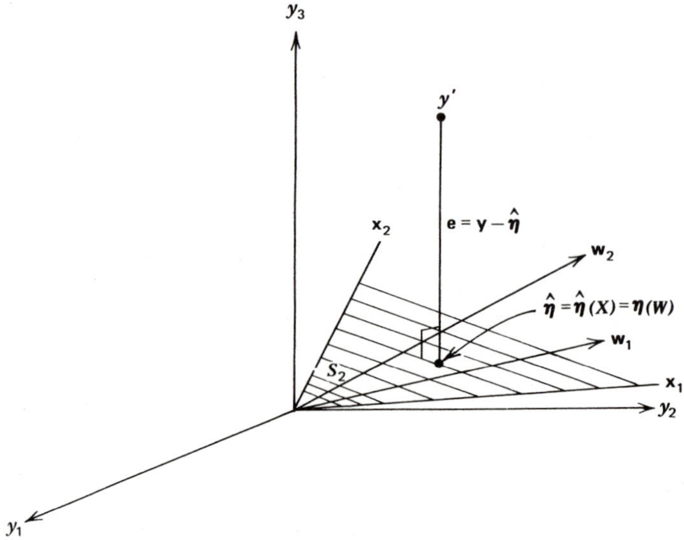

FIGURE 3.2.3. $W = XH$ recoordinatizes $S_k = S_2$ so that $\mathbf{w}_j, j = 1, 2$, belongs to S_2; that is, $S_2(X) = S_2(W)$. Projection into $S_2(X)$ or $S_2(W)$ is thus the same. $\mathbf{w}_j = X\mathbf{h}_j = [\mathbf{x}_1 \mid \mathbf{x}_2]\begin{pmatrix} h_{1j} \\ h_{2j} \end{pmatrix} = \mathbf{x}_1 h_{1j} + \mathbf{x}_2 h_{2j} \in S_2$. $S_2(X) = S_2(W) = S_2$.

which implies that *the sum of squares due to regression and error are invariant under nonsingular linear transformations of X*.

The above proof has been algebraic; the geometric proof (see Figure 3.2.3) is obvious, since we have just recoordinatized S_k by the transformation $W = XH$. After all, the jth column of W is $\mathbf{w}_j = X\mathbf{h}_j \in S_k$, and

$$\boldsymbol{\eta} = X\boldsymbol{\beta} = W\boldsymbol{\phi} \tag{3.70}$$

so that projection of $\mathbf{y}$ into S_k using X coordinates is accomplished by the matrix $\mathcal{R}(X)$, and using W coordinates by $\mathcal{R}(W)$, since the projection of $\mathbf{y}$ into S_k is the same using either coordinate system. That is,

$$\hat{\boldsymbol{\eta}} = \hat{\boldsymbol{\eta}}(X) = X\hat{\beta} = \mathcal{R}(X)\mathbf{y} = \hat{\boldsymbol{\eta}}(W) = W\hat{\boldsymbol{\phi}} = \mathcal{R}(W)\mathbf{y}, \tag{3.71}$$

since

$$\mathcal{R}(X) = \mathcal{R}(W). \tag{3.71a}$$

Geometrically, since the projection is the same, the squared length of the residual vector is

$$(3.72)\qquad (\mathbf{y} - \hat{\boldsymbol{\eta}}(X))'(\mathbf{y} - \hat{\boldsymbol{\eta}}(X)) = (\mathbf{y} - \hat{\boldsymbol{\eta}}(W))'(\mathbf{y} - \hat{\boldsymbol{\eta}}(W)),$$

and the squared length of the projection vector $\hat{\boldsymbol{\eta}}$ is

$$(3.73)\qquad \hat{\boldsymbol{\eta}}'(X)\hat{\boldsymbol{\eta}}(X) = \hat{\boldsymbol{\eta}}'(W)\hat{\boldsymbol{\eta}}(W).$$

3.3. BREAKDOWN OF THE REGRESSION SUM OF SQUARES: THE QUESTION OF NUISANCE PARAMETERS AND ELIMINATION BY ORTHOGONALIZATION

Often when working with the linear model, interest may focus on only a subset of the regression parameters involved. A simple example has been explored in Chapter 1; there we dealt with the linear model

$$(3.74)\qquad y = \beta_0 + \beta_1 x + \varepsilon = \eta + \varepsilon$$

and interest may be restricted to the rate constant $\beta_1 = d\eta/dx$. In this case we discovered that it was necessary to find the regression sum of squares due solely to the inclusion of $\beta_1 x$ in the model, and, indeed, when this was done, that part of the regression sum of squares due to $\beta_1 x$ was used in conjunction with the error sum of squares for making inference about β_1. The error sum of squares itself is found after fitting the "complete" model (3.74) to the data: It is the squared length of the vector $\mathbf{y} - \hat{\boldsymbol{\eta}}$, where $\hat{\boldsymbol{\eta}}$ is the projection of $\mathbf{y}$ into the estimation space defined as the span of two vectors, $\mathbf{1}$ and $\mathbf{x}$; the vector $\mathbf{1}$ is associated with the overall constant β_0 and the vector $\mathbf{x}$ is associated with the rate constant β_1. (We sometimes say that vector $\mathbf{1}$ *carries* β_0, the mean effect, and the vector $\mathbf{x}$ *carries* β_1, the effect due to the inclusion of the factor x in the model.)

Now consider the general model

$$(3.75)\qquad \mathbf{y} = X\boldsymbol{\beta} + \boldsymbol{\varepsilon} = \boldsymbol{\eta} + \boldsymbol{\varepsilon},$$

where X is of full rank, $r(X) = k < N$, and suppose that interest focuses on certain parameters β_j. Now partition X and $\boldsymbol{\beta}$ as

$$(3.76)\qquad X = (X_1 \mid X_2), \qquad \boldsymbol{\beta}' = (\boldsymbol{\beta}_1' \mid \boldsymbol{\beta}_2'),$$

where X_1 is of order $N \times k_1$, X_2 is $N \times k_2$, $\boldsymbol{\beta}_1$ is $k_1 \times 1$, and $\boldsymbol{\beta}_2$ is $k_2 \times 1$ with $k_1 + k_2 = k$, and suppose that interest is restricted to $\boldsymbol{\beta}_2$. We then say that $\boldsymbol{\beta}_1$ is a vector of *nuisance parameters*.

We may write the model (3.75) as

$$\mathbf{y} = X_1\boldsymbol{\beta}_1 + X_2\boldsymbol{\beta}_2 + \boldsymbol{\varepsilon}. \tag{3.77}$$

Two cases may occur.

CASE A. $\boldsymbol{\beta}_2$ is of interest, and X_1 is orthogonal to X_2; that is

$$X_1'X_2 = \bigcirc \tag{3.78}$$

where $\bigcirc$ stands for a $k_1 \times k_2$ matrix of zeros. Note that (3.78) implies that the k_1 column vectors of X_1 are orthogonal to the k_2 columns of X_2.

It is interesting to note the structure of the least-square estimates and the various regression sum of squares, when the condition of Case A holds. Note that

$$X'X = \begin{pmatrix} X_1' \\ \hline X_2' \end{pmatrix} \left(X_1 \mid X_2 \right) = \left(\begin{array}{c|c} X_1'X_1 & X_1'X_2 \\ \hline X_2'X_1 & X_2'X_2 \end{array}\right). \tag{3.79}$$

Using (3.78),

(3.79a)

$$X'X = \left(\begin{array}{c|c} X_1'X_1 & \bigcirc \\ \hline \bigcirc' & X_2'X_2 \end{array}\right) \quad \text{and} \quad (X'X)^{-1} = \left(\begin{array}{c|c} (X_1'X_1)^{-1} & \bigcirc \\ \hline \bigcirc' & (X_2'X_2)^{-1} \end{array}\right).$$

This in turn implies that

$$\hat{\boldsymbol{\beta}} = (X'X)^{-1}X'\mathbf{y} = \begin{pmatrix} (X_1'X_1)^{-1}X_1'\mathbf{y} \\ \hline (X_2'X_2)^{-1}X_2'\mathbf{y} \end{pmatrix} = \begin{pmatrix} \hat{\boldsymbol{\beta}}_1 \\ \hline \hat{\boldsymbol{\beta}}_2 \end{pmatrix}; \tag{3.79b}$$

that is, we may find the least-square estimates of $\boldsymbol{\beta}_2$ by regressing $\mathbf{y}$ only on X_2. Indeed, the variance–covariance matrix of $\hat{\boldsymbol{\beta}}$ is

$$V(\hat{\boldsymbol{\beta}}) = V\begin{bmatrix} \hat{\boldsymbol{\beta}}_1 \\ \hline \hat{\boldsymbol{\beta}}_2 \end{bmatrix} = \sigma^2(X'X)^{-1} = \sigma^2 \left(\begin{array}{c|c} (X_1'X_1)^{-1} & \bigcirc \\ \hline \bigcirc' & (X_2'X_2)^{-1} \end{array}\right) \tag{3.79c}$$

which is to say that the estimator $\hat{\boldsymbol{\beta}}_1$ is uncorrelated with the estimator $\hat{\boldsymbol{\beta}}_2$. If $\mathbf{y}$ is normal, this further means that $\hat{\boldsymbol{\beta}}_1$ and $\hat{\boldsymbol{\beta}}_2$ are independent of one another.

We recall [see (3.19*c*)] that the orthogonal projection of $\mathbf{y}$ into S_k is

$$\hat{\boldsymbol{\eta}} = \mathcal{R}(X)\mathbf{y}, \qquad \mathcal{R}(X) = X(X'X)^{-1}X' \tag{3.80}$$

and using (3.76) and (3.79*b*), we have

$$\begin{aligned} \mathcal{R}(X) &= X_1(X_1'X_1)^{-1}X_1' + X_2(X_2'X_2)^{-1}X_2' \\ &= \mathcal{R}(X_1) + \mathcal{R}(X_2). \end{aligned} \tag{3.80a}$$

Note that $\mathcal{R}(X_1)$ and $\mathcal{R}(X_2)$ are symmetric and idempotent. Now the regression sum of squares when projecting $\mathbf{y}$ into S_k is $\mathbf{y}'\mathcal{R}(X)\mathbf{y}$ which, in view of (3.80*a*), can be written

$$\hat{\boldsymbol{\eta}}'\hat{\boldsymbol{\eta}}' = \mathbf{y}'\mathcal{R}(X)\mathbf{y} = \mathbf{y}'\mathcal{R}(X_1)\mathbf{y} + \mathbf{y}'\mathcal{R}(X_2)\mathbf{y}. \tag{3.80b}$$

We have arrived at a breakdown of the regression sum of squares $\hat{\boldsymbol{\eta}}'\hat{\boldsymbol{\eta}}$ into the sum of two parts:

(i) $\mathbf{y}'\mathcal{R}(X_1)\mathbf{y}$, due to the inclusion of $X_1\boldsymbol{\beta}_1$ in the model;
(ii) $\mathbf{y}'\mathcal{R}(X_2)\mathbf{y}$, due to the inclusion of $X_2\boldsymbol{\beta}_2$ in the model.

Geometrically, it is very easy to see what has happened. The observational vector $\mathbf{y}$ may be projected (see Figure 3.3.1) into S_{k_1}, the span of the columns of X_1, to give $\hat{\boldsymbol{\eta}}_1 = \mathcal{R}(X_1)\mathbf{y}$ and also into S_{k_2}, the span of the columns of X_2, to give $\hat{\boldsymbol{\eta}}_2 = \mathcal{R}(X_2)\mathbf{y}$. The vectors $\hat{\boldsymbol{\eta}}_1$ and $\hat{\boldsymbol{\eta}}_2$ belong to S_k, since S_{k_1} and S_{k_2} are clearly subsets of S_k, and it is easy to see (Problem 3.12) that

$$\hat{\boldsymbol{\eta}}_1'\hat{\boldsymbol{\eta}}_2 = 0, \tag{3.80c}$$

since X_1 is orthogonal to X_2. The resultant vector of $\hat{\boldsymbol{\eta}}_1$ and $\hat{\boldsymbol{\eta}}_2$ is [see (3.80*a*)]

$$\hat{\boldsymbol{\eta}}_1 + \hat{\boldsymbol{\eta}}_2 = [\mathcal{R}(X_1) + \mathcal{R}(X_2)]\mathbf{y} = \mathcal{R}(X)\mathbf{y} = \hat{\boldsymbol{\eta}}, \tag{3.81}$$

that is, the sum of the projections into S_{k_1} and S_{k_2} is the same as the projection of $\mathbf{y}$ into S_k, where S_k is the sum of the sets S_{k_1} and S_{k_2} (see Figure 3.3.1) and S_{k_1} is orthogonal to S_{k_2}. If we project down first into S_k, arriving at $\hat{\boldsymbol{\eta}}$, and *resolve* $\hat{\eta}$ by projection into S_{k_1} and S_{k_2}, we find (Problem 3.12)

$$\hat{\boldsymbol{\eta}}_i = \mathcal{R}(X_i)\hat{\boldsymbol{\eta}} = \mathcal{R}(X_i)\mathbf{y}. \tag{3.82}$$

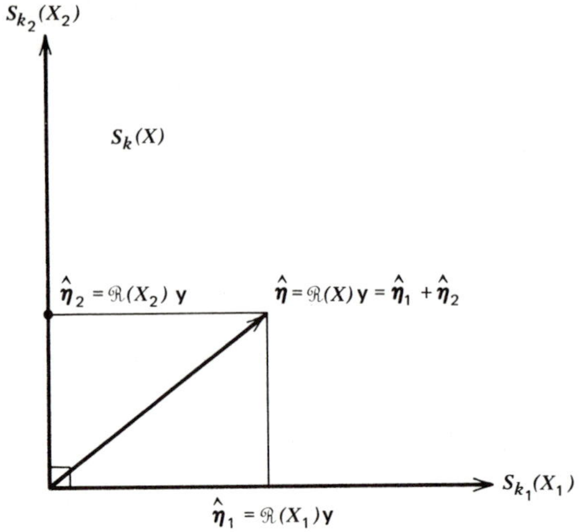

FIGURE 3.3.1. Diagrammatic view of the projection of **y** (**y** not shown) into S_{k_1} (the span of X_1) and S_{k_2} (the span of X_2). $\hat{\boldsymbol{\eta}}'\hat{\boldsymbol{\eta}} = \hat{\boldsymbol{\eta}}_1'\hat{\boldsymbol{\eta}}_1 + \hat{\boldsymbol{\eta}}_2'\hat{\boldsymbol{\eta}}_2$; $\hat{\boldsymbol{\eta}}_1'\hat{\boldsymbol{\eta}}_2 = 0$.

Since X_1 is orthogonal to X_2, we have

$$\hat{\boldsymbol{\eta}}_1 + \hat{\boldsymbol{\eta}}_2 = \hat{\boldsymbol{\eta}}. \tag{3.82a}$$

Again, because X_1 is orthogonal to X_2, we obtain the breakdown (3.80*b*), since in view of the above (see Figure 3.3.1),

$$\hat{\boldsymbol{\eta}}'\hat{\boldsymbol{\eta}} = \hat{\boldsymbol{\eta}}_1'\hat{\boldsymbol{\eta}}_1 + \hat{\boldsymbol{\eta}}_2'\hat{\boldsymbol{\eta}}_2, \tag{3.83}$$

and because $\mathcal{R}(X_i)$ are symmetric and idempotent, as is $\mathcal{R}(X)$, we easily find

$$\hat{\boldsymbol{\eta}}'\hat{\boldsymbol{\eta}} = \mathbf{y}'\mathcal{R}(X)\mathbf{y} = \mathbf{y}'\mathcal{R}(X_1)\mathbf{y} + \mathbf{y}'\mathcal{R}(X_2)\mathbf{y}. \tag{3.83a}$$

We note that the error sum of squares, when regressing **y** on X, is

$$\begin{aligned}\mathbf{y}'\mathbf{y} - \mathbf{y}'\mathcal{R}(X)\mathbf{y} &= \mathbf{y}'\mathbf{y} - [\mathbf{y}'\mathcal{R}(X_1)\mathbf{y} + \mathbf{y}'\mathcal{R}(X_2)\mathbf{y}] \\ &= \mathbf{y}'(I - \mathcal{R})\mathbf{y} = \mathbf{y}'[I - \mathcal{R}(X_1) - \mathcal{R}(X_2)]\mathbf{y}.\end{aligned} \tag{3.84}$$

It is convenient to summarize these results in an analysis of variance table—Table 3.3.1.

TABLE 3.3.1. Analysis of Variance Table for Case A

Source	Sum of Squares	Degrees of Freedom	Mean Square	Expected Mean Square
Due to X_1	$\mathbf{y}'\mathfrak{R}(X_1)\mathbf{y}$	k_1	$\mathbf{y}'\mathfrak{R}(X_1)\mathbf{y}/k_1$	$\sigma^2 + \frac{1}{k_1}\boldsymbol{\beta}_1'X_1'X_1\boldsymbol{\beta}_1$
Due to X_2	$\mathbf{y}'\mathfrak{R}(X_2)\mathbf{y}$	k_2	$\mathbf{y}'\mathfrak{R}(X_2)\mathbf{y}/k_2$	$\sigma^2 + \frac{1}{k_2}\boldsymbol{\beta}_2'X_2'X_2\boldsymbol{\beta}_2$
Residual	$\mathbf{y}'[I - \mathfrak{R}(X_1) - \mathfrak{R}(X_2)]\mathbf{y}$	$N - k = N - k_1 - k_2$	$\mathbf{y}'[I - \mathfrak{R}(X_1) - \mathfrak{R}(X_2)]\mathbf{y}/(N - k)$	σ^2
Total	$\mathbf{y}'\mathbf{y}$	N		

TABLE 3.3.1a. Usual Analysis of Variance Table for Case A
$X_1'X_2 = O$; **Interest in** $\boldsymbol{\beta}_2$

Source	Sum of Squares	Degrees of Freedom	Mean Square	Expected Mean Square
Due to X_2	$\mathbf{y}'\mathfrak{R}(X_2)\mathbf{y}$	k_2	$\mathbf{y}'\mathfrak{R}(X_2)\mathbf{y}/k_2$	$\sigma^2 + \frac{1}{k_2}\boldsymbol{\beta}_2'X_2'X_2\boldsymbol{\beta}_2$
Residual	$\mathbf{y}'[I - \mathfrak{R}(X_1) - \mathfrak{R}(X_2)]\mathbf{y}$	$N - k_1 - k_2$	$\mathbf{y}'[I - \mathfrak{R}(X_1) - \mathfrak{R}(X_2)]\mathbf{y}/(N - k)$	σ^2
Total	$\mathbf{y}'[I - \mathfrak{R}(X_1)]\mathbf{y}$	$N - k_1$		

If interest is restricted to $\boldsymbol{\beta}_2$, the usual analysis of variance table quoted is a modified Table 3.3.1, with the first line missing, and the last line (i.e., the total line) adjusted accordingly, as in Table 3.3.1a.

The two tables just presented are really quite informative. The first states that when interested in all of $\boldsymbol{\beta} = (\boldsymbol{\beta}_1', \boldsymbol{\beta}_2')$, regress $\mathbf{y}$ on $X = (X_1 \mid X_2)$, and the total sum of squares is $\mathbf{y}'\mathbf{y}$. The second table states that, when interest is restricted to $\boldsymbol{\beta}_2$ alone, the total sum of squares is

(3.85) $$\mathbf{y}'[I - \mathfrak{R}(X_1)]\mathbf{y} = \mathbf{y}'[I - \mathfrak{R}(X_1)]'[I - \mathfrak{R}(X_1)]\mathbf{y}$$

(since $[I - \mathfrak{R}(X_1)]$ is idempotent and symmetric); that is, we may write

(3.86) $$\mathbf{y}'[I - \mathfrak{R}(X_1)]\mathbf{y} = \mathbf{e}_1'\mathbf{e}_1,$$

where

(3.86a) $$\mathbf{e}_1 = [I - \mathfrak{R}(X_1)]\mathbf{y} = \mathbf{y} - \hat{\boldsymbol{\eta}}_1.$$

In words, $\mathbf{e}_1$ is the vector of residuals found after regressing $\mathbf{y}$ on X_1 alone [remember, we are dealing with the case $X = (X_1 \mid X_2)$, $X_1'X_2 = \bigcirc$]. This suggests the following two-stage procedure.

Stage (a). Regress $\mathbf{y}$ on X_1 and obtain, in the usual way, $\hat{\boldsymbol{\beta}}_1$, $\mathbf{y}'\mathcal{R}(X_1)\mathbf{y}$, and residuals $\mathbf{e}_1 = \mathbf{y} - X_1\hat{\boldsymbol{\beta}}_1 = [I - \mathcal{R}(X_1)]\mathbf{y}$. We note that (3.86) holds, that is, the residual sum of squares at this point is $\mathbf{e}_1'\mathbf{e}_1 = \mathbf{y}'(I - \mathcal{R}_1)\mathbf{y}$. But the model includes X_2 as well as X_1, so this residual sum of squares cannot serve as an estimator of σ^2 alone, since it does not have a portion which calls for the removal of the effects carried by X_2. Indeed (see Problem 3.13),

$$E[\mathbf{e}_1'\mathbf{e}_1] = (N - k_1)\sigma^2 + \boldsymbol{\beta}_2'X_2'X_2\boldsymbol{\beta}_2. \tag{3.87}$$

Hence, continuing our two-stage procedure, we have the following.

Stage (b). Regress $\mathbf{e}_1$, regarded as a (new) dependent variable vector, on X_2. We find, in the ordinary way, but being careful with our notation, that the least-squares estimator of the regression coefficients is

$$(X_2'X_2)^{-1}X_2'\mathbf{e}_1 = \hat{\boldsymbol{\beta}}_2, \tag{3.88}$$

since the left-hand side of (3.88) is

$$\begin{aligned}(X_2'X_2)^{-1}X_2'[I - \mathcal{R}(X_1)]\mathbf{y} &= (X_2'X_2)^{-1}[X_2' - X_2'\mathcal{R}(X_1)]\mathbf{y} \\ &= (X_2'X_2)^{-1}X_2'\mathbf{y} = \hat{\boldsymbol{\beta}}_2\end{aligned} \tag{3.88a}$$

[see (3.79*b*)] since $X_2'\mathcal{R}(X_1) = X_2'X_1(X_1'X_1)^{-1}X_1' = \bigcirc$.

Recall that when we are regressing $\mathbf{y}$ on X, the regression sum of squares is $\mathbf{y}'\mathcal{R}\mathbf{y}$, with $\mathcal{R} = \mathcal{R}(X) = X(X'X)^{-1}X'$. So for Stage (b), the regression sum of squares (we are regressing $\mathbf{e}_1$ on X_2) is

$$\mathbf{e}_1'\mathcal{R}(X_2)\mathbf{e}_1 = \mathbf{y}'[I - \mathcal{R}(X_1)]'\mathcal{R}(X_2)[I - \mathcal{R}(X_1)]\mathbf{y}, \tag{3.89}$$

and this (see Problem 3.14) is

$$\mathbf{e}_1'\mathcal{R}(X_2)\mathbf{e}_1 = \mathbf{y}'\mathcal{R}(X_2)\mathbf{y}, \tag{3.89a}$$

which is the regression sum of squares due to X_2 being in the model. That is, this two-stage procedure gives the correct regression sum of squares due to X_2.

We further recall that when regressing $\mathbf{y}$ on X, the error sum of squares is $\mathbf{y}'[I - \mathcal{R}(X)]\mathbf{y}$. Here we are regressing $\mathbf{e}_1$ on X_2 so the residual sum of squares for Stage (b) of the procedure is

$$\mathbf{e}_1'[I - \mathcal{R}(X_2)]\mathbf{e}_1 = \mathbf{y}'[I - \mathcal{R}(X_1) - \mathcal{R}(X_2)]\mathbf{y} \tag{3.90}$$

which we know is the correct residual sum of squares when regressing $\mathbf{y}$ on $X = (X_1 \mid X_2)$, with $X_1'X_2 = O$, and of course serves as a basis for estimating the pure error σ^2 (see the EMS column of Table 3.3.1a).

This two-stage procedure is a useful way of implementing a regression analysis when $E(\mathbf{y}) = X_1\boldsymbol{\beta}_1 + X_2\boldsymbol{\beta}_2$, and interest is in $\boldsymbol{\beta}_2$ alone, and either $X_1'X_2 = O$ or $X_1'X_2 \neq O$—the latter case to be discussed below. At this point we emphasize the following five points which hold when $X_1'X_2 = O$.

1. If X_2 had not been included in the model, then the regression sum of squares found after regressing $\mathbf{y}$ on X_1 alone would have been $\mathbf{y}'\mathcal{R}(X_1)\mathbf{y}$ and $\mathbf{y}'[I - \mathcal{R}(X_1)]\mathbf{y}$ would have been the residual sum of squares. We then call $\mathbf{y}'\mathcal{R}(X_2)\mathbf{y}$ the *extra regression sum of squares* due to inclusion of X_2 in the model.

2. We also note that if X_2 had not been included in the model, the residual sum of squares if regressing $\mathbf{y}$ on X_1 alone would have been $\mathbf{y}'[I - \mathcal{R}(X_1)]\mathbf{y}$ and we note that in Case A ($X_1'X_2 = O$),

$$\mathbf{y}'[I - \mathcal{R}(X_1)]\mathbf{y} > \mathbf{y}'[I - \mathcal{R}(X_1) - \mathcal{R}(X_2)]\mathbf{y}. \tag{3.91}$$

Points 1 and 2 may be rephrased as follows: when X_2 is included in the model in addition to X_1, the regression sum of squares *increases* [by the amount $\mathbf{y}'\mathcal{R}(X_2)\mathbf{y}$, the regression sum of squares due to X_2], and the residual sum of squares *decreases* (by the same amount).

3. Note that the residual sum of squares (3.90) may be expressed as

$$\mathbf{e}_1'\mathbf{e}_1 - \mathbf{e}_1'\mathcal{R}(X_2)\mathbf{e}_1 = \mathbf{y}'[I - \mathcal{R}(X_1)]\mathbf{y} - \mathbf{y}'\mathcal{R}(X_2)\mathbf{y}. \tag{3.91a}$$

In terms of our two-step procedure, this may be described as follows. We regress $\mathbf{y}$ on a matrix X_1 and form the residual sum of squares $\mathbf{e}_1'\mathbf{e}_1$, where $\mathbf{e}_1 = [I - \mathcal{R}(X_1)]\mathbf{y}$. From this residual sum of squares, we subtract $\mathbf{e}_1'\mathcal{R}(X_2)\mathbf{e}_1 = \mathbf{y}'\mathcal{R}(X_2)\mathbf{y}$, a regression sum of squares found by regressing $\mathbf{e}_1$ on the *orthogonal* part to X_1; namely, in this case X_2 ($X_1'X_2 = O$). Note that $\mathcal{R}(X_1)$ is orthogonal to $\mathcal{R}(X_2)$, as is easily verified.

4. The extra regression sum of squares due to X_2, $\mathbf{y}'\mathcal{R}(X_2)\mathbf{y}$, may be calculated using (remember $X_1'X_2 = O$ here)

$$\mathbf{y}'[I - \mathcal{R}(X_1)]\mathbf{y} - \mathbf{y}'[I - \mathcal{R}(X_1) - \mathcal{R}(X_2)]\mathbf{y} = \mathbf{y}'\mathcal{R}(X_2)\mathbf{y}, \tag{3.91b}$$

which is the difference between the residual sum of squares found after regressing $\mathbf{y}$ on X_1 and the residual sum of squares found after regressing $\mathbf{y}$ on $X = (X_1 \mid X_2)$, or

$$\mathbf{y}'\mathcal{R}(X)\mathbf{y} - \mathbf{y}'\mathcal{R}(X_1)\mathbf{y} = \mathbf{y}'\mathcal{R}(X_2)\mathbf{y}, \tag{3.91c}$$

that is, by subtracting the regression sum of squares found after regressing $\mathbf{y}$ on X_1 from the (total) regression sum of squares found by regressing $\mathbf{y}$ on $X = (X_1 \mid X_2)$.

Note that we may write the extra regression sum of squares due to $\boldsymbol{\beta}_2$ as [using (3.91b); compare with (1.59f) of Chapter 1]

$$SS_r(\boldsymbol{\beta}_2) = SS_e(\boldsymbol{\beta}_1) - SS_e(\boldsymbol{\beta}_1, \boldsymbol{\beta}_2), \tag{3.91d}$$

where of course

$$\begin{aligned} SS_e(\boldsymbol{\beta}_1) &= SS_e(X_1) = \mathbf{y}'[I - \mathcal{R}(X_1)]\mathbf{y} \\ SS_e(\boldsymbol{\beta}_1, \boldsymbol{\beta}_2) &= SS_e(X_1, X_2) = \mathbf{y}'[I - \mathcal{R}(X_1) - \mathcal{R}(X_2)]\mathbf{y} \\ SS_r(\boldsymbol{\beta}_2) &= \mathbf{y}'\mathcal{R}(X_2)\mathbf{y}. \end{aligned} \tag{3.91e}$$

In fact, if we do some simple algebra on the left-hand side of (3.91b), we find

$$SS_r(\boldsymbol{\beta}_2) = SS_r(\boldsymbol{\beta}_1, \boldsymbol{\beta}_2) - SS_r(\boldsymbol{\beta}_1) \tag{3.91f}$$

with

$$\begin{aligned} SS_r(\boldsymbol{\beta}_1, \boldsymbol{\beta}_2) &= \mathbf{y}'\mathcal{R}(X)\mathbf{y} = \mathbf{y}'[\mathcal{R}(X_1) + \mathcal{R}(X_2)]\mathbf{y}, \\ SS_r(\boldsymbol{\beta}_1) &= \mathbf{y}'\mathcal{R}(X_1)\mathbf{y}. \end{aligned} \tag{3.91g}$$

Equation (3.91d) states that if we regress $\mathbf{y}$ on X_1 and find the error sum of squares $SS_e(\boldsymbol{\beta}_1)$, and then regress $\mathbf{y}$ on $X = (X_1 \mid X_2)$, with $X_1'X_2 = \mathrm{O}$, finding $SS_e(X) = SS_e(X_1 \mid X_2) = SS_e(\boldsymbol{\beta}_1, \boldsymbol{\beta}_2)$, then the difference of these two error sum of squares is the extra regression sum of squares due to $\boldsymbol{\beta}_2$. A similar description in terms of the regression sum of squares results in (3.91f). Equations (3.91d) and (3.91f) are quite general and hold even when $X_1'X_2 \neq \mathrm{O}$—see below. An example of (3.91d) is given in Chapter 1; see (1.59f) of Chapter 1.

5. A *cautionary* remark for the reader is warranted at this point. One might suppose that if interest is in $\boldsymbol{\beta}_2$ alone, and that if we wish to calculate $\hat{\boldsymbol{\beta}}_2$ and the contribution to the regression sum of squares due to X_2 (which

carries $\boldsymbol{\beta}_2$), then we should proceed by regressing $\mathbf{y}$ on X_2 alone. This produces $\hat{\boldsymbol{\beta}}_2 = (X_2'X_2)^{-1}X_2'\mathbf{y}$, but the residual sum of squares would be $\mathbf{y}'[I - \mathfrak{R}(X_2)]\mathbf{y}$, and it can be shown (Problem 3.15) that

$$E\{\mathbf{y}'[I - \mathfrak{R}(X_2)]\mathbf{y}\} = (N - k_2)\sigma^2 + \boldsymbol{\beta}_1'X_1'X_1\boldsymbol{\beta}_1. \tag{3.92}$$

That is, since X_1 is in the model and is used to generate $\mathbf{y}$, $\mathbf{y}'[I - \mathfrak{R}(X_2)]\mathbf{y}$ does not lead to information about σ^2 alone.

Finally, our two-step procedure gives rise to the following analysis of variance table, Table 3.3.2, which of course is equivalent to Tables 3.3.1 and 3.3.1a.

Case B. $\boldsymbol{\beta}_2$ is of interest, and $X_1'X_2 \neq O$. We again assume

$$\begin{aligned} E(\mathbf{y}) &= X\boldsymbol{\beta} = (X_1 \mid X_2)(\boldsymbol{\beta}_1' \mid \boldsymbol{\beta}_2')' \\ &= X_1\boldsymbol{\beta}_1 + X_2\boldsymbol{\beta}_2, \end{aligned} \tag{3.93}$$

where interest focuses on $\boldsymbol{\beta}_2$, and X_1 is of order $N \times k_1$, X_2 is $N \times k_2$, $\boldsymbol{\beta}_2$ is $k_2 \times 1$, and so on. We may reduce this to the orthogonal case by a process which we shall call *orthogonalization*.

TABLE 3.3.2. Analysis of Variance For $E(\mathbf{y}) = X_1\boldsymbol{\beta}_1 + X_2\boldsymbol{\beta}_2$, $X_1'X_2 = O$; Interest in $\boldsymbol{\beta}_2$

Source	Sum of Squares	Degrees of Freedom	Expected Mean Square
Due to X_1	$\mathbf{y}'\mathfrak{R}(X_1)\mathbf{y}$	k_1	$\sigma^2 + \frac{1}{k_1}\boldsymbol{\beta}_1'X_1'X_1\boldsymbol{\beta}_1$
Extra, due to X_2	$\mathbf{e}_1'\mathfrak{R}(X_2)\mathbf{e}_1 = \mathbf{y}'\mathfrak{R}(X_2)\mathbf{y}$	k_2	$\sigma^2 + \frac{1}{k_2}\boldsymbol{\beta}_2'X_2'X_2\boldsymbol{\beta}_2$
Residual	$\mathbf{e}_1'[I - \mathfrak{R}(X_2)]\mathbf{e}_1 =$ $\mathbf{y}'[I - \mathfrak{R}(X_1) - \mathfrak{R}(X_2)]\mathbf{y}$	$N - k$	σ^2
Total	$\mathbf{y}'\mathbf{y}$	N	

Line 1: Regress $\mathbf{y}$ on X_1, find residuals $\mathbf{e}_1$ and $\mathbf{y}'\mathfrak{R}(X_1)\mathbf{y}$.
Line 2: Regress $\mathbf{e}_1$ on X_2, find residuals $\mathbf{e}_2 = [I - \mathfrak{R}(X_2)]\mathbf{e}_1$, extra regression sum of squares $\mathbf{e}_1'\mathfrak{R}(X_2)\mathbf{e}_1$, and residual sum of squares $\mathbf{e}_2'\mathbf{e}_2 = \mathbf{y}'[I - \mathfrak{R}(X_1) - \mathfrak{R}(X_2)]\mathbf{y}$.

The term *orthogonalization* means the process of regressing each column of X_2 on X_1 and constructing a matrix of residuals, where, for example the jth column of this matrix is the residual vector found when regressing the jth column of X_2 on X_1, for $j = 1,\ldots,k_2$. This matrix of residuals, denoted by $X_{2\cdot 1}$ (the notation stands for the fact that columns of X_2 are regressed on X_1), is given by

$$X_{2\cdot 1} = X_2 - X_1 A, \qquad A = (X_1'X_1)^{-1}X_1'X_2, \tag{3.94}$$

where A is of order $k_1 \times k_2$ and X_1 is orthogonal to $X_{2\cdot 1}$, since

$$X_1'X_{2\cdot 1} = X_1'X_2 - X_1'X_1A = X_1'X_2 - X_1'X_2 = \bigcirc, \tag{3.94a}$$

where $\bigcirc$ is a $k_1 \times k_2$ matrix of zeros. The $k_1 \times k_2$ matrix A is called in the literature the *alias* matrix [see the remarks after (3.95b) and (3.200a)]. We note that vectors belonging to S_k are contained in the space spanned by the columns of X_1 and $X_{2\cdot 1}$, and vice versa. Indeed, we have recoordinatized S_k so that we are referring not to the columns of X_1 and X_2, which are not orthogonal to each other, but to columns of X_1 and $X_{2\cdot 1}$, which are orthogonal to each other (see Figure 3.3.2). Indeed, we now may write

$$\begin{aligned} E(\mathbf{y}) &= X_1\boldsymbol{\beta}_1 + X_2\boldsymbol{\beta}_2 \\ &= (X_1\boldsymbol{\beta}_1 + X_1A\boldsymbol{\beta}_2) + (X_2\boldsymbol{\beta}_2 - X_1A\boldsymbol{\beta}_2) \\ &= X_1(\boldsymbol{\beta}_1 + A\boldsymbol{\beta}_2) + (X_2 - X_1A)\boldsymbol{\beta}_2; \end{aligned} \tag{3.95}$$

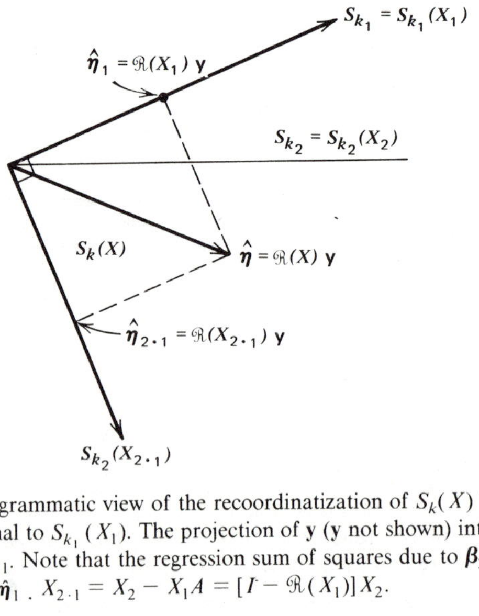

FIGURE 3.3.2. Diagrammatic view of the recoordinatization of $S_k(X)$ by orthogonalization. $S_{k_2}(X_{2\cdot 1})$ is orthogonal to $S_{k_1}(X_1)$. The projection of $\mathbf{y}$ ($\mathbf{y}$ not shown) into S_k may be resolved so that $\hat{\boldsymbol{\eta}} = \hat{\boldsymbol{\eta}}_1 + \hat{\boldsymbol{\eta}}_{2\cdot 1}$. Note that the regression sum of squares due to $\boldsymbol{\beta}_2$ is $\hat{\boldsymbol{\eta}}_{2\cdot 1}'\hat{\boldsymbol{\eta}}_{2\cdot 1}$ and that $\hat{\boldsymbol{\eta}}_{2\cdot 1}'\hat{\boldsymbol{\eta}}_{2\cdot 1} = \hat{\boldsymbol{\eta}}'\hat{\boldsymbol{\eta}} - \hat{\boldsymbol{\eta}}_1'\hat{\boldsymbol{\eta}}_1$. $X_{2\cdot 1} = X_2 - X_1A = [I - \mathcal{R}(X_1)]X_2$.

that is,

$$E(\mathbf{y}) = X_1\boldsymbol{\phi} + X_{2\cdot 1}\boldsymbol{\beta}_2 \tag{3.95a}$$

where $X_1'X_{2\cdot 1} = \bigcirc$, and the $k_1 \times 1$ vector $\boldsymbol{\phi}$ is given by

$$\boldsymbol{\phi} = \boldsymbol{\beta}_1 + A\boldsymbol{\beta}_2. \tag{3.95b}$$

Note that $\boldsymbol{\beta}_1$ differs from $\boldsymbol{\phi}$ by a quantity that depends on A, and this is one reason for calling A the "alias" matrix. We are now in the same position as in Case A—interest focuses on $\boldsymbol{\beta}_2$, and $X_{2\cdot 1}'X_1 = \bigcirc$, a $k_2 \times k_1$ matrix of zeros. We know from Case A that

$$\hat{\boldsymbol{\beta}}_2 = (X_{2\cdot 1}'X_{2\cdot 1})^{-1}X_{2\cdot 1}'\mathbf{y}, \tag{3.96}$$

and we may construct an analysis of variance table as shown in Table 3.3.3.

We note, *using results of Case A*, that

(3.97)

$$\begin{pmatrix} \hat{\boldsymbol{\phi}} \\ \hat{\boldsymbol{\beta}}_2 \end{pmatrix} = \begin{pmatrix} (X_1'X_1)^{-1} & \bigcirc \\ \bigcirc' & (X_{2\cdot 1}'X_{2\cdot 1})^{-1} \end{pmatrix} \begin{pmatrix} X_1' \\ X_{2\cdot 1}' \end{pmatrix} \mathbf{y} = \begin{pmatrix} (X_1'X_1)^{-1}X_1'\mathbf{y} \\ (X_{2\cdot 1}'X_{2\cdot 1})^{-1}X_{2\cdot 1}'\mathbf{y} \end{pmatrix}$$

and that the regression sum of squares when regressing $\mathbf{y}$ on X is

$$\mathbf{y}'\mathfrak{R}(X)\mathbf{y} = \mathbf{y}'\mathfrak{R}(X_1)\mathbf{y} + \mathbf{y}'\mathfrak{R}(X_{2\cdot 1})\mathbf{y} \tag{3.97a}$$

TABLE 3.3.3. $E(\mathbf{y}) = X_1\boldsymbol{\beta}_1 + X_2\boldsymbol{\beta}_2$; **Interest in** $\boldsymbol{\beta}_2$; $X_1'X_2 \neq \bigcirc$

Source	Sum of Squares	Degrees of Freedom	Expected Mean Square
Due to X_1	$\hat{\boldsymbol{\phi}}'X_1'X_1\hat{\boldsymbol{\phi}} = \mathbf{y}'\mathfrak{R}(X_1)\mathbf{y}$	k_1	$\sigma^2 + \frac{1}{k_1}\boldsymbol{\phi}'X_1'X_1\boldsymbol{\phi}$
Extra due to $X_{2\cdot 1}$	$\hat{\boldsymbol{\beta}}_2'X_{2\cdot 1}'X_{2\cdot 1}\hat{\boldsymbol{\beta}}_2 = \mathbf{y}'\mathfrak{R}(X_{2\cdot 1})\mathbf{y}$	k_2	$\sigma^2 + \frac{1}{k_2}\boldsymbol{\beta}_2'X_{2\cdot 1}'X_{2\cdot 1}\boldsymbol{\beta}_2$
Residual	$\mathbf{y}'[I - \mathfrak{R}(X_1) - \mathfrak{R}(X_{2\cdot 1})]\mathbf{y}$	$N - k_1 - k_2$	σ^2
Total	$\mathbf{y}'\mathbf{y}$	N	

$\mathfrak{R}(X_1) = X_1(X_1'X_1)^{-1}X_1'$; $\mathfrak{R}(X_{2\cdot 1}) = X_{2\cdot 1}(X_{2\cdot 1}'X_{2\cdot 1})^{-1}X_{2\cdot 1}'$

so that the residual sum of squares is

$$(3.97b)\quad \mathbf{y}'\mathbf{y} - \mathbf{y}'\mathcal{R}(X_1)\mathbf{y} - \mathbf{y}'\mathcal{R}(X_{2\cdot 1})\mathbf{y} = \mathbf{y}'[I - \mathcal{R}(X_1) - \mathcal{R}(X_{2\cdot 1})]\mathbf{y}.$$

We may use the two-stage procedure of Case A to formulate a three-stage procedure for Case B, as follows.

1. Regress X_2 on X_1 and take residuals to form $X_{2\cdot 1}$; rewrite the model as in (3.95*a*).
2. Regress $\mathbf{y}$ on X_1 and find $\mathbf{e}_1 = [I - \mathcal{R}(X_1)]\mathbf{y}$;
3. Regress $\mathbf{e}_1$ on $X_{2\cdot 1}$.

This three-stage procedure gives rise to the analysis of variance table shown in Table 3.3.3. Omitting the usual first line and adjusting the last line accordingly, we obtain Table 3.3.4 (see also Table 3.3.1a).

When we regress $\mathbf{y}$ on X_1, and obtain residuals $\mathbf{e}_1$ in the above procedure, we often say that we have *eliminated* X_1 (or the effects due to X_1), since the above procedure calls for regressing $\mathbf{e}_1$ on $X_{2\cdot 1}$, where, as we have seen in (3.95*a*), $X_{2\cdot 1}$ carries $\boldsymbol{\beta}_2$ ($\boldsymbol{\beta}_2$ is of interest, with $\boldsymbol{\beta}_1$ nuisance parameters, where X_1 carries $\boldsymbol{\beta}_1$). Indeed (see Problem 3.16), we have

$$(3.98)\qquad E(\mathbf{e}_1) = X_{2\cdot 1}\boldsymbol{\beta}_2,$$

so at this stage $\mathbf{e}_1$ is regressed on $X_{2\cdot 1}$ in the explicit absence of X_1.

Geometrically, what we have done is recoordinatized S_k so that each column vector of $X_{2\cdot 1}$ is orthogonal to X_1 (see Figure 3.3.2). We do this by

TABLE 3.3.4. Analysis of Variance for $E(\mathbf{y}) = X_1\boldsymbol{\beta}_1 + X_2\boldsymbol{\beta}_2$; $X_1'X_2 \neq O$; Interest in $\boldsymbol{\beta}_2$

Source	Sum of Squares	Degrees of Freedom	Expected Mean Square
Extra, due to $X_{2\cdot 1}$ (X_2, given X_1 in model)	$\mathbf{e}_1'\mathcal{R}(X_{2\cdot 1})\mathbf{e}_1 = \mathbf{y}'\mathcal{R}(X_{2\cdot 1})\mathbf{y}$	k_2	$\sigma^2 + \frac{1}{k_2}\boldsymbol{\beta}_2'X_{2\cdot 1}'X_{2\cdot 1}\boldsymbol{\beta}_2$
Residual	$\mathbf{e}_1'[I - \mathcal{R}(X_{2\cdot 1})]\mathbf{e}_1 = \mathbf{y}'[I - \mathcal{R}(X_1) - \mathcal{R}(X_{2\cdot 1})]\mathbf{y}$	$N - k_1 - k_2 = N - k$	σ^2
Total	$\mathbf{e}_1'\mathbf{e}_1 = \mathbf{y}'[I - \mathcal{R}(X_1)]\mathbf{y}$	$N - k_1$	

projecting X_2 on $S_{k_1} = S_{k_1}(X_1)$. The projection, using previous results, is

(3.99) $$\mathcal{R}(X_1)X_2 = X_1(X_1'X_1)^{-1}X_1'X_2 = X_1A, \qquad A = (X_1'X_1)^{-1}X_1'X_2.$$

It is clear that each column of $\mathcal{R}(X_1)X_2 = X_1A$ lies in S_k, and for that matter, the span of X_1 and $X_{2\cdot 1}$ is the span of X_1 and X_2, which is, S_k. We then form the residual

(3.99a) $$X_2 - X_1A = X_{2\cdot 1}$$

and

(3.99b) $$X_1'X_{2\cdot 1} = \bigcirc.$$

We then use X_1 and $X_{2\cdot 1}$ as our basis matrices or coordinates instead of X_1 and X_2. We are in the same situation as in Figure 3.3.1, but here we label the set of axes that are orthogonal to X_1 by $X_{2\cdot 1}$.

A geometrical summary of the above procedure is that we may obtain the extra sum of squares due to $\boldsymbol{\beta}_2$ by projecting $\mathbf{y}$ into $S_{k_2}(X_{2\cdot 1})$, where $S_{k_2}(X_{2\cdot 1})$ is the subspace orthogonal to $S_{k_1}(X_1)$ in $S_k(X) = S_k$, $k_1 + k_2 = k$. We sometimes speak of $S_{k_2}(X_{2\cdot 1})$ as the *orthogonal complement* of X_1 in $S_k(X)$.

(We remark that if we did not use the matrices X_1 and $X_{2\cdot 1}$, and, for example, regressed $\mathbf{y}$ on X_1 and X_2 separately and then took the resolvent of the projections, we would not obtain $\hat{\boldsymbol{\eta}} = \mathcal{R}(X)\mathbf{y}$, and so on. Problem 3.17 gives details.)

That we have obtained the correct sums of squares by the above process is clear, for recoordinatizing means that we have used a nonsingular linear transformation on X: Specifically we have that

$$W = XH = (X_1 \mid X_2)\begin{pmatrix} I_{k_1} & -A \\ \bigcirc & I_{k_2} \end{pmatrix}$$

(3.100) $$= (X_1 \mid -X_1A + X_2)$$

$$= (X_1 \mid X_{2\cdot 1})$$

and Theorem 3.2.1 shows that the regression and residual sum of squares are invariant under such transformations. Furthermore, since X_1 is orthogonal to $X_{2\cdot 1}$, we may break up the regression sum of squares into a part due to X_1 and a part due to $X_{2\cdot 1}$; that is, due to X_2, given X_1 in the model, and we may do this in a manner analogous to that of Case A, since $X_1'X_{2\cdot 1} = O$ and $X_{2\cdot 1}$ carries $\boldsymbol{\beta}_2$.

In fact, recalling Case A, we have

$$\mathbf{y}'\mathcal{R}(X)\mathbf{y} = \mathbf{y}'\mathcal{R}(X_1)\mathbf{y} + \mathbf{y}'\mathcal{R}(X_{2\cdot 1})\mathbf{y} \tag{3.100a}$$

so that the extra regression sum of squares due to X_2, given X_1 in the model, is [see (3.91f)]

$$\mathbf{y}'\mathcal{R}(X_{2\cdot 1})\mathbf{y} = \mathbf{y}'\mathcal{R}(X)\mathbf{y} - \mathbf{y}'\mathcal{R}(X_1)\mathbf{y}. \tag{3.100b}$$

(Geometrically, this is obvious from Figure 3.3.2.) Now after some algebra,

$$\begin{aligned} \mathbf{y}'\mathcal{R}(X_{2\cdot 1})\mathbf{y} &= \mathbf{y}'[I - \mathcal{R}(X_1)]\mathbf{y} - \mathbf{y}'[I - \mathcal{R}(X)]\mathbf{y} \\ &= \mathbf{y}'[I - \mathcal{R}(X_1)]\mathbf{y} - \mathbf{y}'[I - \mathcal{R}(X_1) - \mathcal{R}(X_{2\cdot 1})]\mathbf{y} \end{aligned} \tag{3.100c}$$

[see (3.91d)]. Using (3.100), we may write our model as

$$\begin{aligned} E(\mathbf{y}) &= X\boldsymbol{\beta} = XHH^{-1}\boldsymbol{\beta} \\ &= [XH]\left[\begin{array}{c|c} I_{k_1} & A \\ \hline O & I_{k_2} \end{array}\right]\begin{pmatrix} \boldsymbol{\beta}_1 \\ \boldsymbol{\beta}_2 \end{pmatrix} \\ &= [X_1 \mid X_{2\cdot 1}]\begin{pmatrix} \boldsymbol{\beta}_1 + A\boldsymbol{\beta}_2 \\ \boldsymbol{\beta}_2 \end{pmatrix} \\ &= X_1\boldsymbol{\phi} + X_{2\cdot 1}\boldsymbol{\beta}_2 \end{aligned} \tag{3.101}$$

where $\boldsymbol{\phi} = \boldsymbol{\beta}_1 + A\boldsymbol{\beta}_2$. Since $X_2'X_{2\cdot 1} = O$, regressing $\mathbf{y}$ on $XH = [X_1 \mid X_{2\cdot 1}]$ leads to

$$\begin{pmatrix} \hat{\boldsymbol{\phi}} \\ \hat{\boldsymbol{\beta}}_2 \end{pmatrix} = \begin{pmatrix} (X_1'X_1)^{-1}X_1'\mathbf{y} \\ (X_{2\cdot 1}'X_{2\cdot 1})^{-1}X_{2\cdot 1}'\mathbf{y} \end{pmatrix}. \tag{3.102}$$

Now

$$\begin{pmatrix} \hat{\boldsymbol{\phi}} \\ \hat{\boldsymbol{\beta}}_2 \end{pmatrix} = H^{-1}\begin{pmatrix} \hat{\boldsymbol{\beta}}_1 \\ \hat{\boldsymbol{\beta}}_2 \end{pmatrix} \quad \text{or} \quad \begin{pmatrix} \hat{\boldsymbol{\beta}}_1 \\ \hat{\boldsymbol{\beta}}_2 \end{pmatrix} = H\begin{pmatrix} \hat{\boldsymbol{\phi}} \\ \hat{\boldsymbol{\beta}}_2 \end{pmatrix}. \tag{3.102a}$$

[It may be easily verified, see Problem 3.17(b), that $\hat{\boldsymbol{\beta}}_2$, as given in (3.102), is provided by the last k_2 components of $\hat{\boldsymbol{\beta}} = (X'X)^{-1}X'\mathbf{y}$.] Hence

$$\hat{\boldsymbol{\beta}} = H\begin{pmatrix} \hat{\boldsymbol{\phi}} \\ \hat{\boldsymbol{\beta}}_2 \end{pmatrix} = \left(\begin{array}{c|c} I_{k_1} & -A \\ \hline \bigcirc & I_{k_2} \end{array}\right)\begin{pmatrix} \hat{\boldsymbol{\phi}} \\ \hat{\boldsymbol{\beta}}_2 \end{pmatrix} \tag{3.103}$$

so that in particular

$$\hat{\boldsymbol{\beta}}_1 = \hat{\boldsymbol{\phi}} - A\hat{\boldsymbol{\beta}}_2 \tag{3.103a}$$

which is unbiased for $\boldsymbol{\beta}_1$ and is a linear combination of the y_j's, so that from the Gauss theorem, $\hat{\boldsymbol{\beta}}_1$ is UMV unbiased for $\boldsymbol{\beta}_1$ (see Problem 3.18).

Note that

$$V\begin{pmatrix} \hat{\boldsymbol{\phi}} \\ \hat{\boldsymbol{\beta}}_2 \end{pmatrix} = \sigma^2\left(\begin{array}{c|c} (X_1'X_1)^{-1} & \bigcirc \\ \hline \bigcirc & (X_{2\cdot1}'X_{2\cdot1})^{-1} \end{array}\right) \tag{3.104}$$

so that $\hat{\boldsymbol{\phi}}$ and $\hat{\boldsymbol{\beta}}_2$ are uncorrelated. It is left as an exercise for the reader (see Problem 3.19) to determine the covariance of $\hat{\boldsymbol{\beta}}_1$ and $\hat{\boldsymbol{\beta}}_2$ in terms of A and $X_{2\cdot1}$.

Note too that the residual sum of squares (see Table 3.3.4) is formed by subtracting from $\mathbf{e}_1'\mathbf{e}_1$, where $\mathbf{e}_1 = [I - \mathcal{R}(X_1)]\mathbf{y}$, a regression sum of squares found by regressing $\mathbf{e}_1$ on $X_{2\cdot1}$, where X_1 is orthogonal to $X_{2\cdot1}$.

It should be noted that the above work has the following interesting implication. Suppose $\mathbf{y}$ is regressed on X_1, that is, the model

$$E(\mathbf{y}) = X_1\boldsymbol{\beta}_1, \qquad V(\mathbf{y}) = \sigma^2 I_N \tag{3.105}$$

is tacitly assumed, and assume further that after $\mathbf{y}$ is regressed on X_1, it is desired to take into account additional factors which have been used and summarized in the matrix X_2, so that we now wish to find estimators of $\boldsymbol{\beta}_2$, where

$$E(\mathbf{y}) = X_1\boldsymbol{\beta}_1 + X_2\boldsymbol{\beta}_2. \tag{3.105a}$$

This is of course easily accomplished, because we may write

$$E(\mathbf{y}) = X_1\boldsymbol{\phi} + X_{2\cdot1}\boldsymbol{\beta}_2, \tag{3.105b}$$

yielding

$$\hat{\boldsymbol{\beta}}_2 = (X_{2\cdot1}'X_{2\cdot1})^{-1}X_{2\cdot1}'\mathbf{y}, \tag{3.105c}$$

that is, (3.105*b*) and (3.105*c*) state that to add additional factors, we need only do so by adding the additional factors in orthogonally as in (3.105*b*), since $X_{2\cdot 1}'X_1 = O$. Furthermore, the extra regression sum of squares due to $\boldsymbol{\beta}_2$ is $\hat{\boldsymbol{\beta}}_2' X_{2\cdot 1}' X_{2\cdot 1}\hat{\boldsymbol{\beta}}_2$ or $\mathbf{y}'\mathcal{R}(X_{2\cdot 1})\mathbf{y}$, while $\mathrm{SS}_e = \mathbf{y}'[I - \mathcal{R}(X_1)]\mathbf{y} - \mathbf{y}'\mathcal{R}(X_{2\cdot 1})\mathbf{y}$, and so on.

EXAMPLE 3.3.1.

This is an "elimination of the mean" example.

In many situations, the appropriate model is of the form

$$E(\mathbf{y}) = \beta_0\mathbf{1} + \beta_1\mathbf{x}_1 + \cdots + \beta_{k-1}\mathbf{x}_{k-1}, \tag{3.106}$$

where $\mathbf{1}$ is a $N \times 1$ vector of ones, and in which the parameters $(\beta_1,\ldots,\beta_{k-1})$ are of interest. We note that if $(\beta_1,\ldots,\beta_{k-1}) = (0,\ldots,0)$ then β_0 would be the common mean value of the y_j's, so that β_0 is often referred to as a general mean, and, as stated above, it often is the case that there is no intrinsic interest in β_0, but rather in the effects of the k factors $x_1,\ldots,x_{k-1}$ as measured by $(\beta_1,\ldots,\beta_{k-1})$.

Using notation similar to that of Case B, we have

$$E(\mathbf{y}) = \left(X_0 \,\vert\, X_1\right)\begin{pmatrix}\beta_0\\ \hline \boldsymbol{\beta}_1\end{pmatrix}, \tag{3.107}$$

where $X_0 = \mathbf{x}_0 = \mathbf{1}$, $X_1 = (\mathbf{x}_1 \,\vert\, \mathbf{x}_2 \,\vert\, \cdots \,\vert\, \mathbf{x}_{k-1})$; that is, X_0 is of order $N \times k_0 = N \times 1$, X_1 is $N \times k_1 = N \times (k-1)$, and interest is in the $k_1 \times 1$ vector $\boldsymbol{\beta}_1 = (\beta_1,\ldots,\beta_{k-1})'$. We wish to find an estimate of $\boldsymbol{\beta}_1$ and the contribution to the regression sum of squares of X_1, given that there is a general mean in the model and an appropriate term on which to base an estimate of the "error", that is, of σ^2, where we are assuming that

$$V(\mathbf{y}) = \sigma^2 I, \tag{3.107a}$$

that is, we wish to extract the correct residual sum of squares. We proceed as in Case B. We first *orthogonalize* by regressing X_1 on $X_0 = \mathbf{1}$ and taking residuals. Let $\bar{x}_j = (1/N)\Sigma_{u=1}^N x_{uj}$,

$$A = (\mathbf{x}_0'\mathbf{x}_0)^{-1}\mathbf{x}_0' X_1 = N^{-1}\mathbf{1}'X_1 = (\bar{x}_1,\ldots,\bar{x}_{k-1}), \tag{3.108}$$

which is to say that A is of order $k_0 \times k_1 = 1 \times (k-1)$. Thus,

(3.108*a*)

$$\begin{aligned} X_{1\cdot 0} &= X_1 - \mathbf{x}_0 A = \left(\mathbf{x}_1 \,\vert\, \mathbf{x}_2 \,\vert\, \cdots \,\vert\, \mathbf{x}_{k-1}\right) - \mathbf{1}\left(\bar{x}_1 \,\vert\, \cdots \,\vert\, \bar{x}_{k-1}\right) \\ &= \left(\mathbf{x}_1 - \bar{x}_1\mathbf{1} \,\vert\, \mathbf{x}_2 - \bar{x}_2\mathbf{1} \,\vert\, \cdots \,\vert\, \mathbf{x}_{k-1} - \bar{x}_{k-1}\mathbf{1}\right). \end{aligned}$$

We note that columns of $X_{1\cdot 0}$ are "ordinary" residuals—the jth column has

uth entry $x_{uj} - \bar{x}_j$, and we have

$$\sum_{u=1}^{N} (x_{uj} - \bar{x}_j) = \mathbf{1}'(\mathbf{x}_j - \bar{x}_j\mathbf{1}) = 0. \tag{3.108b}$$

Thus, we have that

$$\mathbf{1}'X_{1\cdot 0} = \mathbf{0}'_{k-1}, \tag{3.108c}$$

that is, here $X_0 = \mathbf{x}_0 = \mathbf{1}$ is orthogonal to $X_{1\cdot 0}$. Paralleling (3.101), we have, from (3.106), (3.107), and (3.108), that

$$\mathbf{y} = \phi\mathbf{x}_0 + X_{1\cdot 0}\boldsymbol{\beta}_1 + \boldsymbol{\varepsilon}, \tag{3.109}$$

where

$$E(\boldsymbol{\varepsilon}) = \mathbf{0}, \quad V(\boldsymbol{\varepsilon}) = \sigma^2 I, \quad \phi = \beta_0 + A\boldsymbol{\beta}_1 = \beta_0 + \sum_1^{k-1} \beta_j\bar{x}_j. \tag{3.109a}$$

Using (3.108c),

$$\begin{aligned} \mathbf{1}'\mathbf{y} &= \phi\mathbf{1}'\mathbf{1} + \mathbf{1}'X_{1\cdot 0}\boldsymbol{\beta}_1 + \mathbf{1}'\boldsymbol{\varepsilon} \\ &= N\phi + N\bar{\varepsilon} \end{aligned} \tag{3.110}$$

so that

$$\bar{y} = \phi + \bar{\varepsilon}. \tag{3.110a}$$

Using (3.109a),

$$E(\bar{y}) = \phi \tag{3.110b}$$

so that $\hat{\phi} = \bar{y}$ is, by the Gauss theorem, linear UMV unbiased for ϕ, since by (3.108c), (3.109), and (3.109a), we have

$$\begin{aligned} \begin{pmatrix} \hat{\phi} \\ \hat{\boldsymbol{\beta}}_1 \end{pmatrix} &= \left(\begin{array}{c|c} (\mathbf{1}'\mathbf{1})^{-1} & \mathbf{0}' \\ \hline \mathbf{0} & (X'_{1\cdot 0}X_{1\cdot 0})^{-1} \end{array}\right) \begin{pmatrix} \mathbf{1}'\mathbf{y} \\ X'_{1\cdot 0}\mathbf{y} \end{pmatrix} \\ &= \begin{pmatrix} \bar{y} \\ (X'_{1\cdot 0}X_{1\cdot 0})^{-1}X'_{1\cdot 0}\mathbf{y} \end{pmatrix}. \end{aligned} \tag{3.111}$$

We note that

$$V(\hat{\phi}) = \sigma^2(\mathbf{1}'\mathbf{1})^{-1} = \frac{\sigma^2}{N} \tag{3.112}$$

and

$$V(\hat{\boldsymbol{\beta}}_1) = \sigma^2(X'_{1\cdot 0}X_{1\cdot 0})^{-1}, \qquad \text{cov}(\hat{\phi}, \hat{\boldsymbol{\beta}}_1) = \mathbf{0}'. \tag{3.112a}$$

Now interest is not in ϕ, but in $\boldsymbol{\beta}_1$, and so, continuing our procedure, we regress $\mathbf{y}$ on $X_0 = \mathbf{x}_0 = \mathbf{1}$ and take residuals. But regressing $\mathbf{y}$ on $\mathbf{1}$ means [see Problem 3.20(d) and (3.108a)] that residuals are given by

$$\mathbf{e}_0 = \mathbf{y} - \bar{y}\mathbf{1} = [I - \mathcal{R}(\mathbf{1})]\mathbf{y}, \qquad \mathcal{R}(\mathbf{1}) = N^{-1}\mathbf{1}\mathbf{1}', \tag{3.113}$$

and we regress $\mathbf{e}_0$ on $X_{1\cdot 0}$, since we are now considering the model

$$E(\mathbf{e}_0) = X_{1\cdot 0}\boldsymbol{\beta}_1. \tag{3.113a}$$

Writing the original model (3.106) in the form (3.113a) eliminates ϕ from further consideration, and we refer to this as having *eliminated the (general) mean*. We may now find the regression sum of squares and, in particular, the extra regression sum of squares due to X_1, given that β_0 (or ϕ) is in the model, and imitating Table 3.3.4, we construct the relevant analysis of variance table shown in Table 3.3.5.

Problem 3.21 asks the reader to verify the entries of the sum of squares column and the expected mean square column of Table 3.3.5.

The reader will recall that in the work on the simple linear model $E(y) = \alpha + \beta x$ in Chapter 1, we defined the sample correlation coefficient R between x and y (see Section 1.5) and found that we may write R^2 as

$$\begin{aligned} R^2 &= \frac{\text{SS}_r(\beta)}{\Sigma(y_i - \bar{y})^2} \\ &= \frac{\text{SS}_e(\alpha) - \text{SS}_e(\alpha, \beta)}{\text{SS}_e(\alpha)}, \end{aligned} \tag{3.114}$$

and indeed that

$$\text{SS}_e = \text{SS}_e(\alpha, \beta) = \Sigma(y_i - \bar{y})^2(1 - R^2). \tag{3.114a}$$

Similarly, in working with the model of this example, given in (3.106), we

TABLE 3.3.5. Analysis of Variance for (3.106) and (3.107); Interest in $\boldsymbol{\beta}_1$

Source	Sum of Squares	Degrees of Freedom	Expected Mean Square
Regression: (Due to $X_{1\cdot 0}$, i.e., due to $\mathbf{x}_1, \ldots, \mathbf{x}_{k-1}$, given **1** in model)	$\hat{\boldsymbol{\beta}}_1' X_{1\cdot 0}' X_{1\cdot 0} \hat{\boldsymbol{\beta}}_1$ $= \mathbf{y}' \mathfrak{R}(X_{1\cdot 0}) \mathbf{y}$	$k - 1$	$\sigma^2 + \frac{1}{k-1} \boldsymbol{\beta}_1' X_{1\cdot 0}' X_{1\cdot 0} \boldsymbol{\beta}_1$
Residual	$\mathbf{y}\{[I - \mathfrak{R}(\mathbf{1})] - \mathfrak{R}(X_{1\cdot 0})\}\mathbf{y}$ $= \mathbf{y}'[I - \mathfrak{R}(\mathbf{1})]\mathbf{y} - \mathbf{y}'\mathfrak{R}(X_{1\cdot 0})\mathbf{y}$ $= \mathbf{e}_0'\mathbf{e}_0 - \mathbf{y}'\mathfrak{R}(X_{1\cdot 0})\mathbf{y}$	$N - k$	σ^2
Total	$\mathbf{e}_0'\mathbf{e}_0 = \Sigma(y_i - \bar{y})^2$	$N - 1$	
Notation	$\mathfrak{R}(\mathbf{1}) = \frac{1}{N}\mathbf{1}\mathbf{1}'$; $\mathfrak{R}(X_{1\cdot 0}) = X_{1\cdot 0}(X_{1\cdot 0}' X_{1\cdot 0})^{-1} X_{1\cdot 0}'$		

may define the *multiple* correlation coefficient between $x_1, \ldots, x_{k-1}$ and y by

$$(3.114b)\qquad R^2 = \frac{SS_r(\boldsymbol{\beta}_1)}{\Sigma(y_i - \bar{y})^2} = \frac{SS_e(\beta_0) - SS_e(\beta_0, \boldsymbol{\beta}_1)}{SS_e(\beta_0)}.$$

Now, from the analysis of variance table, Table 3.3.5, we have

$$(3.114c)\qquad \sum_{i=1}^{n} (y_i - \bar{y})^2 = SS_r(\boldsymbol{\beta}_1) + SS_e(\beta_0, \boldsymbol{\beta}_1),$$

so that in using (3.114*b*) in (3.114*c*) we find

$$(3.114d)\qquad SS_e(\beta_0, \boldsymbol{\beta}_1) = (1 - R^2)SS_e(\beta_0),$$

and we note that $0 \le R^2 \le 1$.

Equation (3.114*d*) shows that R^2 is a measure of how good the fit of $\mathbf{y}$ is to $x_1, \ldots, x_{k-1}$; a value of R^2 near 1 implies that the fit to all of $x_1, \ldots, x_{k-1}$ is desirable, and a value of R^2 near 0 shows that a fit of $\mathbf{y}$ to the model $E(\mathbf{y}) = \mathbf{1}\beta_0$ is all that is necessary. It turns out that R, as defined by (3.114*b*), is the correlation coefficient between the y_i and $\hat{\eta}_i$; that is, it may be proved that

$$(3.114e)\qquad R = \frac{\Sigma(y_i - \bar{y})(\hat{\eta}_i - \bar{\hat{\eta}})}{\left[\Sigma(y_i - \bar{y})^2 \times \Sigma(\hat{\eta}_i - \bar{\hat{\eta}})^2\right]^{1/2}},$$

$-1 \le R \le 1$, where $\bar{\hat{\eta}} = (1/N)\Sigma_{i=1}^{N} \hat{\eta}_i$. Problem 3.21(b) asks for details.

3.4. TWO SPECIAL EXAMPLES: ONE-WAY ANALYSIS OF VARIANCE AND ONE-WAY ANALYSIS OF COVARIANCE

Many experimental situations are variants of the *one-way analysis of variance* or *one-way analysis of covariance* experimental setups. We shall see that the analysis for each may be viewed as an example of elimination of certain effects, for which we have available the orthogonalization procedure of the previous section.

EXAMPLE 3.4.1. (One-Way Analysis of Variance)

The one-way analysis of variance usually arises in such a way that we may state the following assumptions:

> There are N observations generated from k populations (or processes), where we have n_j observations generated from population j, $j = 1,\ldots,k$, and

$$\sum_{j=1}^{k} n_j = N \tag{3.115}$$

where *all observations are independent*. The k populations, for example, could be k different fertilizers, and the observations y_{ij} could be yields of a certain type of wheat in bushels from the ith (standard sized) plot that has fertilizer j applied to it, $i = 1,\ldots,n_j$ and $j = 1,\ldots,k$. The observations may be conveniently tabulated as in Table 3.4.1. Note that an observation is classified only by which fertilizer generated it, and it is for this reason that we speak of a table such as Table 3.4.1 as a *one-way classification*, and when we find the corresponding analysis of variance table for such a set of data, we call the process a *one-way analysis of variance*. We assume

$$E(y_{ij}) = \beta_j, \qquad i = 1,\ldots,n_j \tag{3.116}$$

where $j = 1,\ldots,k$. We further assume that

$$V(y_{ij}) = \sigma^2, \qquad \text{for all } i,\ j, \tag{3.116a}$$

TABLE 3.4.1. Fertilizers

1	2	$\cdots$	j	$\cdots$	k
y_{11}	y_{12}	$\cdots$	y_{1j}	$\cdots$	y_{1k}
y_{21}	y_{22}	$\cdots$	$\vdots$		
	$\vdots$		y_{ij}		$\vdots$
$\vdots$	$y_{n_2 2}$				$y_{n_k k}$
$y_{n_1 1}$			$\vdots$		
			$y_{n_j j}$		
$\sum_{i=1}^{n_1} y_{i1}$	$\sum_{i=1}^{n_2} y_{i2}$	$\cdots$	$\sum_{i=1}^{n_j} y_{ij}$	$\cdots$	$\sum_{i=1}^{n_k} y_{ik}$

and for any two observations, y_{ij} and y_{rs}, we have

$$\operatorname{cov}(y_{ij}, y_{rs}) = 0, \qquad (i, j) \neq (r, s). \tag{3.116b}$$

These assumptions can be summarized by using the general form of the linear model.

(a) Let

$$\begin{aligned} \mathbf{y} &= \left(y_{11}, y_{21}, \ldots, y_{n_1 1}, \ldots, y_{1j}, \ldots, y_{n_j j}, \ldots, y_{1k}, \ldots, y_{n_k k}\right)' \\ &= \left(y_1, y_2, \ldots, y_{n_1}, \ldots, y_N\right)'. \end{aligned} \tag{3.117}$$

where $N = \Sigma_1^k n_j$.

(b) Define x_{uj} as

$$x_{uj} = \begin{cases} 1, & \text{if } y_u \text{ is an observation obtained from the } j\text{th group} \\ 0, & \text{if } y_u \text{ is not from the } j\text{th group,} \end{cases} \tag{3.117a}$$

where $u = 1, \ldots, N$, $j = 1, \ldots, k$, so that there are Nk such x's.

(c) We may summarize (3.116)–(3.117a) by writing

$$y_u = x_{u1}\beta_1 + x_{u2}\beta_2 + \cdots + x_{uk}\beta_k + \varepsilon_u, \tag{3.117b}$$

where

$$\begin{aligned} E(\varepsilon_u) = 0, \quad V(\varepsilon_u) = \sigma^2, \qquad &\text{for all } u, \\ \operatorname{cov}(\varepsilon_u, \varepsilon_v) = 0, \qquad &u \neq v. \end{aligned} \tag{3.117c}$$

That is,

$$\mathbf{y} = X\boldsymbol{\beta} + \boldsymbol{\varepsilon} \tag{3.118}$$

where the $N \times 1$ vector $\mathbf{y}$ is given in (3.117),

$$\boldsymbol{\beta} = (\beta_1, \ldots, \beta_k)', \tag{3.118a}$$

$$E(\boldsymbol{\varepsilon}) = 0, \qquad V(\boldsymbol{\varepsilon}) = \sigma^2 I, \tag{3.118b}$$

and where X is the $N \times k$ matrix given by

$$
(3.118c)\quad X = \left(\begin{array}{ccccc}
1 & 0 & 0 & \cdots & 0 \\
1 & 0 & 0 & \cdots & 0 \\
\vdots & \vdots & \vdots & & \vdots \\
1 & 0 & 0 & \cdots & 0 \\
\hline
0 & 1 & 0 & \cdots & 0 \\
\vdots & \vdots & \vdots & & \vdots \\
0 & 1 & 0 & \cdots & 0 \\
\hline
0 & 0 & 1 & \cdots & 0 \\
\vdots & \vdots & \vdots & & \vdots \\
0 & 0 & 1 & \cdots & 0 \\
\hline
 & & 0 & \cdots & 0 \\
\vdots & \vdots & \vdots & & \vdots \\
 & & 0 & \cdots & 0 \\
\hline
0 & 0 & 0 & \cdots & 1 \\
\vdots & \vdots & \vdots & & \vdots \\
0 & 0 & 0 & \cdots & 1
\end{array}\right)
\begin{array}{c}
\uparrow \\ n_1 \text{ rows} \\ \downarrow \\
\hline
\uparrow \\ n_2 \text{ rows} \\ \downarrow \\
\hline
\uparrow \\ n_3 \text{ rows} \\ \downarrow \\
\hline
\vdots \\
\hline
\uparrow \\ n_k \text{ rows} \\ \downarrow
\end{array}
= [\mathbf{x}_1 \mid \mathbf{x}_2 \mid \cdots \mid \mathbf{x}_k].
$$

[The reader should satisfy himself or herself that $X\boldsymbol{\beta}$ does give the required values of $E(y_u)$.] A matrix X of the form (3.118c) is often referred to as an *incidence* or *design* matrix, since it tells the experimenter to take n_1 observations from group 1, n_2 from group 2, and so on. The typical value x_{uj} is one or zero, often referred to as *presence* or *absence* values. The variables x_{uj} are often called *dummy variables*, because they only tell us from which group an observation y_u comes. For example, in the agricultural situation described above, it could be that the different populations correspond to different amounts of nitrogen used in the fertilizer and x_{uj} simply tells us whether or not y_u has been generated using the jth fertilizer. We note that the ith column of X is orthogonal to $\mathbf{x}_j$, for all $i \neq j$, and that X is of full rank.

It is easy to verify that $X'X$ is diagonal, and indeed

$$
(3.119)\qquad X'X = \begin{pmatrix} n_1 & & & \bigcirc \\ & n_2 & & \\ & & \ddots & \\ \bigcirc & & & n_k \end{pmatrix}
$$

which of course means that

$$(3.119a) \qquad (X'X)^{-1} = \begin{pmatrix} 1/n_1 & & \\ & \ddots & \\ & & 1/n_k \end{pmatrix}.$$

The least-squares estimate is

$$(3.120) \qquad \hat{\boldsymbol{\beta}} = \begin{pmatrix} 1/n_1 & & \\ & \ddots & \\ & & 1/n_k \end{pmatrix} X'\mathbf{y}$$

or

$$(3.120a) \quad \hat{\boldsymbol{\beta}} = \begin{pmatrix} 1/n_1 & & \\ & \ddots & \\ & & 1/n_k \end{pmatrix} \begin{pmatrix} \sum_1^{n_1} y_{i1} \\ \vdots \\ \sum_1^{n_k} y_{ik} \end{pmatrix} = \begin{pmatrix} \bar{y}_1 \\ \vdots \\ \bar{y}_k \end{pmatrix}.$$

There are two reasons why we might have expected this. First, we have from the basic setup of the problem that

$$(3.121) \qquad E(y_{ij}) = \beta_j, \qquad i = 1,\ldots,n_j$$

so that $\bar{y}_j = n_j^{-1}\Sigma_{i=1}^{n_j} y_{ij}$,

$$(3.121a) \qquad E(\bar{y}_j) = \beta_j,$$

which is to say that $\bar{y}_j$ is unbiased for β_j, and it is of course linear in the y_{ij}. Second, we note from (3.118) and the following that

$$(3.122) \qquad \mathbf{y} = \mathbf{x}_1\beta_1 + \cdots + \mathbf{x}_k\beta_k + \boldsymbol{\varepsilon},$$

where the $\mathbf{x}_j$ are orthogonal, so that the work of the previous section implies (see Problem 3.22) that the β_j can be estimated one at a time.

The least-squares estimator of $\boldsymbol{\eta}$ is

$$\hat{\boldsymbol{\eta}} = X\hat{\boldsymbol{\beta}} = \mathbf{x}_1\hat{\beta}_1 + \cdots + \mathbf{x}_k\hat{\beta}_k, \tag{3.123}$$

whose uth component is

$$\hat{\eta}_u = \bar{y}_1 x_{u1} + \cdots + \bar{y}_j x_{uj} + \cdots + \bar{y}_k x_{uk} = \bar{y}_j, \tag{3.123a}$$

where, using (3.117a), u corresponds to ij. The uth component of the residual $\mathbf{y} - \hat{\boldsymbol{\eta}}$ is

$$y_u - \hat{\eta}_u = y_{ij} - \bar{y}_j. \tag{3.123b}$$

Again, this is to be expected. The sample of size n_j is such that $E(\bar{y}_j) = \beta_j$ and $y_{ij} - \bar{y}_j$ give information about $V(y_{ij}) = \sigma^2$, in particular, through $S_j^2 = [\Sigma_{i=1}^{n}(y_{ij} - \bar{y}_j)^2]/(n_j - 1)$. Indeed, $E(y_{ij} - \bar{y}_j) = 0$.

Returning to the full model we have that the regression sum of squares $\mathbf{y}'\mathcal{R}(X)\mathbf{y}$ is

$$\text{SS}_r = \hat{\boldsymbol{\beta}}'X'X\hat{\boldsymbol{\beta}} = (\bar{y}_1,\ldots,\bar{y}_k)\begin{pmatrix} n_1 & & \\ & \ddots & \\ & & n_k \end{pmatrix}\begin{pmatrix} \bar{y}_1 \\ \vdots \\ \bar{y}_k \end{pmatrix}, \tag{3.124}$$

that is,

$$\text{SS}_r = \sum_{j=1}^{k} n_j \bar{y}_j^2 \tag{3.124a}$$

and the residual sum of squares is

$$\begin{aligned} \text{SS}_e &= (\mathbf{y} - \hat{\boldsymbol{\eta}})'(\mathbf{y} - \hat{\boldsymbol{\eta}}) = \mathbf{y}'\mathbf{y} - \mathbf{y}'\mathcal{R}\mathbf{y} \\ &= \sum_{j=1}^{k}\sum_{i=1}^{n_j} y_{ij}^2 - \sum_{j=1}^{k} n_j\bar{y}_j^2 \\ &= \sum_{j=1}^{k}\left(\sum_{i=1}^{n_j} y_{ij}^2 - n_j\bar{y}_j^2\right) \end{aligned} \tag{3.124b}$$

or

$$\text{SS}_e = \sum_{j=1}^{k}\sum_{i=1}^{n_j}(y_{ij} - \bar{y}_j)^2. \tag{3.124c}$$

TABLE 3.4.2. Preliminary Analysis of Variance for One-Way Situation

Source	Degrees of Freedom	Sum of Squares	Expected Mean Square
Regression	k	$\sum_j n_j \bar{y}_j^2$	$\sigma^2 + \frac{1}{k}\sum_{j=1}^{k} n_j \beta_j^2$
Residual	$N-k$	$\sum_j \sum_i (y_{ij} - \bar{y}_j)^2$	σ^2
Total	N	$\sum_j \sum_i y_{ij}^2$	

We may now state the "usual" analysis of variance table as in Table 3.4.2.

It is clear under the hypothesis $\beta_1 = \cdots = \beta_k = 0$ that $\sum_j n_j \bar{y}_j^2/k$ and $\sum\sum(y_{ij} - \bar{y}_j)^2/(N-k)$ are both unbiased estimates of σ^2. If we now assume that the y_{ij} are normally distributed in addition to the assumptions previously made, then it can be shown that the ratio of these random quantities is distributed as an $F_{k,\,N-k}$ variable. Hence, a level γ test of

$$\beta_1 = \cdots = \beta_k = 0 \tag{3.125}$$

is

$$\text{Reject if} \quad F = \frac{\left[\sum n_j \bar{y}_j^2\right]/k}{\left[\sum\sum(y_{ij} - \bar{y}_j)^2\right]/(N-k)} > F_{k,\,N-k;\,\gamma}; \tag{3.125a}$$

Accept otherwise.

(Problem 3.23 does assume normality of the y_{ij} and asks for all the necessary details.)

Often in problems of this type, interest focuses on the comparative values of the parameters, and very frequently it may be that the question of whether or not

$$\beta_1 = \beta_2 = \cdots = \beta_k \tag{3.125b}$$

is true is of prime interest. [In our agricultural example, if (3.125*b*) held, we would be saying that the fertilizers are as effective as one another.] The common value of the β_j's need not be specified and certainly need not be zero. [In our agricultural example, if (3.125*b*) held, there is no reason to believe that the expected yield when using any one of the fertilizers is zero.] How to examine the statement (3.125*b*) is certainly not clear from the analysis of variance table given in Table 3.4.2.

To test (3.125*b*) we may proceed as follows. Since the model is of the form (3.122), and interest is in the statement $\beta_1 = \beta_2 = \cdots = \beta_k$, we may rewrite (3.122) as [see (3.118*c*)]

$$\text{(3.126)} \qquad \mathbf{y} = \beta_1 \mathbf{1}_N + \theta_2 \mathbf{x}_2 + \cdots + \theta_k \mathbf{x}_k + \boldsymbol{\varepsilon},$$

where

$$\text{(3.126}a\text{)} \qquad \theta_j = \beta_j - \beta_1, \qquad j = 2,\ldots,k,$$

and $\mathbf{1}_N$ is a $N \times 1$ vector of ones. We have, of course, that (3.125*b*) holds if and only if $\theta_2 = \theta_3 = \cdots = \theta_k = 0$. We are thus in the classical "elimination of the mean" situation, since we may write (3.126) as

$$\text{(3.127)} \qquad \begin{aligned} \mathbf{y} &= (\mathbf{1}_N \mid \mathbf{x}_2,\ldots,\mathbf{x}_k)(\beta_1 \mid \theta_2,\ldots,\theta_k)' + \boldsymbol{\varepsilon} \\ &= (\mathbf{1}_N \mid X_2)(\beta_1 \mid \boldsymbol{\theta}')' + \boldsymbol{\varepsilon}, \end{aligned}$$

where $X_2 = (\mathbf{x}_2 \mid \cdots \mid \mathbf{x}_k)$ and where $\boldsymbol{\theta} = (\theta_2,\ldots,\theta_k)' = (\beta_2 - \beta_1,\ldots,\beta_k - \beta_1)'$ is of interest. Using the appropriate procedures of Section 3.3, we orthogonalize by regressing X_2 on $\mathbf{1}_N$, taking residuals $X_2 - \mathbf{1}_N A$, regressing $\mathbf{y}$ on $\mathbf{1}_N$, and denoting the residuals by $\mathbf{e}_1 = \mathbf{y} - \bar{y}\mathbf{1}_N$, then regressing $\mathbf{e}_1$ on $X_2 - \mathbf{1}_N A$. Problem 3.24 asks the reader to complete the details and to show that the analysis of variance table (with all lines present) is as given in Table 3.4.3.

TABLE 3.4.3 One-Way Analysis of Variance; Interest in the Statement $\beta_1 = \cdots = \beta_k$

Source	Sum of Squares	Degrees of Freedom	Expected Mean Square
Regression: Mean	$N\bar{y}^2$	1	$\sigma^2 + N\bar{\beta}^2,\ \bar{\beta} = \Sigma n_j\beta_j/N$
Regression: Between	$\Sigma n_j(\bar{y}_j - \bar{y})^2$	$k-1$	$\sigma^2 + \dfrac{1}{k-1}\sum_{j=1}^{k} n_j(\beta_j - \bar{\beta})^2$
Error (within)	$\Sigma\Sigma(y_{ij} - \bar{y}_j)^2$	$N-k$	σ^2
Total	$\Sigma\Sigma y_{ij}^2$	N	

It is of interest to note that we may obtain Table 3.4.3 as follows. We have that, for any nonsingular H,

$$E(\mathbf{y}) = (XH)(H^{-1}\boldsymbol{\beta}), \tag{3.128}$$

where X is as in (3.118*c*). In view of (3.126) and (3.126*a*), that is, since we wish to examine the statement

$$\beta_1 = \beta_2 = \cdots = \beta_k, \text{ which is equivalent to } \theta_2 = \cdots = \theta_k = 0, \tag{3.129}$$

we take H^{-1} as

(3.130)

$$H^{-1} = \left[\begin{array}{c|cccccc} 1 & 0 & & \cdots & & & 0 \\ \hline -1 & 1 & 0 & & \cdots & & 0 \\ -1 & 0 & 1 & 0 & \cdots & & 0 \\ -1 & 0 & 0 & 1 & 0 & & 0 \\ \vdots & \vdots & \vdots & 0 & 1 & & \vdots \\ & & & \vdots & & \ddots & \\ -1 & 0 & 0 & 0 & \cdots & & 1 \end{array}\right] = \left[\begin{array}{c|c} 1 & \mathbf{0}'_{k-1} \\ \hline -\mathbf{1}_{k-1} & I_{k-1} \end{array}\right]$$

in order that

$$\begin{aligned} H^{-1}\boldsymbol{\beta} &= (\beta_1, \theta_2, \ldots, \theta_k)', \qquad \theta_j = \beta_j - \beta_1 \\ &= (\beta_1 \mid \boldsymbol{\theta}')', \end{aligned} \tag{3.130a}$$

with $\boldsymbol{\theta} = (\theta_2, \ldots, \theta_k)'$.

Now the inverse of H^{-1} is

$$H = \left[\begin{array}{c|c} 1 & \mathbf{0}'_{k-1} \\ \hline \mathbf{1}_{k-1} & I_{k-1} \end{array}\right] \tag{3.130b}$$

and it is easy to see that

$$XH = (\mathbf{1}_N \mid \mathbf{x}_2 \mid \cdots \mid \mathbf{x}_k). \tag{3.130c}$$

Substituting (3.130a) and (3.130c) into (3.128) yields

$$E(\mathbf{y}) = (\mathbf{1}_N \mid \mathbf{x}_2,\ldots,\mathbf{x}_k)(\beta_1 \mid \boldsymbol{\theta}')' \tag{3.130d}$$
$$= \mathbf{1}_N\beta_1 + \mathbf{x}_2\theta_2 + \cdots + \mathbf{x}_k\theta_k.$$

By Theorem 3.2.1, the regression sum of squares and the residual sum of squares are invariant. In particular,

$$\mathbf{y}'\mathcal{R}(X)\mathbf{y} = \mathbf{y}'\mathcal{R}(XH)\mathbf{y}. \tag{3.131}$$

If we were to orthogonalize in (3.130d), with $X_1 = \mathbf{1}_N$, then eliminating the mean leads to the regression sum of squares because the mean is given by $\mathbf{y}'\mathcal{R}(\mathbf{1})\mathbf{y} = (\Sigma\, y_j)^2/N = N\bar{y}^2$. But

$$\mathbf{y}'\mathcal{R}(X)\mathbf{y} - N\bar{y}^2 = \mathbf{y}'\mathcal{R}(XH)\mathbf{y} - N\bar{y}^2; \tag{3.132}$$

that is, the extra regression sum of squares due to $\beta_2,\ldots,\beta_k$, given that β_1 is in the model, is thus invariant. But this in turn means, by using Table 3.4.2, that the extra regression sum of squares, due to $\theta_2,\ldots,\theta_k$ and given β_1 is in the model, is

$$\mathbf{y}'\mathcal{R}(XH)\mathbf{y} - N\bar{y}^2 = \Sigma\, n_j\bar{y}_j^2 - N\bar{y}^2 \tag{3.133}$$
$$= \Sigma\, n_j(\bar{y}_j - \bar{y})^2,$$

where, of course

$$\bar{y} = N^{-1}\mathbf{1}'\mathbf{y} = N^{-1}\sum_{j=1}^{k} n_j\bar{y}_j. \tag{3.133a}$$

Needless to say, we have from (3.133) that the regression sum of squares has been broken down into

$$\mathbf{y}'\mathcal{R}(X)\mathbf{y} = N\bar{y}^2 + \Sigma\, n_j(\bar{y}_j - \bar{y})^2. \tag{3.133b}$$

The first term $N\bar{y}^2 = (\Sigma\Sigma y_{ij})^2/N$ is, using the notation of the previous section, due to the general mean level ϕ (Problem 3.24 asks the reader to find $\phi = \beta_1 + A\boldsymbol{\theta}$) and indeed it can be shown (Problem 3.24) that $N\bar{y}^2$

gives information, apart from error, about ϕ; that is

$$E(N\bar{y}^2) = \sigma^2 + N\phi^2, \tag{3.133c}$$

where $\phi = \bar{\beta} = N^{-1}\Sigma_{j=1}^{k} n_j\beta_j$. The second term of (3.133*b*), usually denoted by SS_B, standing for the *between* sum of squares, arises due to $\theta_2,\ldots,\theta_k$ being present in the model (3.130*d*), and in fact gives information, apart from error, about $\theta_2,\ldots,\theta_k$; that is,

$$E[SS_B] = (k-1)\sigma^2 + \sum_{j=1}^{k} n_j(\beta_j - \bar{\beta})^2. \tag{3.133d}$$

As mentioned, the reader is asked to show that $\phi = \bar{\beta}$ and to prove (3.133*c*) and (3.133*d*) in Problem 3.24.

We call (3.133) the *between sum of squares*; this terminology is given in all first courses in Analysis of Variance through the fundamental identity for observations generated as in Table 3.4.1; namely,

(3.134)

$$\sum_{j=1}^{k}\sum_{i=1}^{n_j}(y_{ij}-\bar{y})^2 = \sum_{j=1}^{k}\sum_{i=1}^{n_j}(y_{ij}-\bar{y}_j)^2 + \sum_{j=1}^{k}\sum_{i=1}^{n_j}(\bar{y}_j-\bar{y})^2$$

$$= \sum_{j=1}^{k}\sum_{i=1}^{n_j}(y_{ij}-\bar{y}_j)^2 + \sum_{j=1}^{k} n_j(\bar{y}_j-\bar{y})^2$$

or $$SS_T = SS_W + SS_B.$$

(In words, the *total* sum of squares of deviations SS_T is the sum of squares of the *within* sum of squares of deviations SS_W plus the *between* sum of squares of deviations SS_B.) We may now tabulate the elements of (3.134) in Table 3.4.3.

It is easy to see from the Expected Mean Square column that a comparison of the between mean square with the residual mean square will provide a test of the statement $\beta_1 = \cdots = \beta_k$. Problem 3.25(a) asks for details when we may assume the y_{ij}'s to be normal and so on. The reader is asked to analyze a set of data in Problem 3.25(b).

EXAMPLE 3.4.2. Analysis of Covariance

Before discussing analysis of covariance, it will be helpful to emphasize some of the results of Section 3.3 in a slightly different way and to examine a special regression model. Firstly, the reader is asked to reexamine Table 3.3.4 and the work that led to it. Of interest to us here will be the last two lines of Table 3.3.4. We recall that $\mathbf{e}_1$ was found by regressing $\mathbf{y}$ on X_1 and indeed acting as if our model was now

$$E(\mathbf{y}) = X_1\boldsymbol{\beta}_1. \tag{3.135}$$

We recall from previous work that the residual sum of squares, after regressing $\mathbf{y}$ on X_1, would be

$$\mathbf{y}'[I - \mathcal{R}(X_1)]\mathbf{y} = \mathbf{e}_1'\mathbf{e}_1, \tag{3.135a}$$

since, from Step 2 of the procedure outlined before Table 3.3.4, we have that

$$\mathbf{e}_1 = [I - \mathcal{R}(X_1)]\mathbf{y}. \tag{3.135b}$$

Now for Table 3.3.4, *we are eliminating* X_1, *and so, the sum of squares* $\mathbf{e}_1'\mathbf{e}_1$ *becomes the entry of the total line*. We note that if we are now asked what the residual sum of squares is due to the fact that $X_{2\cdot 1}$ is in the model—that is, (3.135) does not apply—but it is the case that

$$\begin{aligned} E(\mathbf{y}) &= X_1\boldsymbol{\beta}_1 + X_2\boldsymbol{\beta}_2 \\ &= X_1\boldsymbol{\phi} + X_{2\cdot 1}\boldsymbol{\beta}_2, \end{aligned} \tag{3.136}$$

then we note that the answer is supplied (see Table 3.3.4) by

$$\mathbf{e}_1'\mathbf{e}_1 - \mathbf{y}'\mathcal{R}(X_{2\cdot 1})\mathbf{y} = \mathbf{y}'[I - \mathcal{R}(X_1) - \mathcal{R}(X_{2\cdot 1})]\mathbf{y}. \tag{3.136a}$$

In words, we obtained the answer by subtracting a regression sum of squares, a quadratic form in $\mathbf{y}$, whose matrix is a projection matrix $\mathcal{R}$ defined in terms of the orthogonal part of X_2 to X_1, that is, $\mathbf{y}'\mathcal{R}(X_{2\cdot 1})\mathbf{y}$ is subtracted from $\mathbf{e}_1'\mathbf{e}_1$.

Also, we wish to emphasize the particular form the regression sum of squares takes when the model is

$$E(\mathbf{y}) = \mathbf{x}\beta \tag{3.137}$$

(that is, the expected value of y is a linear function βx through the origin).

We have (X is now replaced by $\mathbf{x}$ in our general model, and so on)

$$\hat{\beta} = (\mathbf{x}'\mathbf{x})^{-1}\mathbf{x}'\mathbf{y} = \frac{\Sigma_{i=1}^N x_i y_i}{\Sigma x_i^2} \tag{3.138}$$

so that

$$\hat{\boldsymbol{\eta}} = \mathbf{x}\hat{\beta}. \tag{3.138a}$$

Then the regression sum of squares is

$$\mathrm{SS}_r = \hat{\boldsymbol{\eta}}'\hat{\boldsymbol{\eta}} = \frac{(\Sigma_1^N x_i y_i)^2}{\Sigma x_i^2} = \mathbf{y}'\mathcal{R}(\mathbf{x})\mathbf{y}, \tag{3.138b}$$

where $\mathcal{R}(\mathbf{x}) = \mathbf{x}(\mathbf{x}'\mathbf{x})^{-1}\mathbf{x}' = \mathbf{x}\mathbf{x}'/\Sigma x_i^2$, and the residual sum of squares is

$$\begin{aligned} \mathrm{SS}_e &= \mathbf{y}'\mathbf{y} - \mathbf{y}'\mathcal{R}(\mathbf{x})\mathbf{y} = \mathbf{y}'\mathbf{y} - \frac{(\mathbf{x}'\mathbf{y})^2}{\mathbf{x}'\mathbf{x}} \\ &= \Sigma y_i^2 - \frac{(\Sigma x_i y_i)^2}{\Sigma x_i^2}. \end{aligned} \tag{3.138c}$$

We also draw attention to the fact that if

$$\begin{aligned} S_{12e} &= \text{residual sum of squares due to regressing } \mathbf{y} \text{ on } (X_1 \mid X_2), \\ S_{1e} &= \text{residual sum of squares due to regressing } \mathbf{y} \text{ on } X_1, \end{aligned} \tag{3.139}$$

then

$$\begin{aligned} S_{2\cdot 1r} &= \text{extra regression sum of squares due to } X_2, \\ &\quad \text{given } X_1 \text{ is in the model} \\ &= S_{1e} - S_{12e} \end{aligned} \tag{3.139a}$$

[see (3.100c)]. The above is important and may be emphasized by constructing Table 3.4.4.

We have

$$\begin{aligned} S_{1e} &= \mathbf{e}_1'\mathbf{e}_1 = \mathbf{y}'[I - \mathcal{R}(X_1)]\mathbf{y}, \\ \mathbf{e}_1 &= [I - \mathcal{R}(X_1)]\mathbf{y}, \end{aligned} \tag{3.140}$$

TABLE 3.4.4 Determination of the (Extra) Regression Sum of Squares due to $X_{2\cdot 1}$ (X_2, given X_1)

Source	Sum of Squares	Degrees of Freedom
Residual, due to X_1	S_{1e}	$N - k_1$
Residual, due to X_1 and X_2 (or X_1 and $X_{2\cdot 1}$)	S_{12e}	$N - (k_1 + k_2)$
Regression, due to X_2 given X_1 in model ($X_{2\cdot 1}$)	$S_{2\cdot 1r} = S_{1e} - S_{12e}$	k_2

and

$$S_{12e} = \mathbf{e}_{12}'\mathbf{e}_{12} = \mathbf{y}'[I - \mathcal{R}(X_1) - \mathcal{R}(X_{2\cdot 1})]\mathbf{y}$$

(3.140a)
$$= \mathbf{e}_1'\mathbf{e}_1 - \mathbf{y}'\mathcal{R}(X_{2\cdot 1})\mathbf{y}$$

$$\mathbf{e}_{12} = [I - \mathcal{R}(X_1) - \mathcal{R}(X_{2\cdot 1})]\mathbf{y} = \mathbf{e}_1 - \mathcal{R}(X_{2\cdot 1})\mathbf{y}.$$

Recall too that in regressing $\mathbf{e}_1$ on $X_{2\cdot 1}$, the regression sum of squares is

(3.140b)
$$\mathbf{e}_1'\mathcal{R}(X_{2\cdot 1})\mathbf{e}_1.$$

Using (3.140) we have that the regression sum of squares due to $X_{2\cdot 1}$ is

(3.140c)
$$\mathbf{y}'[I - \mathcal{R}(X_1)]'\mathcal{R}(X_{2\cdot 1})[I - \mathcal{R}(X_1)]\mathbf{y}$$

and as discussed in Section 3.3—see (3.89) and (3.89a)—this equals

(3.140d)
$$\mathbf{y}'\mathcal{R}(X_{2\cdot 1})\mathbf{y}.$$

Hence we may write (3.140a) as

(3.140e)
$$S_{12e} = \mathbf{e}_1'\mathbf{e}_1 - \mathbf{e}_1'\mathcal{R}(X_{2\cdot 1})\mathbf{e}_1$$

or

(3.140f)
$$S_{12e} = \mathbf{e}_1'[I - \mathcal{R}(X_{2\cdot 1})]\mathbf{e}_1,$$

a form that suggests again that $\mathbf{e}_1$ has been regressed on $X_{2\cdot 1}$, residuals $\mathbf{e}_{12}$

computed, where

$$\begin{aligned}\mathbf{e}_{12} &= [I - \mathcal{R}(X_1) - \mathcal{R}(X_{2\cdot 1})]\mathbf{y} \\ &= \mathbf{e}_1 - \mathcal{R}(X_{2\cdot 1})\mathbf{y},\end{aligned} \tag{3.140g}$$

and we may write

$$\begin{aligned}\mathbf{e}_{12} &= \mathbf{e}_1 - \mathcal{R}(X_{2\cdot 1})\mathbf{e}_1 \\ &= [I - \mathcal{R}(X_{2\cdot 1})]\mathbf{e}_1\end{aligned} \tag{3.140h}$$

(see Problem 3.26). Hence, using (3.140f),

$$\begin{aligned}S_{12e} &= \mathbf{e}_{12}'\mathbf{e}_{12} = \mathbf{y}'[I - \mathcal{R}(X_1) - \mathcal{R}(X_{2\cdot 1})]\mathbf{y} \\ &= \mathbf{e}_1'\mathbf{e}_1 - \mathbf{e}_1'\mathcal{R}(X_{2\cdot 1})\mathbf{e}_1.\end{aligned} \tag{3.140i}$$

When $X_1 = \mathbf{x}_1$, it is easy to see that

$$S_{1e} = \mathbf{y}'\mathbf{y} - \frac{(\mathbf{x}_1'\mathbf{y})^2}{\mathbf{x}_1'\mathbf{x}_1} \tag{3.140j}$$

so that, in particular, if $\mathbf{x}_1 = \mathbf{1}$, we have

$$S_{1e} = \mathbf{y}'\mathbf{y} - \frac{(\Sigma\, y_u)^2}{N} = \Sigma\,(y_u - \bar{y})^2. \tag{3.140k}$$

Consider the following situation, where interest lies in $\boldsymbol{\beta}_2$, and the model

$$E(\mathbf{y}) = X_1\boldsymbol{\beta}_1 + X_2\boldsymbol{\beta}_2 \tag{3.141}$$

is thought to be correct, so that X_1 is eliminated in usual fashion, and we construct Table 3.4.4. It is then determined that the effects of some factors

TABLE 3.4.5. Determination of (Extra) Regression Sum of Squares When X_1, X_2, and X_0 Are in the Model

Source	Sum of Squares	Degrees of Freedom
Residual, due to X_0, X_1	S_{01e}	$N - k_0 - k_1$
Residual, due to X_0, X_1, and X_2	S_{012e}	$N - k_0 - k_1 - k_2$
Regression, due to X_2, given X_0 and X_1 are in the model	$S_{2\cdot 01r} = S_{01e} - S_{012e}$	k_2

have been left out, indeed that the model should be

$$E(\mathbf{y}) = X_0\boldsymbol{\beta}_0 + X_1\boldsymbol{\beta}_1 + X_2\boldsymbol{\beta}_2, \qquad (3.141a)$$

where interest still lies in $\boldsymbol{\beta}_2$, and where we assume that the rank of $[X_0 \mid X_1 \mid X_2] = \mathcal{X}$ is full; that is, if X_0 is $N \times k_0$, then $\mathcal{X}$ is $N \times (k_0 + k_1 + k_2)$, and we assume

$$r(\mathcal{X}) = k_0 + k_1 + k_2 < N. \qquad (3.141b)$$

We now wish to find (i) the correct (extra) regression sum of squares due to X_2, given that X_1 *and* X_0 are in the model, and (ii) the correct residual sum of squares, given (3.141*a*). Of course, we could ignore our work that led to Table 3.4.4 and start all over. We would take $\mathcal{X}_1 = [X_0 \mid X_1]$, $\mathcal{X}_2 = X_2$, orthogonalize $\mathcal{X}_2$ with respect to $\mathcal{X}_1$, regress $\mathbf{y}$ on $\mathcal{X}_1$ and take residuals, and regress the residuals on the matrix

$$\mathcal{X}_{2\cdot 1} = \mathcal{X}_2 - \mathcal{X}_1\mathcal{A}, \qquad \mathcal{A} = (\mathcal{X}_1'\mathcal{X}_1)^{-1}\mathcal{X}_1'\mathcal{X}_2; \qquad (3.141c)$$

that is, go through the procedure of Case B in Section 3.3.

But the work leading to Table 3.4.4 may be used to deduce a similar procedure for eliminating X_0, resulting in Table 3.4.5.

The question now is how to find S_{01e} and S_{012e}, given that the results in Table 3.4.4 are available. Let us examine the entries of Table 3.4.5. We have that S_{01e} is the residual sum of squares found after regressing $\mathbf{y}$ on *both* X_0 and X_1. This, as we have often seen, is equivalent to regressing $\mathbf{e}_1$, given in (3.140), on $X_{0\cdot 1}$, where

$$X_1'X_{0\cdot 1} = O \quad \text{and} \quad X_{0\cdot 1} = X_0 - X_1(X_1'X_1)^{-1}X_1'X_0. \qquad (3.142)$$

Paralleling (3.140*e*), it is clear that

$$S_{01e} = \mathbf{e}_1'\mathbf{e}_1 - \mathbf{e}_1'\mathcal{R}(X_{0\cdot 1})\mathbf{e}_1 = S_{1e} - \mathbf{e}_1'\mathcal{R}(X_{0\cdot 1})\mathbf{e}_1, \qquad (3.143)$$

where $\mathcal{R}(X_{0\cdot 1}) = X_{0\cdot 1}(X_{0\cdot 1}'X_{0\cdot 1})^{-1}X_{0\cdot 1}'$. Problem 3.27 asks the reader to show that (3.143) is indeed the sum of squares of residuals if X_1 is eliminated and residuals are taken and regressed on $X_{0\cdot 1}$, the orthogonal part of X_0.

In the special case $X_0 = \mathbf{x}_0$, it is easy to see (Problem 3.27) that

$$S_{01e} = S_{1e} - \frac{(\mathbf{x}_{0\cdot 1}'\mathbf{e}_1)^2}{(\mathbf{x}_{0\cdot 1}'\mathbf{x}_{0\cdot 1})} \qquad (3.143a)$$

since $\mathcal{R}(X_{0\cdot 1}) = \mathbf{x}_{0\cdot 1}\mathbf{x}_{0\cdot 1}'/\mathbf{x}_{0\cdot 1}'\mathbf{x}_{0\cdot 1}$. The form of (3.143*a*) is suggestive—see

(3.138c)—since we are in effect regressing a column vector $\mathbf{e}_1$ on $\mathbf{x}_{0\cdot 1}$, and so on.

In a similar way we find

$$
\begin{aligned}
S_{012e} &= S_{12e} - \mathbf{e}'_{12}\mathcal{R}(X_{0\cdot 12})\mathbf{e}_{12} \\
&= \mathbf{e}'_{12}\mathbf{e}_{12} - \mathbf{e}'_{12}\mathcal{R}(X_{0\cdot 12})\mathbf{e}_{12},
\end{aligned}
\tag{3.143b}
$$

where

$$
\begin{aligned}
X_{0\cdot 12} &= X_0 - X_1(X'_1X_1)^{-1}X'_1X_0 - X_{2\cdot 1}(X'_{2\cdot 1}X_{2\cdot 1})^{-1}X'_{2\cdot 1}X_0 \\
&= [I - \mathcal{R}(X_1) - \mathcal{R}(X_{2\cdot 1})]X_0
\end{aligned}
\tag{3.143c}
$$

and $X'_1X_{2\cdot 1} = \text{O}$, so that

$$
X'_1X_{0\cdot 12} = \text{O}, \qquad X'_{2\cdot 1}X_{0\cdot 12} = \text{O}
\tag{3.143d}
$$

(see Problem 3.27). The point here is that we are regressing $\mathbf{e}_{12}$ on $X_{0\cdot 12}$, where $X_{0\cdot 12}$ is orthogonal to X_1 and $X_{2\cdot 1}$, and $\mathbf{e}_{12} = [I - \mathcal{R}(X_1) - \mathcal{R}(X_{2\cdot 1})]\mathbf{y}$, so that the form of (3.143b) is the same as (3.143). Again, in the special case where $X_0 = \mathbf{x}_0$, we find

$$
X_{0\cdot 12} = [I - \mathcal{R}(X_0) - \mathcal{R}(X_{2\cdot 1})]\mathbf{x}_0 = \mathbf{x}_{0\cdot 12}
\tag{3.144}
$$

and, from (3.143b),

$$
S_{012e} = S_{12e} - \frac{(\mathbf{x}'_{0\cdot 12}\mathbf{e}_{12})^2}{\mathbf{x}'_{0\cdot 12}\mathbf{x}_{0\cdot 12}}.
\tag{3.144a}
$$

In summary, having constructed Table 3.4.4 using (3.140) and (3.140a) or (3.140f) and (3.140h), we may, if we wish to take into account factors carried by X_0, find the needed entries for Table 3.4.5 by calculating (3.143) and (3.143b), which makes use of Table 3.4.4.

After doing the calculations necessary for (3.143b), the reader should note that we are able to test, if so desired, the statement $\boldsymbol{\beta}_0 = \mathbf{0}$. After all, the extra regression sum of squares due to $\boldsymbol{\beta}_0$ is $S_{12e} - S_{012e}$, and from (3.143b) this is $\mathbf{e}'_{12}\mathcal{R}(X_{0\cdot 12})\mathbf{e}_{12}$. Furthermore, it is straightforward to verify that under normality, $\mathbf{e}'_{12}\mathcal{R}(X_{0\cdot 12})\mathbf{e}_{12}$ is independent of S_{012e}. (The proof is most easily accomplished if we first write out the relevant sums of squares as quadratic forms in terms of $\mathbf{y}$.) For the case $X_0 = \mathbf{x}_0$, Problem 3.49(d) asks for the necessary details in constructing the test of $\boldsymbol{\beta}_0 = \mathbf{0}$ at level γ.

We can now discuss situations that are called *analysis of covariance* in the literature. In terms of a related example, we refer the reader back to the agricultural example, mentioned at the beginning of this section. There, y_{ij} was the yield from the ith plot using the jth fertilizer, $i=1,\dots,n_j$, $j=1,\dots,k$. There were k fertilizers used, k sets of plots, and n_j plots with fertilizer j (see Table 3.4.1). We recall too that if interested in the statement that the fertilizers were equally effective, the model was written, in the notation of (3.126) and (3.126a),

$$\mathbf{y} = \beta_1 \mathbf{1}_N + \theta_2 \mathbf{x}_2 + \cdots + \theta_k \mathbf{x}_k + \boldsymbol{\varepsilon}, \tag{3.145}$$

where

$$\theta_j = \beta_j - \beta_1, \qquad j = 2,\dots,k. \tag{3.145a}$$

Now the parameters $(\theta_2,\dots,\theta_k) = \boldsymbol{\theta}'$ are of interest and we wish to examine, in particular, the statement

$$\beta_1 = \cdots = \beta_k \quad \text{or} \quad \theta_2 = \cdots = \theta_k = 0. \tag{3.145b}$$

By elimination of the mean, we were finally led to Table 3.4.3, or equivalently, Table 3.4.4.

However, we may now wish to take into account the acidity of the soil of the various plots used, and which as a matter of fact were observed, recorded, and available before the seeding of the plots. We will let x_{0ij} denote the acidity observed (at the outset of the experiment) of the ith plot that is to receive the jth fertilizer, $i=1,\dots,n_j$, $j=1,\dots,k$. This implies that instead of the model

$$E(y_u) = \beta_1 + \theta_2 x_{u2} + \cdots + \theta_k x_{uk}, \tag{3.146}$$

we should now use

$$E(y_u) = \beta_0 x_{u0} + \beta_1 + \theta_2 x_{u2} + \cdots + \theta_k x_{uk}, \tag{3.146a}$$

where x_{u0} is defined by

$$x_{u0} = \text{acidity of the plot from which } y_u \text{ is observed}, \tag{3.146b}$$

and where for $j \geq 1$, the x_{uj} terms are defined as in (3.117a). In vector notation, we use the model

$$E(\mathbf{y}) = \mathbf{x}_0\beta_0 + \mathbf{1}\beta_1 + \mathbf{x}_2\theta_2 + \cdots + \mathbf{x}_k\theta_k \tag{3.147}$$

TABLE 3.4.6. One-Way Analysis of Variance for Model (3.147*a*), with β_1 Eliminated

Source	Degrees of Freedom	Sum of Squares
Between	$k-1$	$\Sigma n_j(\bar{y}_{ij} - \bar{y})^2$
Residual	$N-k$	$\Sigma\Sigma(y_{ij} - \bar{y}_j)^2 \quad [= S_{12e}]$
Total	$N-1$	$\Sigma\Sigma(y_{ij} - \bar{y})^2 \quad [= S_{1e}]$

instead of

$$E(\mathbf{y}) = \mathbf{1}\beta_1 + \mathbf{x}_2\theta_2 + \cdots + \mathbf{x}_k\theta_k. \tag{3.147a}$$

We have already constructed Table 3.4.3 for the model (3.147*a*). Since we are interested in $\boldsymbol{\theta} = (\theta_2, \dots, \theta_k)'$, we eliminate the mean β_1 and construct Table 3.4.6.

Now we know [see the discussion centering on (3.113) and Table 3.3.5] that when eliminating the general mean, we regress $\mathbf{y}$ on $\mathbf{1}$ and find the residual $\mathbf{e}_1 = \mathbf{y} - \bar{y}\mathbf{1}$, so that the residual sum of squares after regressing $\mathbf{y}$ on $\mathbf{1}$ is

$$\begin{aligned} \mathbf{e}_1'\mathbf{e}_1 &= \Sigma(y_u - \bar{y})^2 = \sum_{j=1}^{k}\sum_{i=1}^{n_j}(y_{ij} - \bar{y})^2 \\ &= S_{1e}. \end{aligned} \tag{3.148}$$

This is the total line of Table 3.4.6, mentioned in the first line of Table 3.4.4. Now when we regress $\mathbf{y}$ on $(\mathbf{1}, \mathbf{x}_2, \dots, \mathbf{x}_k)$ in (3.147*a*), which, as we have seen, is equivalent to regressing $\mathbf{y}$ on $(\mathbf{x}_1, \mathbf{x}_2, \dots, \mathbf{x}_k) = X$, then the residual sum of squares [see (3.123*b*)], from Table 3.4.6, is

$$\Sigma\Sigma(y_{ij} - \bar{y}_j)^2 = S_{12e}, \tag{3.149}$$

the entry of the second line of Table 3.4.4. We have recorded facts (3.148) and (3.149) in brackets in Table 3.4.6. Now we wish to find the counterpart of Table 3.4.5. From (3.143*a*), we obtain

$$S_{01e} = S_{1e} - \frac{(\mathbf{x}_{0\cdot 1}'\mathbf{e}_1)^2}{\mathbf{x}_{0\cdot 1}'\mathbf{x}_{0\cdot 1}}. \tag{3.150}$$

We have $\mathbf{e}_1 = \mathbf{y} - \bar{\mathbf{y}}\mathbf{1}$ and recall that $\mathbf{x}_{0\cdot 1}$ is the residual obtained after regressing $\mathbf{x}_0$ on $\mathbf{1}$, so that paralleling $\mathbf{e}_1$ (the residual obtained after regressing $\mathbf{y}$ on $\mathbf{1}$), we have

$$\text{(3.151)} \qquad \mathbf{x}_{0\cdot 1} = \mathbf{x}_0 - \bar{x}_0\mathbf{1}.$$

Then (3.150) becomes

$$\text{(3.152)} \qquad S_{01e} = \Sigma\Sigma\,(y_{ij} - \bar{y})^2 - \frac{\left[\Sigma\Sigma\,(x_{0ij} - \bar{x}_0)(y_{ij} - \bar{y})\right]^2}{\Sigma\Sigma\,(x_{0ij} - \bar{x}_0)^2},$$

where

$$\bar{x}_0 = N^{-1}(\mathbf{1}'\mathbf{x}_0) = N^{-1}\Sigma\Sigma\,x_{0ij}.$$

We now wish to determine S_{012e}, and since $X_0 = \mathbf{x}_0$, $X_1 = \mathbf{1}$, we find [see (3.144a)] that

$$\text{(3.153)} \qquad S_{012e} = S_{12e} - \frac{(\mathbf{x}'_{0\cdot 12}\mathbf{e}_{12})^2}{\mathbf{x}'_{0\cdot 12}\mathbf{x}_{0\cdot 12}},$$

where $\mathbf{e}_{12}$ has components $e_{12ij} = y_{ij} - \bar{y}_j$ and S_{12e} is recorded in Table 3.4.6. Problem 3.28 asks for direct verification that $\mathbf{e}_{12}$ is as just given. It remains to find $\mathbf{x}_{0\cdot 12}$, the residual obtained after regressing $\mathbf{x}_0$ on $\mathbf{x}_1 = \mathbf{1}$ and $X_2 = (\mathbf{x}_2, \ldots, \mathbf{x}_k)$. By analogy with $\mathbf{e}_{12}$, the components of $\mathbf{x}_{0\cdot 12}$ are

$$\text{(3.153}a\text{)} \qquad x_{0\cdot 12ij} = x_{0ij} - \bar{x}_{0\cdot j},$$

where $\bar{x}_{0\cdot j} = n_j^{-1}\Sigma_{i=1}^{n_j}\,x_{0ij}$ is the mean of the n_j x_0 values that correspond (belong) to the jth group: In our agricultural example, $\bar{x}_{0\cdot j}$ is the mean of the acidity values of the n_j plots that received fertilizer j. Hence

$$\text{(3.153}b\text{)} \qquad S_{012e} = \Sigma\Sigma\,(y_{ij} - \bar{y}_j)^2 - \frac{\left[\Sigma\Sigma\,(x_{0ij} - \bar{x}_{0\cdot j})(y_{ij} - \bar{y}_j)\right]^2}{\Sigma\Sigma\,(x_{0ij} - \bar{x}_{0\cdot j})^2}.$$

By (3.150) and (3.153b), the correct regression sum of squares due to X_2, given that *both* $\mathbf{x}_0$ and $\mathbf{1}$ are in the model, is

$$\text{(3.153}c\text{)} \qquad S_{01e} - S_{012e}.$$

With these results we may now construct the relevant analysis of covariance table shown as Table 3.4.7.

TABLE 3.4.7. One-Way Analysis of Covariance for Model (3.147); $\mathbf{x}_0(\beta_0)$ and $\mathbf{1}(\beta_1)$ Eliminated

Source	Degrees of Freedom	Sum of Squares
Extra regression sum of squares due to $\theta_2,\ldots,\theta_k$	$k-1$	$S_{01e} - S_{012e}$
Residual, after regressing $\mathbf{y}$ on $\mathbf{x}_0, \mathbf{1}, \mathbf{x}_2,\ldots,\mathbf{x}_k$, or $\mathbf{x}_0, \mathbf{x}_1,\ldots,\mathbf{x}_k$	$N-k-1$	S_{012e}
Total	$N-2$	S_{01e}

Note the total degrees of freedom. We have eliminated two effects, one due to acidity and one due to a general mean, leaving $N-2$ degrees of freedom. Also, the residual degrees of freedom is $N-(k+1)$, since in the full model there are now $k+1$ effects, $\beta_0, \beta_1, \beta_2,\ldots,\beta_k$. Furthermore, the one degree of freedom for β_0 may be isolated and inference made about the statement "$\beta_0 = 0$"—see Problem 3.49(d). We close this discussion with two related points.

Remark 1. *We do not need to go through the above procedure but we can simply orthogonalize by regressing* $X_2 = (\mathbf{x}_2 \mid \cdots \mid \mathbf{x}_k)$ *on* $X_1 = (\mathbf{x}_0 \mid \mathbf{1})$, *regressing* $\mathbf{y}$ *on* X_1, *computing the residual* $\mathbf{y} - \mathcal{R}(X_1)\mathbf{y}$, *and regress the latter on* $X_{2\cdot 1}$, *as discussed earlier. Indeed, if* X_1 *is* $N \times m$, *where* $m > 2$, *that is, there are many* concomitant *variables or factors to include at the outset, then the general procedure is more efficient.*

Remark 2. *The work in Remark* 1 *for* $X_1 = (\mathbf{x}_0 \mid \mathbf{1})$ *is long and tedious, and indeed we can obtain all the components of Table* 3.4.7 *by meshing Table* 3.4.6, *a one-way analysis of variance for the* y_{ij}, *with a similar one-way analysis of variance on the* x_{0ij}, *and a parallel one-way analysis of variance on the cross products, as in Table* 3.4.8. *The table is simply constructed and makes use of* (3.134). *Problem* 3.29 *asks for direct verification that the first two entries in each column of Table* 3.4.8 *add to the third entry of that column. From Tables* 3.4.7 *and* 3.4.8, *the analysis of covariance table takes the form of Table* 3.4.8*a*.

Incidentally, since we are doing a series of one-way analyses of variances in Table 3.4.8, we remind the reader that an easy computational scheme for

TABLE 3.4.8. One-Way Analysis of Variance on the y_{ij}, x_{ij}, and Cross Products[a]

	y_{ij}	$x_{ij}y_{ij}$	x_{ij}
Between	$\Sigma n_j(\bar{y}_j - \bar{y})^2 = q_{y1}$	$\Sigma n_j(\bar{y}_j - \bar{y})(\bar{x}_{0\cdot j} - \bar{x}_0) = q_{xy1}$	$\Sigma n_j(\bar{x}_{0\cdot j} - \bar{x}_0)^2 = q_{x1}$
Within	$\Sigma\Sigma(y_{ij} - \bar{y}_j)^2 = q_{y2}$	$\Sigma\Sigma(y_{ij} - \bar{y}_j)(x_{0ij} - \bar{x}_{0\cdot j}) = q_{xy2}$	$\Sigma\Sigma(x_{0ij} - \bar{x}_{0\cdot j})^2 = q_{x2}$
Total	$\Sigma\Sigma(y_{ij} - \bar{y})^2 = q_y$	$\Sigma\Sigma(y_{ij} - \bar{y})(x_{0ij} - \bar{x}_0) = q_{xy}$	$\Sigma\Sigma(x_{0ij} - \bar{x}_0)^2 = q_x$
		$\bar{x}_0 = \frac{1}{N}\sum_i\sum_j x_{0ij}$	$\bar{x}_{0\cdot j} = n_j^{-1}\sum_i x_{0ij}$

[a] See (3.134).

q_{y1}, q_{y2}, and q_y is

$$q_{y1} = \sum_{j=1}^{k} \frac{T_{\cdot j}(y)}{n_j} - \frac{G^2(y)}{N}, \qquad N = \sum_{j=1}^{k} n_j$$

(3.154)
$$q_{y2} = \sum_{j=1}^{k} \sum_{i=1}^{n_j} y_{ij}^2 - \sum_{j=1}^{k} \frac{[T_{\cdot j}(y)]^2}{n_j}$$

$$q_y = \sum\sum y_{ij}^2 - \frac{G^2(y)}{N},$$

where

(3.154a)
$$G(y) = \sum_{j=1}^{k}\sum_{i=1}^{n_j} y_{ij} = \sum_{j=1}^{k} T_{\cdot j}(y) \quad \text{and} \quad T_{\cdot j}(y) = \sum_{i=1}^{n_j} y_{ij}.$$

Similar expressions exist for q_{x1}, q_{x2}, and q_x. The quantities in (3.154) and,

TABLE 3.4.8a. Analysis of Covariance (One-Way, One Concomitant Variable)

Source	Degrees of Freedom	Sum of Squares
Regression due to X_2, given $\mathbf{x}_0$, $\mathbf{1}$ in model	$k - 1$	$(q_y - q_{xy}^2/q_x) - (q_{y2} - q_{xy2}^2/q_{x2})$
Residual	$N - k - 1$	$q_{y2} - q_{xy2}^2/q_{x2}$
Total	$N - 2$	$q_y - q_{xy}^2/q_x$

of course, q_{x1}, q_{x2}, and q_x are all positive. However q_{xy1}, q_{xy2}, and q_{xy} need not be positive, and is can be shown that an easy computational scheme for these quantities is given by

$$q_{xy1} = \sum_{j=1}^{k} \frac{T_{\cdot j}(x)T_{\cdot j}(y)}{n_j} - \frac{G(x)G(y)}{N}$$

(3.154*b*)
$$q_{xy2} = \sum_{j=1}^{k} \sum_{i=1}^{n_j} y_{ij}x_{0ij} - \sum_{j=1}^{k} \frac{T_{\cdot j}(x)T_{\cdot j}(y)}{n_j}$$

$$q_{xy} = \sum_{j=1}^{k} \sum_{i=1}^{n_j} y_{ij}x_{0ij} - \frac{G(x)G(y)}{N}.$$

The reader is asked to analyze a set of data using the above in Problem 3.49.

3.5. ORTHOGONAL POLYNOMIALS

As we have stated previously, a polynomial model is very often an appropriate description of the expected behavior of a response variable y. We now assume that

(3.155)
$$E(y) = \sum_{j=0}^{m} \beta_j x^j,$$

where $x = \phi(z)$ is a function of the independent variable z.

In many instances, we wish to fit a polynomial of order "as low as possible," so that we may well want to look at the contribution of a first-degree term, *separately* from the contribution of a second-degree term, and these separately from a third-degree term, and so on. But from previous work in this chapter, we know that we may look at certain effects separately if we have orthogonality of the relevant parts of the "X matrix." A short summary is as follows. Suppose

(3.156)
$$E(\mathbf{y}) = X_1\boldsymbol{\beta}_1 + X_2\boldsymbol{\beta}_2 = X\boldsymbol{\beta}, \qquad X = (X_1 \mid X_2), \quad \boldsymbol{\beta} = (\boldsymbol{\beta}_1' \mid \boldsymbol{\beta}_2')'$$

and that interest lies in $\boldsymbol{\beta}_2$. We have seen that we may write (3.156) as

(3.156*a*)
$$E(\mathbf{y}) = X_1\boldsymbol{\phi} + X_{2\cdot 1}\boldsymbol{\beta}_2, \qquad \boldsymbol{\phi} = \boldsymbol{\beta}_1 + A\boldsymbol{\beta}_2,$$

where X_1 is orthogonal to $X_{2\cdot 1} = X_2 - X_1 A$, $A = (X_1'X_1)^{-1}X_1'X_2$. This fact allows us to write down the (extra) regression sum of squares due to $\boldsymbol{\beta}_2$, given that X_1 is in the model. In fact, because X_1 is orthogonal to $X_{2\cdot 1}$, we may estimate $\boldsymbol{\beta}_2$ by regressing $\mathbf{y}$ on $X_{2\cdot 1}$, that is, acting as if the model were $E(\mathbf{y}) = X_{2\cdot 1}\boldsymbol{\beta}_2$ (see Case A of Section 3.3). Then the least-squares estimator of $\boldsymbol{\beta}_2$ is $\hat{\boldsymbol{\beta}}_2 = (X_{2\cdot 1}'X_{2\cdot 1})^{-1}X_{2\cdot 1}'\mathbf{y}$, and indeed the extra regression sum of squares due to $\boldsymbol{\beta}_2$ is $\hat{\boldsymbol{\beta}}_2'X_{2\cdot 1}'X_{2\cdot 1}\hat{\boldsymbol{\beta}}_2 = \mathbf{y}'\mathcal{R}(X_{2\cdot 1})\mathbf{y}$. We may view the above as a linear transformation on X,

(3.157) $$X\boldsymbol{\beta} = XHH^{-1}\boldsymbol{\beta} = W\boldsymbol{\theta},$$

where

(3.157*a*) $$W = XH, \qquad \boldsymbol{\theta} = H^{-1}\boldsymbol{\beta}$$

and

(3.157*b*) $$H = \left(\begin{array}{c|c} I & -A \\ \hline \bigcirc & I \end{array}\right).$$

Then

(3.157*c*) $$W = (X_1 \mid X_2 - X_1 A) = (X_1 \mid X_{2\cdot 1})$$

and

(3.157*d*) $$\boldsymbol{\theta} = H^{-1}\boldsymbol{\beta} = \left(\begin{array}{c|c} I & A \\ \hline \bigcirc & I \end{array}\right)\left(\begin{array}{c} \boldsymbol{\beta}_1 \\ \hline \boldsymbol{\beta}_2 \end{array}\right) = \left(\begin{array}{c} \boldsymbol{\beta}_1 + A\boldsymbol{\beta}_2 \\ \hline \boldsymbol{\beta}_2 \end{array}\right) = \left(\begin{array}{c} \boldsymbol{\phi} \\ \hline \boldsymbol{\beta}_2 \end{array}\right).$$

Note from (3.157*b*) that H is upper triangular and H orthogonalizes X; specifically, columns of X_1 are orthogonal to columns of $X_{2\cdot 1}$. This result can be generalized, since we can find an upper triangular matrix H such that *all* columns of $W = XH$ are orthogonal. This process is called the *Gram–Schmidt process*, and we will outline it now and then use it for the fitting of polynomials.

Suppose we represent the $N \times k$ matrix X as

(3.158) $$X = (\mathbf{x}_1 \mid \cdots \mid \mathbf{x}_k).$$

1. Keep $\mathbf{x}_1, \mathbf{x}_3, \ldots, \mathbf{x}_k$ fixed and regress $\mathbf{x}_2$ on $\mathbf{x}_1$, finding the residual

(3.159) $$\mathbf{x}_{2\cdot 1} = \mathbf{x}_2 - \mathbf{x}_1(\mathbf{x}_1'\mathbf{x}_1)^{-1}\mathbf{x}_1'\mathbf{x}_2$$

which may be written

$$(3.159a)\qquad \mathbf{x}_{2\cdot 1} = \mathbf{x}_2 - \alpha_{21}\mathbf{x}_1, \qquad \alpha_{21} = \frac{\mathbf{x}_2'\mathbf{x}_1}{\mathbf{x}_1'\mathbf{x}_1} = \frac{\mathbf{x}_1'\mathbf{x}_2}{\mathbf{x}_1'\mathbf{x}_1}.$$

We now let

$$(3.159b)\qquad W_1 = \left(\mathbf{x}_1 \mid \mathbf{x}_{2\cdot 1} \mid \mathbf{x}_3 \mid \cdots \mid \mathbf{x}_k\right)$$

and we may easily verify that

$$(3.159c)\quad W_1 = XH_1, \qquad H_1 = \begin{pmatrix} 1 & -\alpha_{21} & 0 & \cdots & 0 \\ 0 & 1 & 0 & \cdots & 0 \\ 0 & 0 & 1 & & \\ \vdots & \vdots & & \ddots & 0 \\ 0 & 0 & \cdots & 0 & 1 \end{pmatrix}.$$

In W_1, note that $\mathbf{x}_{2\cdot 1}$ is orthogonal to $\mathbf{x}_1$, that is,

$$(3.159d)\qquad \mathbf{x}_1'\mathbf{x}_{2\cdot 1} = 0.$$

2. In W_1, keep $\mathbf{x}_1, \mathbf{x}_{2\cdot 1}, \mathbf{x}_4, \ldots, \mathbf{x}_k$ fixed and regress $\mathbf{x}_3$ on $\mathbf{x}_1$ and $\mathbf{x}_{2\cdot 1}$. The residual (see Problem 3.30) is

$$(3.160)\qquad \mathbf{x}_{3\cdot 21} = \mathbf{x}_3 - \alpha_{31}\mathbf{x}_1 - \alpha_{32}\mathbf{x}_{2\cdot 1},$$

where

$$(3.160a)\qquad \alpha_{31} = \frac{\mathbf{x}_3'\mathbf{x}_1}{\mathbf{x}_1'\mathbf{x}_1}, \qquad \alpha_{32} = \frac{\mathbf{x}_3'\mathbf{x}_{2\cdot 1}}{\mathbf{x}_{2\cdot 1}'\mathbf{x}_{2\cdot 1}}.$$

Furthermore, we have that $\mathbf{x}_1$ and $\mathbf{x}_{2\cdot 1}$ are orthogonal to $\mathbf{x}_{3\cdot 21}$; recalling (3.159d), we have

$$(3.160b)\qquad \mathbf{x}_1'\mathbf{x}_{2\cdot 1} = 0; \quad \mathbf{x}_1'\mathbf{x}_{3\cdot 21} = 0; \quad \mathbf{x}_{2\cdot 1}'\mathbf{x}_{3\cdot 21} = 0.$$

Now let

$$(3.160c)\qquad W_2 = \left(\mathbf{x}_1 \mid \mathbf{x}_{2\cdot 1} \mid \mathbf{x}_{3\cdot 21} \mid \mathbf{x}_4 \mid \cdots \mid \mathbf{x}_k\right)$$

so that

$$(3.160d) \quad W_2 = W_1 H_2, \qquad H_2 = \begin{bmatrix} 1 & 0 & -\alpha_{31} & 0 & \cdots & 0 \\ 0 & 1 & -\alpha_{32} & 0 & \cdots & 0 \\ 0 & 0 & 1 & 0 & \cdots & 0 \\ 0 & 0 & 0 & 1 & \cdots & 0 \\ \vdots & \vdots & \vdots & \vdots & \ddots & \vdots \\ 0 & 0 & 0 & 0 & \cdots & 1 \end{bmatrix}.$$

Note that

$$(3.160e) \qquad W_2 = XH_1H_2.$$

Problem 3.30 asks for the form of H_1H_2.

3. After performing this procedure $k-1$ times, we arrive at

$$\begin{aligned} W = W_{k-1} &= (\mathbf{w}_1 \mid \cdots \mid \mathbf{w}_k) = XH_1H_2 \cdots H_{k-1} \\ (3.161) \qquad &= XH \\ &= (\mathbf{x}_1 \mid \mathbf{x}_{2\cdot 1} \mid \mathbf{x}_{3\cdot 21} \mid \cdots \mid \mathbf{x}_{k\cdot \overline{k-1}\,\overline{k-2}\cdots 21}) \end{aligned}$$

with $\mathbf{w}_i'\mathbf{w}_j = 0$, for all $i \neq j$ and where

$$(3.161a) \qquad H = H_1H_2 \cdots H_{k-1}$$

is of upper triangular form,

$$(3.161b) \qquad H = \begin{pmatrix} 1 & h_{12} & \cdots & & h_{1k} \\ 0 & 1 & h_{23} & \cdots & h_{2k} \\ 0 & 0 & 1 & & \vdots \\ \vdots & \vdots & & \ddots & h_{k-1,k} \\ 0 & 0 & \cdots & 0 & 1 \end{pmatrix},$$

which implies that H^{-1} is also upper triangular,

$$(3.161c) \qquad H^{-1} = \begin{bmatrix} 1 & c_{12} & \cdots & & c_{1,k-1} & c_{1k} \\ 0 & 1 & c_{23} & \cdots & c_{2,k-1} & c_{2k} \\ 0 & 0 & 1 & \cdots & & c_{3k} \\ \vdots & \vdots & & \ddots & \vdots & \vdots \\ 0 & 0 & \cdots & & 1 & c_{k-1,1} \\ 0 & 0 & \cdots & & 0 & 1 \end{bmatrix}.$$

Thus, for any model,

$$E(\mathbf{y}) = X\boldsymbol{\beta}, \tag{3.162}$$

we may write

$$\begin{aligned} E(\mathbf{y}) &= XHH^{-1}\boldsymbol{\beta} \\ &= W\boldsymbol{\theta} \\ &= \theta_1\mathbf{w}_1 + \cdots + \theta_k\mathbf{w}_k, \end{aligned} \tag{3.162a}$$

where

$$W = W_{k-1} = XH_1 \cdots H_{k-1} = \left(\mathbf{x}_1 \mid \mathbf{x}_{2\cdot 1} \mid \cdots \mid \mathbf{x}_{k\cdot \overline{k-1}\cdots 1}\right) \tag{3.162b}$$

and

$$\boldsymbol{\theta} = H^{-1}\boldsymbol{\beta} = \begin{bmatrix} 1 & c_{12} & \cdots & & c_{1k} \\ 0 & 1 & & & \vdots \\ 0 & 0 & \ddots & & \\ \vdots & \vdots & & & \\ 0 & \cdots & & 1 & c_{k-1,k} \\ 0 & \cdots & & & 1 \end{bmatrix} \begin{pmatrix} \beta_1 \\ \beta_2 \\ \vdots \\ \beta_k \end{pmatrix}. \tag{3.162c}$$

Note that $\theta_k = \beta_k$ and indeed we may work "from the bottom" to successively find expressions for the θ_i's in terms of the β_i's, and use $\boldsymbol{\beta} = H\boldsymbol{\theta}$ in the same way to find expressions for the β_i's in terms of the θ_i's, and so on. Also note that in (3.162*a*) we have

$$\mathbf{w}_i'\mathbf{w}_j = 0, \qquad \text{for all } i \neq j, \tag{3.162d}$$

that is, columns of W are orthogonal to each other. Thus,

$$\hat{\boldsymbol{\theta}} = (W'W)^{-1}W'\mathbf{y} = \begin{pmatrix} (\mathbf{w}_1'\mathbf{w}_1)^{-1}\mathbf{w}_1'\mathbf{y} \\ \vdots \\ (\mathbf{w}_k'\mathbf{w}_k)^{-1}\mathbf{w}_k'\mathbf{y} \end{pmatrix}, \tag{3.163}$$

that is, $\hat{\theta}_i = \mathbf{w}_i'\mathbf{y}/\mathbf{w}_i'\mathbf{w}_i$, which is the least-squares estimator of θ_i found when

TABLE 3.5.1. Analysis of Variance Following the Gram–Schmidt Procedure

Source	Degrees of Freedom	Sum of Squares	Expected Mean Square
Regression, due to $\mathbf{x}_1$	1	$\dfrac{(\mathbf{w}_1'\mathbf{y})^2}{\mathbf{w}_1'\mathbf{w}_1}$	
Regression (extra), due to $\mathbf{x}_2$, given $\mathbf{x}_1$ in the model	1	$\dfrac{(\mathbf{w}_2'\mathbf{y})^2}{\mathbf{w}_2'\mathbf{w}_2}$	
⋮	⋮	⋮	
Regression (extra), due to $\mathbf{x}_k$, given $\mathbf{x}_{k-1}, \mathbf{x}_{k-2}, \ldots, \mathbf{x}_1$ in the model	1	$\dfrac{(\mathbf{w}_k'\mathbf{y})^2}{\mathbf{w}_k'\mathbf{w}_k}$	
Residual	$N-k$	$\mathbf{y}'\mathbf{y} - \sum_{j=1}^{k} \dfrac{(\mathbf{w}_j'\mathbf{y})^2}{\mathbf{w}_j'\mathbf{w}_j} = \mathbf{y}'[(I - \mathcal{R}(X)]\mathbf{y}$	
Total	N	$\mathbf{y}'\mathbf{y}$	

regressing $\mathbf{y}$ on $\mathbf{w}_i$ alone, and in view of (3.162d), we also have

$$\mathcal{R}(X) = \mathcal{R}(W) = \mathcal{R}(\mathbf{w}_1) + \cdots + \mathcal{R}(\mathbf{w}_k). \tag{3.163a}$$

The sum of squares of regression of $\mathbf{y}$ on $\mathbf{w}_i$ is

$$\hat{\theta}_i \mathbf{w}_i' \mathbf{w}_i \hat{\theta}_i = \frac{(\mathbf{w}_i'\mathbf{y})^2}{\mathbf{w}_i'\mathbf{w}_i}. \tag{3.163b}$$

Furthermore, because W is a linear transformation of X,

$$\begin{aligned} \mathbf{y}'\mathcal{R}(\mathbf{w}_j)\mathbf{y} = \frac{(\mathbf{w}_j'\mathbf{y})^2}{\mathbf{w}_j'\mathbf{w}_j} &= \mathbf{y}'\mathcal{R}(\mathbf{x}_1, \ldots, \mathbf{x}_j)\mathbf{y} - \mathbf{y}'\mathcal{R}(\mathbf{x}_1, \ldots, \mathbf{x}_{j-1})\mathbf{y} \\ &= \mathbf{y}'\mathcal{R}(\mathbf{w}_1, \ldots, \mathbf{w}_j)\mathbf{y} - \mathbf{y}'\mathcal{R}(\mathbf{w}_1, \ldots, \mathbf{w}_{j-1})\mathbf{y}, \end{aligned} \tag{3.164}$$

since $W_j = XH_1 \cdots H_{j-1}$, and so on. Thus the extra regression sum of squares due to $\mathbf{x}_j$, given $\mathbf{x}_1, \ldots, \mathbf{x}_{j-1}$ are in the model, is $\mathbf{y}'\mathcal{R}(\mathbf{w}_j)\mathbf{y}$. We may think of these results in terms of an analysis of variance table which now has the form of Table 3.5.1.

Problem 3.31 asks for the entries in the Expected Mean Square column of Table 3.5.1.

It is useful to recall the geometry involved in the Gram–Schmidt process. From the basis $\{\mathbf{x}_1,\ldots,\mathbf{x}_k\}$ of S_k, we have constructed an orthogonal basis $\{\mathbf{w}_1,\ldots,\mathbf{w}_k\}$ of S_k. This allows us to regress $\mathbf{y}$ on the $\mathbf{w}_j$ one at a time in order to determine the regression sum of squares due to $\mathbf{w}_j$, then to sum (as in Section 3.3) the vectors so obtained and determine the total regression sum of squares,

$$\mathbf{y}'\mathcal{R}(X)\mathbf{y} = \mathbf{y}'\mathcal{R}(W)\mathbf{y}, \qquad W = XH \tag{3.165}$$

and the residual sum of squares,

$$\mathbf{y}'[I - \mathcal{R}(X)]\mathbf{y} = \mathbf{y}'[I - \mathcal{R}(W)]\mathbf{y}. \tag{3.166}$$

We may now apply the Gram–Schmidt procedure outlined above to our problem of fitting orthogonal polynomials. We have not as yet defined orthogonal polynomials and will do so shortly, but the reader may regard the term "fitting orthogonal polynomials" as a synonym for the Gram–Schmidt procedure in the polynomial fitting case for now. Suppose we are using the model described in (3.155). If we take N observations y_u "at the places $x = x_u$," $u = 1,\ldots,N$, then

$$E(y_u) = \sum_{j=0}^{m} \beta_j x_u^j \tag{3.167}$$

so that, in usual notation,

$$E(\mathbf{y}) = X\boldsymbol{\beta}, \tag{3.168}$$

where the $(j+1)$st column of X, $\mathbf{x}_j$, is given by

$$\mathbf{x}_j = \left(x_1^j, x_2^j,\ldots,x_N^j\right)', \qquad j = 0, 1,\ldots,m, \tag{3.168a}$$

with X a $N \times k$ matrix, $k = m + 1$, and $\boldsymbol{\beta} = (\beta_0, \beta_1, \beta_2,\ldots,\beta_m)'$, a $(m+1) \times 1$ vector of regression constants. We will assume that $N > m + 1$. The second column of X, denoted here as $\mathbf{x}_1$, is of course the "design matrix," in that it records the places x_u at which we observe y_u. The completion of the Gram–Schmidt process and the regression of $\mathbf{y}$ on $W = XH$, where X is specified by (3.168a), would eventually yield an analysis of variance table of the form of Table 3.5.2.

For the entries of the EMS column, see Problem 3.31. The procedure here is to test the contribution of x^m, namely, $\text{SS}_{r_m} = (\mathbf{w}_m'\mathbf{y})^2/\mathbf{w}_m'\mathbf{w}_m$ against the residual mean square $\text{SS}_e/(N-m-1) = \mathbf{y}'[(I - \mathcal{R}(W)]\mathbf{y}/(N-m-1)$. If SS_{r_m} is significantly large, we would reject the hypothesis that θ_m $(=\beta_m)$ is zero. (On the assumption of normality, Problem 3.31 also asks for the appropriate test.) If SS_{r_m} is not significantly large, we would then act as

TABLE 3.5.2. Analysis of Variance for the Gram–Schmidt Process Applied to (3.167) and (3.168)

Source	Degrees of Freedom	Sum of Squares	Expected Mean Square
Due to x^0 (mean)	1	$\dfrac{(\mathbf{w}_0'\mathbf{y})^2}{\mathbf{w}_0'\mathbf{w}_0}$	
Extra, due to $x^1 = x$ (linear)	1	$\dfrac{(\mathbf{w}_1'\mathbf{y})^2}{\mathbf{w}_1'\mathbf{w}_1}$	
Extra, due to x^2 (quadratic)	1	$\dfrac{(\mathbf{w}_2'\mathbf{y})^2}{\mathbf{w}_2'\mathbf{w}_2}$	
$\vdots$	$\vdots$	$\vdots$	
Extra, due to x^m	1	$\dfrac{(\mathbf{w}_m'\mathbf{y})^2}{\mathbf{w}_m'\mathbf{w}_m}$	
Residual	$N - m - 1$	$\mathbf{y}'[I - \mathcal{R}(W)]\mathbf{y}$	
Total	N	$\mathbf{y}'\mathbf{y}$	

if the response is not a polynomial function in x of order at most m, and proceed in a similar way to test whether θ_{m-1} is zero or not, and so on.

To gain insight into how this may proceed, and to see how orthogonal polynomials get into the act, suppose we consider the case where the x's used are equally spaced. Some simplification is possible. To illustrate, suppose $N = 5$ and that the five places used are equispaced by one integer apart,

$$\mathbf{x} = (1, 2, 3, 4, 5)', \tag{3.169}$$

and that the expected response is a polynomial in x, of degree at most three. We note that by a location–scale transformation, the places $(a, a + b, a + 2b, a + 3b, a + 4b)$ may always be mapped to $\mathbf{x}$ as above. If we do fit a third-order model, then in matrix form we have

(3.170)

$$E(\mathbf{y}) = X\boldsymbol{\beta} = \begin{bmatrix} 1 & 1 & 1 & 1 \\ 1 & 2 & 4 & 8 \\ 1 & 3 & 9 & 27 \\ 1 & 4 & 16 & 64 \\ 1 & 5 & 25 & 125 \end{bmatrix} \begin{bmatrix} \beta_0 \\ \beta_1 \\ \beta_2 \\ \beta_3 \end{bmatrix} = \beta_0\mathbf{x}_0 + \beta_1\mathbf{x}_1 + \beta_2\mathbf{x}_2 + \beta_3\mathbf{x}_3,$$

that is, $\mathbf{x}_j$ has as its ith component, $x_i^j = i^j$. We denote $\mathbf{x}_1$ by $\mathbf{x}$. Then

$$E(y) = \beta_0 + \beta_1 x + \beta_2 x^2 + \beta_3 x^3. \tag{3.171}$$

We now go through the Gram–Schmidt procedure, calculate the α's, and [Problem 3.31(b)] arrive at

$$E(\mathbf{y}) = W\boldsymbol{\theta} = (XH)(H^{-1}\boldsymbol{\beta}), \tag{3.172}$$

where

(3.172a)

$$W = (\mathbf{w}_0 \mid \mathbf{w}_1 \mid \mathbf{w}_2 \mid \mathbf{w}_3) = (\mathbf{1} \mid \mathbf{x}_{1\cdot 0} \mid \mathbf{x}_{2\cdot 10} \mid \mathbf{x}_{3\cdot 210}), \qquad \boldsymbol{\theta} = H^{-1}\boldsymbol{\beta}$$

with

$$\begin{aligned} \mathbf{w}_0 = \mathbf{x}_0 = \mathbf{1}; \quad \mathbf{w}_1 &= \mathbf{x}_{1\cdot 0} = \mathbf{x} - 3\mathbf{1} = \mathbf{x}_1 - 3\mathbf{1}; \\ \mathbf{w}_2 &= \mathbf{x}_{2\cdot 10} = \mathbf{x}_2 - (11)\mathbf{1} - 6\mathbf{x}_{1\cdot 0}; \\ \mathbf{w}_3 &= \mathbf{x}_{3\cdot 210} = \mathbf{x}_3 - (45)\mathbf{1} - 30.4\mathbf{x}_{1\cdot 0} - 9\mathbf{x}_{2\cdot 10}; \end{aligned} \tag{3.172b}$$

so that

(3.172c)

$$H = \begin{pmatrix} 1 & -3 & 7 & -16.8 \\ 0 & 1 & -6 & 23.6 \\ 0 & 0 & 1 & -9 \\ 0 & 0 & 0 & 1 \end{pmatrix}; \qquad H^{-1} = \begin{pmatrix} 1 & 3 & 11 & 45 \\ 0 & 1 & 6 & 30.4 \\ 0 & 0 & 1 & 9 \\ 0 & 0 & 0 & 1 \end{pmatrix}.$$

The reader may note the similarity of the elements of H^{-1} with those appearing in (3.172b). This is quickly explained by noting, for example, that the last line of (3.172b) gives

$$\mathbf{x}_3 = 45\mathbf{w}_0 + 30.4\mathbf{w}_1 + 9\mathbf{w}_2 + \mathbf{w}_3$$

or

$$\mathbf{x}_3 = W\begin{pmatrix} 45 \\ 30.4 \\ 9 \\ 1 \end{pmatrix}.$$

That is, the fourth column of X (labeled $\mathbf{x}_3$ for our purposes) may be found by using that linear combination of the columns of $W = (\mathbf{w}_0, \mathbf{w}_1, \mathbf{w}_2, \mathbf{w}_3)$ whose weights are 45, 30.4, 9, and 1, respectively. This is so because of the tautology $X = (XH)H^{-1} = WH^{-1}$, and so on. Once the α's are computed we may state H^{-1} and find H by inversion of H^{-1}; we then do not need to compute $H_1 H_2 H_3$.

We note the following interesting fact. Let typical elements of $\mathbf{w}_j$, the jth column of $W = XH$, be denoted by w_j. Then from (3.172b) we have

$$\begin{aligned} w_0 &= (x-3)^0 = & &1 \\ w_1 &= (x-3)^1 = & &(x-3) \\ w_2 &= x^2 - 6w_1 - 11 & & \\ &= x^2 - 6(x-3) - 11 = & &(x-3)^2 - 2 \\ w_3 &= x^3 - 9w_2 - 30.4w_1 - 45 & & \\ &= x^3 - 9[x^2 - 6(x-3) - 11] - 30.4(x-3) - 45 = & & \\ & & &(x-3)^3 - 3.4(x-3). \end{aligned} \tag{3.173}$$

We note that

$$\sum_{x=1}^{5} w_r(x) w_s(x) = 0, \qquad \text{if } r \neq s, \tag{3.174}$$

(Problem 3.31), and we say that the four polynomials,

$$\begin{aligned} &[w_0(x), w_1(x), w_2(x), w_3(x)] \\ &\qquad = [1, x-3, (x-3)^2 - 2, (x-3)^3 - 3.4(x-3)], \end{aligned} \tag{3.174a}$$

are *orthogonal* polynomials since, as we know from (3.174),

$$\mathbf{w}_i'\mathbf{w}_j = \sum_{x=1}^{5} w_i(x) w_j(x) = 0. \tag{3.174b}$$

In fact, these are the polynomials that are orthogonal when the x_t's are equispaced by one integer: $x_t = t$, $t = 1, \ldots, 5$ when fitting a polynomial of degree three (using five points). The general expression for fitting a polynomial of degree three on the basis of N points, that is, using $x = x_t = t$, $t = 1,\ldots,N$, are (see Problem 3.32)

$$w_0 = 1; \qquad w_1 = x - \bar{x}, \quad \bar{x} = \frac{N+1}{2};$$

$$(3.175) \qquad w_2 = (x - \bar{x})^2 - \frac{N^2 - 1}{12};$$

$$w_3 = (x - \bar{x})^3 - \frac{(3N^2 - 7)(x - \bar{x})}{20}.$$

We note that when we are involved with the use of orthogonal polynomials (3.175), we are implying that

$$E(y) = \theta_0 + \theta_1(x - \bar{x}) + \theta_2\left[(x - \bar{x})^2 - \frac{N^2 - 1}{12}\right]$$

$$(3.176) \qquad + \theta_3\left[(x - \bar{x})^3 - \frac{3N^2 - 7}{20}(x - \bar{x})\right]$$

$$= \beta_0 + \beta_1 x + \beta_2 x^2 + \beta_3 x^3,$$

so that

$$\beta_3 = \theta_3; \qquad \beta_2 = -3\bar{x}\theta_3 + \theta_2;$$

$$(3.176a) \quad \beta_1 = \theta_3\left(3\bar{x}^2 - \frac{3N^2 - 7}{20}\right) - 2\bar{x}\theta_2 + \theta_1;$$

$$\beta_0 = \theta_3\left(-\bar{x}^3 + \frac{3N^2 - 7}{20}\bar{x}\right) + \theta_2\left(\bar{x}^2 - \frac{N^2 - 1}{12}\right) - \bar{x}\theta_1 + \theta_0.$$

For $N = 5$, this merely states that

$$(3.176b) \qquad \boldsymbol{\beta} = \begin{pmatrix} 1 & -3 & 7 & -16.8 \\ 0 & 1 & -6 & 23.6 \\ 0 & 0 & 1 & -9 \\ 0 & 0 & 0 & 1 \end{pmatrix} \begin{pmatrix} \theta_0 \\ \theta_1 \\ \theta_2 \\ \theta_3 \end{pmatrix} = H\boldsymbol{\theta}$$

which was found before. Thus, by regressing $\mathbf{y}$ on W, we easily find $\hat{\boldsymbol{\theta}}$ and by the Gauss theorem, we can then trivially find $\hat{\boldsymbol{\beta}}$, using (3.176b); that is

$$(3.176c) \qquad \hat{\boldsymbol{\beta}} = H\hat{\boldsymbol{\theta}}.$$

TABLE 3.5.3. Orthogonal Polynomial Values $w_j(x), j = 0, 1, \dots, 8, N = 9$, $x = 1, \dots, 9$. $w_j(x) = w_j^{(1)}(x)/\lambda_j$.[a]

x	$w_0^{(1)}$	$w_1^{(1)}$	$w_2^{(1)}$	$w_3^{(1)}$	$w_4^{(1)}$	$w_5^{(1)}$	$w_6^{(1)}$	$w_7^{(1)}$	$w_8^{(1)}$
1	1	−4	28	−14	14	−4	4	−1	1
2	1	−3	7	7	−21	11	−17	6	−8
3	1	−2	−8	13	−11	−4	22	−14	28
4	1	−1	−17	9	9	−9	1	14	−56
5	1	0	−20	0	18	0	−20	0	70
6	1	1	−17	−9	9	9	1	−14	−56
7	1	2	−8	−13	−11	4	22	14	28
8	1	3	7	−7	−21	−11	−17	−6	−8
9	1	4	28	14	14	4	4	1	1
λ	1	1	3	5/6	7/12	3/20	11/60	143/1680	429/1344

[a] The λ_j given at the bottom of the table, are the least common multiple of the components of $\mathbf{w}_j(x)$ that allows components of $\mathbf{w}_j^{(1)}(x)$ to take integer values.

Now it turns out that in many applications N and the degree m are such that we do not have to go through the Gram–Schmidt process each time—the reason being that orthogonal polynomials for *equispaced data* $(x, y(x))$ where x takes on the values $(1, 2, \dots, N)$, perhaps after a suitable transformation, are extensively tabulated when fitting polynomials of any degree $m \leq N - 1$: See, for example, Abramowitz and Stegun (1965) and DeLury (1950). An expanded extract from Table IV of the set of Tables given in the Appendix at the end of this book is given in Table 3.5.3 for the case $N = 9$ and $m = 8$.

We remark that if fitting orthogonal polynomials to a cubic ($m = 3$) using $N = 9$ equispaced points, then we need only use the first four columns in Table 3.5.3 as our X matrix, and so on.

Now, in general, when fitting a polynomial on the basis of N points ($N > m + 1$), it turns out [see DeLury (1950) for discussion] that the orthogonal polynomials $w_j(x)$ are such that

(3.177)
$$w_0(x) \equiv 1, \qquad w_1(x) = x - \bar{x}$$
$$w_{r+1}(x) = w_r(x)w_1(x) - \frac{r^2(N^2 - r^2)w_{r-1}(x)}{4(4r^2 - 1)}$$

for $r = 1, \dots, N - 2$.

We note that the columns of Table 3.5.3, that is, the $\mathbf{w}_j^{(1)}, j = 0,\ldots,8$, are mutually orthogonal to each other and indeed may be used in other contexts to provide (linear) single-degree of freedom contrasts $\mathbf{w}_j^{(1)'}\mathbf{y}$, which are mutually uncorrelated. [These are standardized in practice by dividing by $(\mathbf{w}_j^{(1)'}\mathbf{w}_j^{(1)})^{1/2}$.] But for our purposes, when estimating the vector $\boldsymbol{\theta}$—see (3.162*a*)—it is necessary to use the $\mathbf{w}_j$ vectors, where, as stated before, $\mathbf{w}_j = \mathbf{w}_j^{(1)}/\lambda_j$. However, we note that if simply interested in testing (successively if need be) whether $\theta_m, \theta_{m-1},\ldots,$ are zero, then we need only use the $\mathbf{w}_j^{(1)}$. We have that

$$E(\mathbf{y}) = W\boldsymbol{\theta} = WDD^{-1}\boldsymbol{\theta} = W^{(1)}\boldsymbol{\gamma} \tag{3.178}$$

with

$$D = \begin{pmatrix} \lambda_1^{-1} & & \\ & \ddots & \\ & & \lambda_k^{-1} \end{pmatrix} \tag{3.178a}$$

so that

$$\boldsymbol{\gamma} = D^{-1}\boldsymbol{\theta} = \mathbf{0} \qquad \text{if and only if } \boldsymbol{\theta} = \mathbf{0}. \tag{3.178b}$$

Thus, we have a transformation on the W matrix and regression and error sums of squares are invariant, so that for testing purposes either form of $E(\mathbf{y})$ mentioned in (3.178) may be used. To translate back to the estimated regression function, we need to find $\hat{\boldsymbol{\beta}}$ and since $\hat{\boldsymbol{\beta}} = H\hat{\boldsymbol{\theta}}$, with $\hat{\boldsymbol{\theta}} = D\hat{\boldsymbol{\gamma}}$, we need to know the λ_j for estimation purposes to be able to state $\hat{\eta} = (1, x,\ldots,x^p)\hat{\boldsymbol{\beta}}$. Various tables do different things about this point. DeLury (1950), for example, tabulates the $w_j^{(1)}$ and gives a formula for the λ_j, while Fisher and Yates (1974) provide $\mathbf{w}_j^{(1)}$, $\mathbf{w}_j^{(1)'}\mathbf{w}_j^{(1)}$, and the λ_j but for the most part only do so for $j = 1,\ldots,5$.

A point to be emphasized here is a restatement of the discussion centered around (3.105*b*) for the case of orthogonal polynomials. Indeed, if we are concerned with the model

$$E(\mathbf{y}) = W\boldsymbol{\theta} = \mathbf{w}_0\theta_0 + \mathbf{w}_1\theta_1 + \cdots + \mathbf{w}_d\theta_d \tag{3.179}$$

where $\mathbf{w}_0, \mathbf{w}_1,\ldots,\mathbf{w}_d$ are the vectors of values of orthogonal polynomials up to degree d, and if we wish now to investigate whether a polynomial function of degree $d+1$ is appropriate to denote $E(\mathbf{y})$, we need only add $\mathbf{w}_{d+1}\theta_{d+1}$ since $\mathbf{w}_{d+1}$ is orthogonal to W, that is, to each of $\mathbf{w}_0 = \mathbf{1}, \mathbf{w}_1,\ldots,\mathbf{w}_d$, and so on. (We are assuming that $N > d + 2$.) Thus,

$$\hat{\theta}_{d+1} = (\mathbf{w}'_{d+1}\mathbf{w}_{d+1})^{-1}\mathbf{w}'_{d+1}\mathbf{y}, \tag{3.179a}$$

and the extra regression sum of squares would be

$$\mathrm{SS}_r(\theta_{d+1}) = \hat{\theta}^2_{d+1}(\mathbf{w}'_{d+1}\mathbf{w}_{d+1}) \tag{3.179b}$$

and the error sum of squares takes the form

$$\mathrm{SS}_e = \mathbf{y}'[I - \mathcal{R}(W)]\mathbf{y} - \mathrm{SS}_r(\theta_{d+1}) \tag{3.179c}$$

and so on.

3.6. LACK OF FIT

Very often, a model is postulated on tentative grounds, and data are then collected for two reasons—first, to examine whether the tentative model is appropriate, and second, if the model assumed seems appropriate, to estimate the unknown parameters, and so on. If the tentative model is not deemed appropriate, a visit to the "drawing board" is envisaged, and a new tentative model postulated.

Suppose for the ensuing discussion that the data $\mathbf{y}$ are such that $AV(\mathbf{y})$ holds, that is, $V(\mathbf{y}) = \sigma^2 I_N$, and that the experimenter assumes at the outset that

$$E(\mathbf{y}) = X\boldsymbol{\beta} \tag{3.180}$$

and indeed, after the data are collected, that $\mathbf{y}$ is regressed on X. We assume X is $N \times k$, $k < N$, and of full rank. However, suppose it is actually the case that

$$E(\mathbf{y}) = \boldsymbol{\gamma} \neq X\boldsymbol{\beta} = \boldsymbol{\eta}. \tag{3.181}$$

Now since $\boldsymbol{\gamma} \neq \boldsymbol{\eta}$, that is, $\boldsymbol{\gamma}$ is not a linear combination of the column vectors of X, then $\boldsymbol{\gamma}$ is not a "point" of the subspace S_k defined in (3.12). The reader will recall that S_k is the span of the columns of X. If we pretend for the moment that we know $\boldsymbol{\gamma}$, we may then project $\boldsymbol{\gamma}$ (see Figure 3.6.1) into S_k. When a space S_k is defined in terms of column vectors of X, we have seen that the projection into S_k is accomplished by pre-multiplication

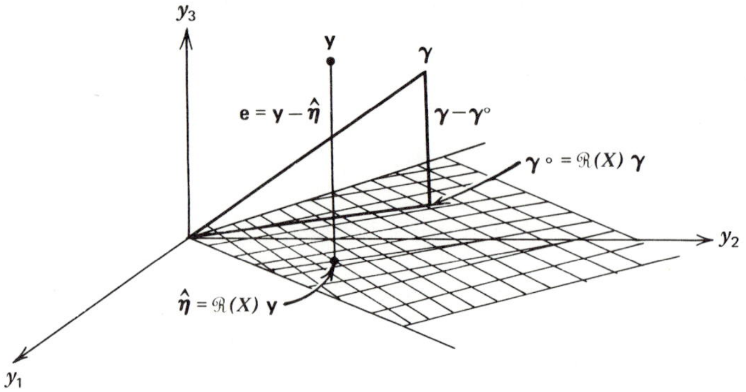

FIGURE 3.6.1. Projection of γ into the (wrongly) assumed solution space S_k. $(\gamma - \gamma^\circ)'(\gamma - \gamma^\circ) + \gamma^{\circ\prime}\gamma^\circ = \gamma'\gamma$.

by $\mathcal{R}(X) = X(X'X)^{-1}X'$. Denote the foot of the projection of γ in S_k by γ°; then

(3.182) $$\gamma^\circ = \mathcal{R}(X)\gamma$$

and the vector

(3.182*a*) $$\gamma - \gamma^\circ = [I - \mathcal{R}(X)]\gamma$$

may be viewed as a *residual model vector* that gives information on the discrepancy between the actual expectation and that assumed. We denote the squared length of $\gamma - \gamma^\circ$ by Λ^2, that is,

(3.182*b*) $$\Lambda^2 = (\gamma - \gamma^\circ)'(\gamma - \gamma^\circ).$$

We have that if $\gamma = \boldsymbol{\eta}$, then $\gamma = \gamma^\circ$ so that

(3.182*c*) $$\Lambda^2 = 0 \qquad \text{if } \gamma = \boldsymbol{\eta}\ (= \gamma^\circ).$$

Now if we project the observation vector $\mathbf{y}$ into S_k, we obtain the residual vector

(3.183) $$\mathbf{e} = \mathbf{y} - \hat{\boldsymbol{\eta}} = \mathbf{y} - \mathcal{R}(X)\mathbf{y} = [I - \mathcal{R}(x)]\mathbf{y}.$$

We note that

(3.183*a*) $$E(\mathbf{e}) = E(\mathbf{y} - \hat{\boldsymbol{\eta}}) = [I - \mathcal{R}(X)]\gamma = \gamma - \gamma^\circ$$

so that the residual vector $(\mathbf{y} - \hat{\boldsymbol{\eta}})$ gives us information about the residual model vector, that is, information on how far off the true model is from the model assumed. Note that (3.183a) also implies that

$$E(\hat{\boldsymbol{\eta}}) = \mathcal{R}(X)\boldsymbol{\gamma} = \boldsymbol{\gamma}^{\circ} \tag{3.183b}$$

so that $\hat{\boldsymbol{\eta}}$ tells us nothing directly about $\boldsymbol{\gamma}$ itself. Now we have, of course, the fact that, irrespective of the true value of $E(\mathbf{y})$,

$$\begin{aligned} V(\hat{\boldsymbol{\eta}}) &= \sigma^2\mathcal{R}(X) \\ V(\mathbf{e}) &= V(\mathbf{y} - \hat{\boldsymbol{\eta}}) = \sigma^2[I - \mathcal{R}(X)] \end{aligned} \tag{3.184}$$

and using all these results we may now construct the usual analysis of variance table, which at this point has the form given in Table 3.6.1. (Problem 3.34 asks for the derivation of the expected mean square column.)

We note that

$$E\left(\frac{\mathbf{e}'\mathbf{e}}{N-k}\right) = \sigma^2 + \frac{\Lambda^2}{N-k}, \tag{3.185}$$

which is to say that the residual mean square no longer reflects just pure error but also a *lack of fit* component through Λ^2.

Now if we were in the situation where we have an independent estimate of σ^2, s'^2, where

$$s'^2 = \frac{\sigma^2\chi_v^2}{v}, \tag{3.186}$$

then we could compare $s^2 = \mathbf{e}'\mathbf{e}/(N-k)$ with s'^2 (if the y's are normal,

TABLE 3.6.1. Analysis of Variance when y is Regressed on X but $E(\mathbf{y}) = \boldsymbol{\gamma} \neq X\boldsymbol{\beta} = \boldsymbol{\eta}$ (X of Full Rank)

Source	Sum of Squares	Degrees of Freedom	Expected Mean Square
Regression	$\hat{\boldsymbol{\eta}}'\hat{\boldsymbol{\eta}} = \mathbf{y}'\mathcal{R}(X)\mathbf{y}$	k	$[\boldsymbol{\gamma}'\mathcal{R}(X)\boldsymbol{\gamma} + k\sigma^2]/k$ $= \sigma^2 + \boldsymbol{\gamma}^{\circ\prime}\boldsymbol{\gamma}^{\circ}/k$
Residual	$\mathbf{e}'\mathbf{e} = (\mathbf{y} - \hat{\boldsymbol{\eta}})'(\mathbf{y} - \hat{\boldsymbol{\eta}})$ $= \mathbf{y}'[I - \mathcal{R}(X)]\mathbf{y}$	$N-k$	$\sigma^2 + \Lambda^2/(N-k)$
Total	$\mathbf{y}'\mathbf{y}$	N	$\sigma^2 + \boldsymbol{\gamma}'\boldsymbol{\gamma}/N$
	$\Lambda^2 = (\boldsymbol{\gamma} - \boldsymbol{\gamma}^{\circ})'(\boldsymbol{\gamma} - \boldsymbol{\gamma}^{\circ})$		

and so on) and examine the statement

(3.187) $$\Lambda^2 = 0 \Leftrightarrow \boldsymbol{\gamma} = \boldsymbol{\eta} = X\boldsymbol{\beta}$$

(see Problem 3.35).

Failing the above, there is another way to examine (3.187) and this may be done by proper consideration of the design of the experiment—that is, the choice of the structure of X. If at the outset, the experimenter has an interest in examining (3.187), then by choosing to run the experiment in such a way that there is some replication, it turns out that the residual sum of squares may be split up into two components—one of which gives information only on the error σ^2 and the other giving information on Λ^2, the lack of fit component.

Suppose we assume that $E(\mathbf{y}) = X\boldsymbol{\beta}$ and the experiment is run in such a way that we have:

(3.188)
$$\begin{aligned} &n_1 \text{ observations on } y \text{ taken at } \mathbf{x}'_{1\cdot} = (x_{11}, x_{12}, \ldots, x_{1k}) \\ &n_2 \text{ observations on } y \text{ taken at } \mathbf{x}'_{2\cdot} = (x_{21}, x_{22}, \ldots, x_{2k}) \\ &\qquad\vdots \\ &n_g \text{ observations on } y \text{ taken at } \mathbf{x}'_{g\cdot} = (x_{g1}, x_{g2}, \ldots, x_{gk}). \end{aligned}$$

That is, we have g groups of observations, the rth group having n_r observations, where we assume that

(3.188a) $$g > k, \qquad \mathbf{x}'_{r\cdot} \neq \mathbf{x}'_{s\cdot},$$

for all pairs (r, s), $r \neq s$, chosen from $(1, \ldots g)$. Let

(3.188b) $$\sum_{r=1}^{g} n_r = N.$$

We will label the tth observation of group r by y_{tr}, $r = 1, \ldots, g$ and $t = 1, \ldots, n_r$, and we assume that for at least one r, $n_r > 1$. We let $\mathbf{y}$ be such that

(3.189) $$\mathbf{y}' = (y_{11}, y_{21}, \ldots, y_{n_1 1}; y_{12}, \ldots, y_{n_2 2}; \ldots; y_{1g}, \ldots, y_{n_g g})$$

so in assuming $E(\mathbf{y}) = X\boldsymbol{\beta}$, we have that X is such that

(3.189a)

$$X' = (\mathbf{x}_{1\cdot} \mid \cdots \mid \mathbf{x}_{1\cdot} \mid \mathbf{x}_{2\cdot} \mid \cdots \mid \mathbf{x}_{2\cdot} \mid \cdots \mid \mathbf{x}_{g\cdot} \mid \cdots \mid \mathbf{x}_{g\cdot}).$$

We assume that X is of full rank. Now in the usual way we may regress $\mathbf{y}$ on X and find that

$$\hat{\boldsymbol{\eta}} = X\hat{\boldsymbol{\beta}}, \qquad \hat{\boldsymbol{\beta}} = (X'X)^{-1}X'\mathbf{y}. \tag{3.190}$$

This means, because of the replication n_r times at $\mathbf{x}'_{r.}$, that the fitted value $\hat{\eta}_{tr}$ corresponding to y_{tr} is such that

$$\hat{\eta}_{tr} = \hat{\eta}_{\cdot r} = \mathbf{x}'_{r\cdot}\hat{\boldsymbol{\beta}}, \qquad \text{for all } t = 1,\ldots,n_r. \tag{3.190a}$$

Hence the residual sum of squares is

$$\mathbf{e}'\mathbf{e} = \sum_{r=1}^{g}\sum_{t=1}^{n_r}(y_{tr} - \hat{\eta}_{\cdot r})^2. \tag{3.191}$$

If we let

$$\bar{y}_{\cdot r} = \frac{1}{n_r}\sum_{t=1}^{n_r} y_{tr}, \tag{3.191a}$$

then (3.191) may be written

$$\mathbf{e}'\mathbf{e} = \Sigma\Sigma\left[(y_{tr} - \bar{y}_{\cdot r}) + (\bar{y}_{\cdot r} - \hat{\eta}_{\cdot r})\right]^2 \tag{3.191b}$$

and it is easily verified that

$$\mathbf{e}'\mathbf{e} = \sum_{r=1}^{g}\sum_{t=1}^{n_r}(y_{tr} - \bar{y}_{\cdot r})^2 + \sum_{r=1}^{g} n_r(\bar{y}_{\cdot r} - \hat{\eta}_{\cdot r})^2. \tag{3.191c}$$

Now for each group, the n_r observations are obtained under exactly the same conditions, so that

$$E(y_{tr}) = \mu_r = \mathbf{x}'_{r\cdot}\boldsymbol{\beta} = x_{r1}\beta_1 + \cdots + x_{rk}\beta_k \tag{3.192}$$

for all $t = 1,\ldots,n_r$. Thus, because $V(\mathbf{y}) = \sigma^2 I$

$$E\left[\sum_{t=1}^{n_r}(y_{tr} - \bar{y}_{\cdot r})^2\right] = (n_r - 1)\sigma^2 \tag{3.192a}$$

or

$$E\left[\sum_{r=1}^{g}\sum_{t=1}^{n_r}(y_{tr} - \bar{y}_{\cdot r})^2\right] = \sigma^2\Sigma(n_r - 1) = \sigma^2(N - g), \tag{3.192b}$$

which is to say that the first term on the right-hand side of equation (3.191c) gives information about σ^2, the pure error. Now using the result tabulated in Table 3.6.1 (see Problem 3.34) and the result (3.191b), we have on taking expectations in (3.191c) and solving that

$$E\left[\sum_{r=1}^{g} n_r(\bar{y}_{\cdot r} - \hat{\eta}_{\cdot r})^2\right] = (g-k)\sigma^2 + \Lambda^2. \tag{3.193}$$

Note that we have assumed $g > k$. Indeed if $g < k$, X would be of rank = [minimum of g, k] = g, contrary to the assumption that the rank of X is k. Furthermore, if $g = k$, it turns out that

$$\hat{\eta}_{tr} = \hat{\eta}_{\cdot r} = \bar{y}_{\cdot r}. \tag{3.194}$$

Problem 3.37 asks for the proof. Note that this means that the second term on the right-hand side of (3.191c) is zero. A simple example results when using the model $E(y) = \beta_0 + \beta_1 x$, that is, $k = 2$ and observations are taken at the two points x_1 and x_2—n_1 at x_1 and n_2 at x_2. Then (see Problem 3.38), $\hat{\eta}_{x_1} = \bar{y}_{\cdot 1}$ and $\hat{\eta}_{x_2} = \bar{y}_{\cdot 2}$, so that the fitted least-squares line is the line that passes through $(x_1, \bar{y}_{\cdot 1})$ and $(x_2, \bar{y}_{\cdot 2})$.

Now with $g > k$ holding, we may put the results (3.191c), (3.192b), and (3.193) into a revised analysis of variance table (see Table 3.6.1) shown in Table 3.6.2.

There is one last point worth remarking on here. Under normality, are the quadratic forms SS_L and SS_E independent? Specifically, if $\boldsymbol{\gamma} = \boldsymbol{\eta}$, which of course implies that $\Lambda^2 = 0$ as we have seen, are SS_L and SS_E independent? It turns out that the answer to the latter question is yes. Suppose we

TABLE 3.6.2. Analysis of Variance when $E(\mathbf{y}) = X\boldsymbol{\beta}$ is Assumed, X contains replications, and $E(\mathbf{y}) = \boldsymbol{\gamma} \neq X\boldsymbol{\beta}$ is of Interest

Source	Sum of Squares	Degrees of Freedom	Expected Sum of Squares
Regression	$\hat{\boldsymbol{\eta}}'\hat{\boldsymbol{\eta}} = \mathbf{y}'\mathcal{R}(X)\mathbf{y}$	k	$k\sigma^2 + \boldsymbol{\gamma}^{\circ\prime}\boldsymbol{\gamma}^{\circ}$
Lack of fit	$SS_L = \sum_{r=1}^{g} n_r(\bar{y}_{\cdot r} - \hat{\eta}_{\cdot r})^2$	$g-k$	$(g-k)\sigma^2 + \Lambda^2$
Pure error	$SS_E = \sum_{r=1}^{g}\sum_{t=1}^{n_r}(y_{tr} - \bar{y}_{\cdot r})^2$	$N-g$	$(N-g)\sigma^2$
Total	$\mathbf{y}'\mathbf{y}$	N	$N\sigma^2 + \boldsymbol{\gamma}'\boldsymbol{\gamma}$

first rewrite (3.189) as

$$\mathbf{y}' = (\mathbf{y}_1', \ldots, \mathbf{y}_r', \ldots, \mathbf{y}_g'), \tag{3.195}$$

where

$$\mathbf{y}_r' = (y_{1r}, \ldots, y_{tr}, \ldots, y_{n_r r}). \tag{3.195a}$$

Then it is easy to see (Problem 3.39) that

$$SS_E = (\mathbf{y}_1', \ldots, \mathbf{y}_g')U \begin{pmatrix} \mathbf{y}_1 \\ \vdots \\ \mathbf{y}_g \end{pmatrix}, \tag{3.195b}$$

where the $N \times N$ matrix U is of the form

(3.195c)

$$U = \begin{bmatrix} I_{n_1} - \dfrac{1}{n_1}\mathbf{1}_{n_1}\mathbf{1}_{n_1}' & \bigcirc & \cdots & \bigcirc \\ \bigcirc & I_{n_2} - \dfrac{1}{n_2}\mathbf{1}_{n_2}\mathbf{1}_{n_2}' & & \bigcirc \\ \vdots & & \ddots & \vdots \\ \bigcirc & \cdots & \bigcirc & I_{n_g} - \dfrac{1}{n_g}\mathbf{1}_{n_g}\mathbf{1}_{n_g}' \end{bmatrix}.$$

But from Table 3.6.2, we have that

$$\begin{aligned} SS_L &= \mathbf{y}'\mathbf{y} - \mathbf{y}'\mathcal{R}(X)\mathbf{y} - \mathbf{y}'U\mathbf{y} \\ &= \mathbf{y}'[I - \mathcal{R}(X) - U]\mathbf{y} \\ &= \mathbf{y}'Z\mathbf{y}, \end{aligned} \tag{3.195d}$$

where

$$Z = I - X(X'X)^{-1}X' - U = I - \mathcal{R}(X) - U. \tag{3.195e}$$

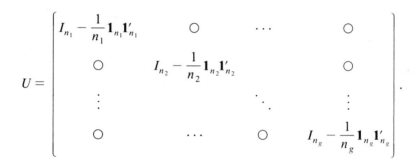

Because of the structure of X (first n_1 rows $\mathbf{x}_{1\cdot}'$, ..., last n_g rows $\mathbf{x}_{g\cdot}'$), it is easy to verify (Problem 3.40) that under the hypothesis $E(\mathbf{y}) = X\boldsymbol{\beta}$, or, equivalently, $\Lambda^2 = 0$, that SS_E and SS_L are two quadratic forms that may be written

$$\text{(3.195}f\text{)} \qquad \begin{aligned} SS_E &= (\mathbf{y} - X\boldsymbol{\beta})'U(\mathbf{y} - X\boldsymbol{\beta}) \\ SS_L &= (\mathbf{y} - X\boldsymbol{\beta})'Z(\mathbf{y} - X\boldsymbol{\beta}) \end{aligned}$$

and such that

$$\text{(3.195}g\text{)} \qquad U[\sigma^2 I]Z = \bigcirc.$$

In other words, SS_E and SS_L are independent. To test the statement (3.187), it is easy to see that (Problem 3.40) an appropriate procedure at significance level α is

Reject the hypothesis $\Lambda^2 = 0$ if the observed value of

$$\text{(3.196)} \qquad F_L = \frac{SS_L/(g-k)}{SS_E/(N-g)} > F_{g-k,\, N-g;\, \alpha}.$$

Accept otherwise.

The implication so far is that to test for *lack of fit* take g places to experiment at, where $g > k$, and select at least one of these, replicating more than once.

Whether or not we have replications at one or more of g places of $(x_1, \ldots, x_k)$, we often accompany our analysis by examining the residuals

$$\text{(3.197)} \qquad e_u = y_u - \hat{\eta}_u.$$

One quick plot is e_u against u, and other plots that may be done are e_u against $\hat{\eta}_u$, or k separate plots of e_u against x_{uj}, $j = 1, \ldots, k$. The scatter of the points should lie in a horizontal band around the line $e = 0$ (see Chapter 1) and if not, then some or all of the assumptions used should be checked, for example, $V(y_u \mid \mathbf{x}'_u) \equiv \sigma^2$, or that the linear model fitted is adequate.

Now at this juncture, we may also ask the question—How are estimates affected if we wrongly assume the model to be such that $E(\mathbf{y}) = X\boldsymbol{\beta}$? Obviously, the answer to this type of question depends heavily on what the nature of the correct model happens to be—γ could be exponential functions evaluated at the places $\mathbf{x}_{u\cdot}$, and so on. It is interesting to investigate this question when it is assumed that

$$E(\mathbf{y}) = X_1\boldsymbol{\beta}_1 = \boldsymbol{\eta}, \tag{3.198}$$

but it is the case that

$$E(\mathbf{y}) = X_1\boldsymbol{\beta}_1 + X_2\boldsymbol{\beta}_2 = \boldsymbol{\gamma}, \tag{3.198a}$$

that is, we are discussing the special case where the model we actually use, (3.198), leaves out some effects $\boldsymbol{\beta}_2$ carried by X_2. Now when we use (3.198) we find

$$\hat{\boldsymbol{\beta}}_1 = (X_1'X_1)^{-1}X_1'\mathbf{y}, \tag{3.199}$$

and we quickly see that

$$E(\hat{\boldsymbol{\beta}}_1) = (X_1'X_1)^{-1}X_1'E(\mathbf{y}) \tag{3.200}$$

which from (3.198*a*) is

$$E(\hat{\boldsymbol{\beta}}_1) = \boldsymbol{\beta}_1 + A\boldsymbol{\beta}_2, \qquad A = (X_1'X_1)^{-1}X_1'X_2 \tag{3.200a}$$

and A is the *alias matrix* introduced in (3.94). Thus, $A\boldsymbol{\beta}_2$ represents the extent of the bias in this case [described by (3.198) and (3.198*a*)]. Also, it is to be noted that the nature of the matrix $(X_1 \mid X_2)$ profoundly affects the estimates. If X_1 is orthogonal to X_2, that is, if

$$X_1'X_2 = \bigcirc, \tag{3.201}$$

then $A = \bigcirc$, so that

$$E(\hat{\boldsymbol{\beta}}_1) = \boldsymbol{\beta}_1 \tag{3.201a}$$

and $\hat{\boldsymbol{\beta}}_1$ is unbiased for $\boldsymbol{\beta}_1$, *even though the model used is wrong*. The message here is clear, of course: If it is not possible to get some factors, which may be judged important, in a controlled way into the experiment, use an experiment that is *orthogonal* to these factors. Of course, there is a problem here since we often don't know what X_2 is really. In that case, so-called randomized designs are used that break the link $A\boldsymbol{\beta}_2$ with $E(\hat{\boldsymbol{\beta}}_1)$. For related reading and discussion see Box and Guttman (1966).

Another interesting point is that if $X_1'X_2 = \bigcirc$, then *maximum* detectability of lack of fit is ensured. We have, when $\mathbf{y}$ is regressed on X_1, that residuals are

(3.202) $$\mathbf{e} = [I - \mathcal{R}(X_1)]\mathbf{y}$$

so that, from (3.198*a*),

(3.202*a*) $$E(\mathbf{e}) = [I - \mathcal{R}(X_1)]E(\mathbf{y}) = [I - \mathcal{R}(X_1)]X_2\boldsymbol{\beta}_2.$$

But recalling that

(3.203) $$\boldsymbol{\gamma} - \boldsymbol{\gamma}^\circ = [I - \mathcal{R}(X_1)]\boldsymbol{\gamma} = [I - \mathcal{R}(X_1)](X_1\boldsymbol{\beta}_1 + X_2\boldsymbol{\beta}_2),$$

we have

(3.203*a*) $$\boldsymbol{\gamma} - \boldsymbol{\gamma}^\circ = [I - \mathcal{R}(X_1)]X_2\boldsymbol{\beta}_2 = E(\mathbf{e}).$$

Also we have that

(3.203*b*) $$\begin{aligned}\Lambda^2 &= (\boldsymbol{\gamma} - \boldsymbol{\gamma}^\circ)'(\boldsymbol{\gamma} - \boldsymbol{\gamma}^\circ) = \boldsymbol{\beta}_2'X_2'[I - \mathcal{R}(X_1)]X_2\boldsymbol{\beta}_2 \\ &= \boldsymbol{\beta}_2'X_2'X_2\boldsymbol{\beta}_2 - \boldsymbol{\beta}_2'X_2'\mathcal{R}(X_1)X_2\boldsymbol{\beta}_2.\end{aligned}$$

But the residual sum of squares is $\mathbf{e}'\mathbf{e}$ and this has expectation, in view of (3.203*a*),

(3.203*c*) $$\begin{aligned}E(\mathbf{e}'\mathbf{e}) &= (\boldsymbol{\gamma} - \boldsymbol{\gamma}^\circ)'(\boldsymbol{\gamma} - \boldsymbol{\gamma}^\circ) + \operatorname{tr}(\{\sigma^2[I - \mathcal{R}(X_1)]\}[I]) \\ &= (N - k_1)\sigma^2 + (\boldsymbol{\gamma} - \boldsymbol{\gamma}^\circ)'(\boldsymbol{\gamma} - \boldsymbol{\gamma}^\circ),\end{aligned}$$

where X_1 is of order $N \times k_1$, X_2 is $N \times k_2$, $(X_1 \mid X_2)$ is $N \times k$, and so on. Thus from (3.203*b*) and (3.203*c*), we see that maximum detectability is achieved when Λ^2 is largest, and this happens when

(3.203*d*) $$\mathcal{R}(X_1)X_2 = \bigcirc,$$

that is, when $X_2'X_1 = \bigcirc$ or $X_1'X_2 = \bigcirc'$, so that a design which is orthogonal to controlled effects not in the experiment gives unbiased estimates and maximum possible detectability of wrong model specification.

3.7. GENERAL LINEAR HYPOTHESES: FULL-RANK CASE

At various points in the preceding sections of this chapter, we have discussed problems having the following general framework. Suppose we are dealing with the normal linear model with the assumptions mentioned in Section 3.1 and labeled $AE(\mathbf{y})$, $AV(\mathbf{y})$, $AD\mathbf{y}$ holding; that is,

$$\mathbf{y} = X\boldsymbol{\beta} + \boldsymbol{\varepsilon}, \qquad \boldsymbol{\varepsilon} = N(\mathbf{0}, \sigma^2 I_N), \tag{3.204}$$

where $\boldsymbol{\beta}$ is of order $k \times 1$, X is $N \times k$, $N > k$, and X is of full rank, $r(X) = k$. Very often, as we have seen in the previous sections of this chapter (see also Chapters 4 and 5), we are interested in making inference about the regression parameters $\boldsymbol{\beta}$ in general and, in particular, examining statements concerning the parameters $\boldsymbol{\beta}$. A typical statement takes the form of *linearly independent constraints* on the parameters $\boldsymbol{\beta}$, and we call such a set a *linear hypothesis*. A linear hypothesis, called H_0, then takes the form

$$H_0\colon B\boldsymbol{\beta} = \mathbf{b}, \tag{3.205}$$

where B is a $q \times k$ matrix of rank q and $\mathbf{b}$ is a $q \times 1$ vector of known constants. Now an important special case that occurs often is where

$$B = \left[\, \bigcirc \mid I_q \,\right], \qquad \mathbf{b} = \mathbf{0}. \tag{3.206}$$

with $\bigcirc$ denoting a $q \times t$ matrix of zeros, $t = k - q$, and $\mathbf{0}$ a $q \times 1$ vector of zeros, so that partitioning $\boldsymbol{\beta}$ as

$$\boldsymbol{\beta}' = \left(\boldsymbol{\beta}_t' \mid \boldsymbol{\beta}_q'\right), \tag{3.206a}$$

where $\boldsymbol{\beta}_t'$ is a $1 \times t$ vector, and so on, (3.205) takes the form, using (3.206) and (3.206a),

$$H_0'\colon \boldsymbol{\beta}_q = \mathbf{0}. \tag{3.207}$$

We have dealt with problems leading to statements of the form (3.207) in this chapter, and the reader should relate the work of the earlier sections of this chapter to the format discussed here.

In examining how to proceed to test statements of the form (3.205), we really need only discuss statements of the form (3.207), as we now show. We recall that the $q \times k$ matrix B, of rank q, may be augmented so that the augmented matrix, A, is a $k \times k$ nonsingular matrix. That is, there exists a matrix of order $t \times k$, $t = k - q$, say C, such that

$$A' = \left[\, C' \mid B' \,\right] \tag{3.208}$$

is of rank k. Hence, we may write

$$(3.209) \qquad E(\mathbf{y}) = X\boldsymbol{\beta} = XA^{-1}A\boldsymbol{\beta} = Z\begin{bmatrix} C \\ \hline B \end{bmatrix}\boldsymbol{\beta}, \qquad Z = XA^{-1},$$

and partitioning $\boldsymbol{\beta}$ as in (3.206a), and Z similarly, we have

$$(3.209a) \qquad E(\mathbf{y}) = [Z_1 \mid Z_2]\begin{bmatrix} \boldsymbol{\theta}_t \\ \hline \boldsymbol{\theta}_q \end{bmatrix} = Z_1\boldsymbol{\theta}_t + Z_2\boldsymbol{\theta}_q,$$

where

$$(3.209b) \qquad Z = XA^{-1}, \qquad Z_1 \text{ is } N \times t \text{ and } Z_2 \text{ is } N \times q$$

and

$$(3.209c) \qquad \boldsymbol{\theta}_t = C\boldsymbol{\beta}, \qquad \boldsymbol{\theta}_q = B\boldsymbol{\beta}.$$

We need now only define

$$(3.210) \qquad \boldsymbol{\gamma}_t = \boldsymbol{\theta}_t, \qquad \boldsymbol{\gamma}_q = \boldsymbol{\theta}_q - \mathbf{b},$$

and letting

$$(3.210a) \qquad \mathbf{u} = \mathbf{y} - Z_2\mathbf{b},$$

we rewrite (3.209a), and thus (3.209), as

$$(3.211) \qquad E(\mathbf{u}) = Z_1\boldsymbol{\gamma}_t + Z_2\boldsymbol{\gamma}_q.$$

We note that the statement H_0 of (3.205) is equivalent to a statement H_0' of the form of (3.207), namely, $\boldsymbol{\gamma}_q = \mathbf{0}$. In summary, we may turn the problem of examining H_0 into one of the form H_0' by the above procedure, and note that this method so far is general, in the sense that X may be of full rank or non-full rank. We now examine a method for testing (3.207) in the full-rank case: the discussion for X not of full rank is given in Chapter 5.

The method that we will use is the likelihood ratio test. Returning to (3.204) and (3.207), we have, perhaps after suitable augmentation, that

$$(3.212) \qquad \mathbf{y} = X_1\boldsymbol{\beta}_t + X_2\boldsymbol{\beta}_q + \boldsymbol{\varepsilon}, \qquad \boldsymbol{\varepsilon} = N(\mathbf{0}, \sigma^2 I_N)$$

where X_1 is of order $N \times t$, X_2 is $N \times q$, $t + q = k$, and so on, and we wish to test

$$H_0': \boldsymbol{\beta}_q = \mathbf{0}. \tag{3.212a}$$

When all parameters are left unspecified, we have that the likelihood function is such that

$$l(\boldsymbol{\beta}, \sigma^2 \mid \mathbf{y}) \propto \frac{1}{(\sigma^2)^{N/2}} \exp\left\{ -\frac{1}{2\sigma^2} [(\mathbf{y} - X\boldsymbol{\beta})'(\mathbf{y} - X\boldsymbol{\beta})] \right\}. \tag{3.213}$$

From previous work it is easy to see that this may be written

(3.213a)

$$l(\boldsymbol{\beta}, \sigma^2 \mid \mathbf{y}) \propto \frac{1}{(\sigma^2)^{N/2}} \exp\left\{ -\frac{1}{2\sigma^2} [\mathrm{SS}_e + (\boldsymbol{\beta} - \hat{\boldsymbol{\beta}})'X'X(\boldsymbol{\beta} - \hat{\boldsymbol{\beta}})] \right\},$$

where the residual (error) sum of squares SS_e is of course given by

$$\mathrm{SS}_e = (\mathbf{y} - X\hat{\boldsymbol{\beta}})'(\mathbf{y} - X\hat{\boldsymbol{\beta}}) \tag{3.213b}$$

and the estimator $\hat{\boldsymbol{\beta}}$ is the least-squares estimator

$$\hat{\boldsymbol{\beta}} = (X'X)^{-1}X'\mathbf{y}. \tag{3.213c}$$

It is straightforward then to see that the maximum likelihood estimators of $\boldsymbol{\beta}$ and σ^2 are given by $\hat{\boldsymbol{\beta}}$ and SS_e/N, respectively, so that

$$\max_{\boldsymbol{\beta}, \sigma^2} l(\boldsymbol{\beta}, \sigma^2 \mid \mathbf{y}) \propto (\mathrm{SS}_e)^{-N/2} \exp\left(-\frac{N}{2}\right). \tag{3.214}$$

Now to arrive at the maximum of the likelihood under H_0', we note first that if H_0' is true then, from (3.212), we have

$$\mathbf{y} = X_1\boldsymbol{\beta}_t + \boldsymbol{\varepsilon}, \qquad \boldsymbol{\varepsilon} = N(\mathbf{0}, \sigma^2 I_N) \tag{3.215}$$

and hence the likelihood under H_0' is

$$l(H_0' \mid \mathbf{y}) \propto \frac{1}{(\sigma^2)^{N/2}} \exp\left\{ -\frac{1}{2\sigma^2} (\mathbf{y} - X_1\boldsymbol{\beta}_t)'(\mathbf{y} - X_1\boldsymbol{\beta}_t) \right\}, \tag{3.216}$$

and clearly the maximum likelihood estimators of $\boldsymbol{\beta}_t$ and σ^2 under H_0' are

$$(3.216a)\quad \tilde{\boldsymbol{\beta}}_t = (X_1'X_1)^{-1}X_1'\mathbf{y}, \qquad \tilde{\sigma}^2 = \frac{(\mathbf{y} - X_1\tilde{\boldsymbol{\beta}}_t)'(\mathbf{y} - X_1\tilde{\boldsymbol{\beta}}_t)}{N}$$

so that

$$(3.217)\quad \max l(H_0' \mid \mathbf{y}) \propto [(\mathbf{y} - X_1\tilde{\boldsymbol{\beta}}_t)'(\mathbf{y} - X_1\tilde{\boldsymbol{\beta}}_t)]^{-N/2}\exp\left(-\frac{N}{2}\right).$$

Thus, taking the ratio of (3.217) to (3.214), we find that the maximum likelihood ratio criterion is

$$(3.218)\qquad \lambda(\mathbf{y}) = K\left[\frac{\mathrm{SS}_e}{(\mathbf{y} - X_1\tilde{\boldsymbol{\beta}}_t)'(\mathbf{y} - X_1\tilde{\boldsymbol{\beta}}_t)}\right]^{N/2}.$$

Now recall from Section 3.3 that we may write, using the notation of that section,

$$(3.219)\qquad \begin{aligned} \mathrm{SS}_e &= \mathbf{y}'[I - \mathcal{R}(X_1)]\mathbf{y} - \mathbf{y}'\mathcal{R}(X_{2\cdot 1})\mathbf{y} \\ &= \mathbf{y}'[I - \mathcal{R}(X_1)]\mathbf{y} - \hat{\boldsymbol{\beta}}_q'X_{2\cdot 1}'X_{2\cdot 1}\hat{\boldsymbol{\beta}}_q, \end{aligned}$$

where

$$(3.219a)\qquad \hat{\boldsymbol{\beta}}_q = (X_{2\cdot 1}'X_{2\cdot 1})^{-1}X_{2\cdot 1}'\mathbf{y}.$$

Furthermore,

$$(3.220)\qquad (\mathbf{y} - X_1\tilde{\boldsymbol{\beta}}_t)'(\mathbf{y} - X_1\tilde{\boldsymbol{\beta}}_t) = \mathbf{y}'[I - \mathcal{R}(X_1)]\mathbf{y}$$

which implies, using (3.219) and (3.220), that

$$(3.220a)\qquad (\mathbf{y} - X_1\tilde{\boldsymbol{\beta}}_t)'(\mathbf{y} - X_1\tilde{\boldsymbol{\beta}}_t) = \mathrm{SS}_e + \hat{\boldsymbol{\beta}}_q'X_{2\cdot 1}'X_{2\cdot 1}\hat{\boldsymbol{\beta}}_q.$$

The reader will recall that $\hat{\boldsymbol{\beta}}_q'X_{2\cdot 1}'X_{2\cdot 1}\hat{\boldsymbol{\beta}}_q$ is the extra regression sum of squares due to $\boldsymbol{\beta}_q$, given $\boldsymbol{\beta}_t$ in the model. Hence, we have that

$$(3.221)\qquad \lambda(\mathbf{y}) = K\left(1 + \frac{\hat{\boldsymbol{\beta}}_q'X_{2\cdot 1}'X_{2\cdot 1}\hat{\boldsymbol{\beta}}_q}{\mathrm{SS}_e}\right)^{-N/2}$$

or, writing

$$F = \frac{(\hat{\boldsymbol{\beta}}_q' X_{2\cdot 1}' X_{2\cdot 1} \hat{\boldsymbol{\beta}}_q)/q}{\mathrm{SS}_e/(N-k)}, \tag{3.222}$$

we have

$$\lambda(\mathbf{y}) = K\left(1 + \frac{q}{N-k}F\right)^{-N/2}; \tag{3.222a}$$

that is, λ is a monotone decreasing function of F, and the critical region for rejecting H_0' is of the form

$$F > c. \tag{3.223}$$

Now because of the properties of $\mathcal{R}(X_1)$ and $\mathcal{R}(X_{2\cdot 1})$, and the fact that X_1 is orthogonal to $X_{2\cdot 1}$, it can be verified that F may be written

$$F = \frac{(\mathbf{y} - X_1\boldsymbol{\beta}_t)'\mathcal{R}(X_{2\cdot 1})(\mathbf{y} - X_1\boldsymbol{\beta}_t)/q}{(\mathbf{y} - X_1\boldsymbol{\beta}_t)'[I - \mathcal{R}(X_1) - \mathcal{R}(X_{2\cdot 1})](\mathbf{y} - X_1\boldsymbol{\beta}_t)/(N-k)}. \tag{3.224}$$

But under H_0', $\mathbf{y} - X_1\boldsymbol{\beta}_t = N(\mathbf{0}, \sigma^2 I_N)$, and since $\mathcal{R}(X_{2\cdot 1})[I - \mathcal{R}(X_1) - \mathcal{R}(X_{2\cdot 1})] = \mathrm{O}$, the quadratic forms in the numerator and denominator are independent and distributed as $\sigma^2\chi_q^2$ and $\sigma^2\chi_{N-k}^2$, respectively. Hence, *under* H_0',

$$F = F_{q,\,N-k} \tag{3.224a}$$

so that, referring to (3.223), the likelihood ratio test *rejects* H_0' at significance level α, if the observed F is such that

$$F > F_{q,\,N-k;\,\alpha}, \tag{3.225}$$

where $F_{q,\,N-k;\,\alpha}$ denotes the point exceeded with probability α when using the F distribution, $(q, N-k)$ degrees of freedom. We have justified the form of this test before by appealing to the examination of EMS columns of relevant analysis of variance tables. We note that this is a similar test; the significance level is α no matter what the values of $\boldsymbol{\beta}_t$ and σ^2 happen to be. The test is also known to be unbiased, that is,

$$P(F > F_{q,\,N-k;\,\alpha} \mid \boldsymbol{\beta}_q \neq \mathbf{0}) > P(F > F_{q,\,N-k;\,\alpha} \mid \boldsymbol{\beta}_q = \mathbf{0}) = \alpha. \tag{3.226}$$

(We defer discussion of the non-full-rank case until Chapter 5.)

Returning to (3.220a), we recall that the right-hand side of (3.220) is the residual sum of squares under H_0', and we often write

$$\mathrm{SS}_e(H_0') = (\mathbf{y} - X_1\tilde{\boldsymbol{\beta}}_t)'(\mathbf{y} - X_1\tilde{\boldsymbol{\beta}}_t) \tag{3.226a}$$

which, for emphasis, using (3.220) is

$$\begin{aligned}\mathrm{SS}_e(H_0') &= \mathbf{y}'[I - \mathcal{R}(X_1)]\mathbf{y} \\ &= \min_{H_0'} (\mathbf{y} - X_1\boldsymbol{\beta}_t - X_2\boldsymbol{\beta}_q)'(\mathbf{y} - X_1\boldsymbol{\beta}_t - X_2\boldsymbol{\beta}_q) \\ &= \min_{\boldsymbol{\beta}_t} (\mathbf{y} - X_1\boldsymbol{\beta}_t)'(\mathbf{y} - X_1\boldsymbol{\beta}_t).\end{aligned} \tag{3.226b}$$

Substituting in (3.222), and using (3.220a) and (3.226a), we find an alternative form for the test statistic F is

$$F = \frac{[\mathrm{SS}_e(H_0') - \mathrm{SS}_e]/q}{\mathrm{SS}_e/(N-k)} \tag{3.227}$$

where SS_e, sometimes referred to as the unrestricted minimum sum of squares under the model (3.212), may be expressed as in (3.219). We turn now to an example.

EXAMPLE 3.7.1.

Suppose $E(\mathbf{y}) = X\boldsymbol{\beta}$, where

$$X = \begin{bmatrix} 1 & 1 \\ 2 & -1 \\ 1 & 2 \end{bmatrix} \quad \text{and} \quad \boldsymbol{\beta} = \begin{bmatrix} \beta_1 \\ \beta_2 \end{bmatrix} \tag{3.228}$$

and that we wish to examine

$$H_0\colon \beta_1 = \beta_2, \tag{3.229}$$

that is, in the language of the foregoing, $B\boldsymbol{\beta} = \mathbf{b}$ with $B = (1, -1)$ and $\mathbf{b} = b = 0$, with $q = 1$, $t = k - q = 2 - 1 = 1$. If we augment for concreteness with $C = (-1, -1)$, we find [see (3.208), and so on]

$$A = \begin{bmatrix} -1 & -1 \\ 1 & -1 \end{bmatrix}, \quad A^{-1} = \frac{1}{2}\begin{bmatrix} -1 & 1 \\ -1 & -1 \end{bmatrix}. \tag{3.230}$$

Thus we can write (since $b = 0$)

$$\begin{aligned} E(\mathbf{y}) &= XA^{-1}A\boldsymbol{\beta} \\ &= Z\boldsymbol{\theta} = \left(\mathbf{z}_1 \mid \mathbf{z}_2\right)\begin{bmatrix}\theta_1 \\ \theta_2\end{bmatrix}, \end{aligned} \tag{3.231}$$

and we take $Z_1 = \mathbf{z}_1 = (-2, -1, -3)', \mathbf{z}_2 = (0, 3, -1)'$, and $\theta_1 = -\frac{1}{2}(\beta_1 + \beta_2)$, $\theta_2 = \frac{1}{2}(\beta_1 - \beta_2)$. We note that H_0 is equivalent to

$$H_0' : \theta_2 = 0. \tag{3.232}$$

We have, with respect to H_0', the problem in the desired form, and we need [see (3.222)] $\hat{\theta}_2 Z_{2\cdot 1}' Z_{2\cdot 1}\hat{\theta}_2$. Now

$$Z_{2\cdot 1} = [I - \mathfrak{R}(Z_1)]Z_2 = (0, 3, -1)', \tag{3.233}$$

and we have

$$\hat{\theta}_2 Z_{2\cdot 1}' Z_{2\cdot 1}\hat{\theta}_2 = \left[0^2 + 3^2 + (-1)^2\right]\hat{\theta}_2^2 = 10\hat{\theta}_2^2. \tag{3.233a}$$

But $\hat{\theta}_2^2 = (\hat{\beta}_1 - \hat{\beta}_2)^2/4$, so that we may write

$$\hat{\theta}_2 Z_{2\cdot 1}' Z_{2\cdot 1}\hat{\theta}_2 = \frac{5(\hat{\beta}_1 - \hat{\beta}_2)^2}{2}, \tag{3.233b}$$

where, of course,

$$\hat{\boldsymbol{\beta}} = (\hat{\beta}_1, \hat{\beta}_2)' = (X'X)^{-1}X'\mathbf{y} = \frac{1}{35}\begin{bmatrix} 5y_1 + 13y_2 + 4y_3 \\ 5y_1 - 8y_2 + 11y_3 \end{bmatrix}. \tag{3.233c}$$

Since $N - k = 3 - 2 = 1$,

$$F = \frac{5(\hat{\beta}_1 - \hat{\beta}_2)^2/2}{\text{MS}_e} \tag{3.234}$$

with

$$\text{MS}_e = \text{SS}_e/1, \tag{3.234a}$$

and where the unrestricted minimum sum of squares SS_e is given by

$$SS_e = \mathbf{y}'\left[I - X(X'X)^{-1}X'\right]\mathbf{y} = \mathbf{y}'\left[I - Z(Z'Z)^{-1}Z'\right]\mathbf{y}$$

$$(3.235) \qquad = \mathbf{y}'\left[\frac{1}{35}\begin{Bmatrix} 25 & -5 & -15 \\ -5 & 1 & 3 \\ -15 & 3 & 9 \end{Bmatrix}\right]\mathbf{y}$$

$$= \tfrac{1}{35}\left[25y_1^2 + y_2^2 + 9y_3^2 - 10y_1y_2 - 30y_1y_3 + 6y_2y_3\right].$$

We could of course determine $SS_e(H_0')$ directly, since we have that

$$(3.236) \qquad \begin{aligned} SS_e(H_0') &= \min_{\theta_1}\left[(\mathbf{y} - \mathbf{z}_1\theta_1)'(\mathbf{y} - \mathbf{z}_1\theta_1)\right] \\ &= (\mathbf{y} - \mathbf{z}_1\hat{\theta}_1)'(\mathbf{y} - \mathbf{z}_1\hat{\theta}_1), \end{aligned}$$

where $\hat{\theta}_1 = \mathbf{z}_1'\mathbf{y}/\mathbf{z}_1'\mathbf{z}_1$, or

$$(3.236a) \qquad \begin{aligned} SS_e(H_0') &= \mathbf{y}'\left(I - \frac{\mathbf{z}_1\mathbf{z}_1'}{\mathbf{z}_1'\mathbf{z}_1}\right)\mathbf{y} \\ &= \tfrac{1}{14}\left[10y_1^2 + 13y_2^2 + 5y_3^2 - 4y_1y_2 - 12y_1y_3 - 6y_2y_3\right]. \end{aligned}$$

The difference between $SS_e(H_0')$ and SS_e is the numerator of the F statistic (3.234), which may be easily checked and which the reader is asked to do in Problem 3.56.

The reader may well wish to ask at this point—What would happen if we use procedure (3.225), derived under the assumption of normality, if the normality assumption does not hold true? Generally speaking, the result (3.224a) remains true, to good approximation when the ε_u's are moderately non-normal. The measure of non-normality that comes into play is the theoretical coefficient of skewness, denoted by γ_2, where

$$(3.237) \qquad \gamma_2 = \frac{\mu_4}{(\sigma^2)^2} - 3, \qquad \mu_4 = E(\varepsilon^4).$$

For normality, $\gamma_2 = 0$. We say that F, as defined by (3.222), is robust under departures from normality, and the above states that use of (3.225) results in a test procedure of significance level α, approximately.

As a word of caution, however, it also turns out that the nature of the matrix X involved plays a role, in particular, through the matrices $\mathcal{R}(X_{2\cdot 1})$

and $I - \mathcal{R}(X_1) - \mathcal{R}(X_{2 \cdot 1})$. However, the effect of X for most of the designs commonly used, such as analysis of variance designs used in this chapter and Chapters 4 and 5, is minimal and the approximation of F by a $F_{q, N-k}$ variable remains valid.

Finally, it should be pointed out that when we know what the non-normal distribution of the ε_u's happens to be, the best criterion to test H_0' may not be the F statistic. For a discussion of this matter, the reader is referred to the discussion of robustness in the books by Box and Tiao (1973, Chapter 3), and Seber (1977, Chapter 6).

3.8. OPTIMUM STATISTICAL DESIGNS: THE POLYNOMIAL CASE

Suppose an experimenter wishes to fit the polynomial model

$$E(y) = \beta_0 + \beta_1 x + \cdots + \beta_d x^d, \tag{3.238}$$

where d is known in advance. Indeed, we are assuming that the experimenter, because of lack of fit considerations and prior experience, believes that *the above model is correct*. As mentioned in Chapter 1, the question arises as to where to experiment, that is, what are the best values x_j at which observations should be generated. (We assume that $x_i < x_j$ for $i < j$, without loss of generality.) We recall that, when we do know the places x_j at which observations are taken,

$$V(\hat{\boldsymbol{\beta}}) = \sigma^2 (X'X)^{-1}, \tag{3.239}$$

so that a criterion that suggests itself is to choose the places x_j such that

$$|V(\hat{\boldsymbol{\beta}})| = (\sigma^2)^{d+1} |X'X|^{-1} \tag{3.239a}$$

is minimized. $|V(\hat{\boldsymbol{\beta}})|$ is called the generalized variance and we see that minimizing (3.238a) is equivalent to maximizing $|X'X|$.

Because our model is a polynomial of dth-degree, which is always fixed by $d + 1$ points, we assume that experimentation will be conducted at $d + 1$ points, in some interval of values for x, which without loss of generality we take to be $(-1, 1)$. (This may always be accomplished since, if originally x' denotes our independent variable and if the experimenter wishes to choose values in $[a, b]$, we let $x = [2x' - (a + b)]/(b - a)$, and clearly $-1 \leq x \leq 1$. Suppose then that n_j observations are to be taken at x_j, $j = 1, \ldots, d + 1$.

Hence, our X matrix takes the form

$$X = \begin{bmatrix} \mathbf{1}_{n_1} & \mathbf{1}_{n_1}x_1 & \mathbf{1}_{n_1}x_1^2 & \cdots & \mathbf{1}_{n_1}x_1^d \\ \vdots & \vdots & \vdots & & \vdots \\ \mathbf{1}_{n_j} & \mathbf{1}_{n_j}x_j & \mathbf{1}_{n_j}x_j^2 & \cdots & \mathbf{1}_{n_j}x_j^d \\ \vdots & \vdots & \vdots & & \vdots \\ \mathbf{1}_{n_{d+1}} & \mathbf{1}_{n_{d+1}}x_{d+1} & \mathbf{1}_{n_{d+1}}x_{d+1}^2 & \cdots & \mathbf{1}_{n_{d+1}}x_{d+1}^d \end{bmatrix} \tag{3.240}$$

with $\Sigma_{j=1}^{d+1} n_j = N$, so that X is of order $N \times k = N \times (d+1)$. Thus [Problem 3.61(a)]

$$X'X = Z'DZ, \tag{3.240a}$$

where D and Z are $(d+1) \times (d+1)$ matrices, D is diagonal,

$$D = \begin{pmatrix} n_1 & & \\ & \ddots & \\ & & n_{d+1} \end{pmatrix}, \tag{3.240b}$$

and

$$Z = \begin{bmatrix} 1 & \cdots & 1 \\ x_1 & \cdots & x_{d+1} \\ \vdots & & \vdots \\ x_1^p & \cdots & x_{d+1}^p \\ \vdots & & \vdots \\ x_1^d & \cdots & x_{d+1}^d \end{bmatrix}. \tag{3.240c}$$

We wish to maximize

$$|X'X| = \prod_{j=1}^{d+1} n_j |Z|^2. \tag{3.240d}$$

Now maximizing $\Pi_{j=1}^{d+1} n_j$ subject to $\Sigma n_j = N$ results in: $n_1 = \cdots = n_{d+1} = N/(d+1)$ minimizing $\Sigma_{j=1}^{d+1} n_j$, whatever the x_j may be. It remains to maximize $|Z|^2$. But, as is easily verified,

$$|Z| = \prod_{\substack{i<j \\ 1}}^{d+1} (x_j - x_i) \tag{3.241}$$

so that

$$|Z|^2 = \prod_{\substack{i<j \\ 1}}^{d+1} (x_j - x_i)^2. \tag{3.241a}$$

This product of squares can always be increased by moving the extreme points to the ends of the interval. (This of course implies that we should always experiment over as wide an interval as possible.)

We may now proceed to find for each d the solution to maximizing (3.241a), with the above borne in mind. For $d = 1$—the linear case—we have

$$|Z|^2 = (x_1 - x_2)^2 \tag{3.242}$$

which of course is maximized by letting $x_1 = -1$, $x_2 = 1$, and observing $N/2$ at each of the end points of the interval.

For $d = 2$—the quadratic case—we have

$$|Z|^2 = (x_1 - x_2)^2(x_2 - x_3)^2(x_1 - x_3)^2 \tag{3.243}$$

so that we need to find x_2 which maximizes Z^2. Because of the above considerations (3.243) can be written

$$\begin{aligned} |Z|^2 &= (-1 - x_2)^2(x_2 - 1)^2(-1 - 1)^2 \\ &= (1 + x_2)^2(x_2 - 1)^2(2^2), \end{aligned} \tag{3.243a}$$

that is,

$$|Z|^2 = 4(1 - x_2^2)^2. \tag{3.243b}$$

This is maximized at $x_2 = 0$, so that the optimal design for the quadratic case is to generate $N/3$ observations at each of the places $x = -1$, 0, and 1.

For the cubic case, $d = 3$, $d + 1 = 4$, we set $x_1 = -1$, $x_4 = 1$, and, because of symmetry considerations, set $x_2 = -x_3$. We then find [Problem

3.61(c)]

$$|Z|^2 = 16x_2^2(1 - x_2^2)^4 \tag{3.244}$$

and this is maximized when $x_2^2 = 1/5$. Thus, we generate $N/4$ observations at each of the places -1, -0.447, 0.447, and 1.

For a fuller discussion of optimum designs, the reader is referred to the book by Kendall and Stuart (1966, Volume 3, Chapter 38) and the references therein.

Problems

3.1. Using the notation of this chapter, show algebraically that $(\hat{\boldsymbol{\eta}} - X\boldsymbol{\theta})'(\mathbf{y} - \hat{\boldsymbol{\eta}}) = 0$.

3.2. Verify that $\mathcal{R} = X(X'X)^{-1}X'$ and $\mathcal{M} = I_N - \mathcal{R}$ are symmetric and idempotent matrices. (I_N is the identity of order N.)

3.3. Invoking $AE(\mathbf{y})$, deduce (i)–(iii) of (3.30).

3.4. Invoking $AE(\mathbf{y})$ and $AV(\mathbf{y})$, deduce (i)–(iii) of (3.31).

3.5. (a) Deduce (3.38*b*). *Hint*: Denoting $2\sigma^2\boldsymbol{\lambda}$ as a vector of Lagrange multipliers, minimize $u/\sigma^2 = \mathbf{c}'\mathbf{c} - 2\boldsymbol{\lambda}'(X'\mathbf{c} - \mathbf{a}) = \mathbf{c}'\mathbf{c} - (\mathbf{c}'X - \mathbf{a}')(2\boldsymbol{\lambda})$, and, denoting the solution by $\hat{\boldsymbol{\lambda}}, \hat{\mathbf{c}}$, show that $\hat{\mathbf{c}}' = \mathbf{a}'(X'X)^{-1}X' = \mathbf{d}'$.

(b) Show that differentiation of (3.42) leads to the derivatives given in (3.43).

3.6. Consider the model $E(\mathbf{y}) = \boldsymbol{\eta} = \beta\mathbf{1}$, $V(\mathbf{y}) = \sigma^2 I_N$.

(a) Describe the estimation space and the error space for this case.

(b) Develop the least-squares estimator for β.

(c) State explicitly the form of the ith component of $\mathbf{y} - \hat{\boldsymbol{\eta}}$ and show that X and $\mathbf{y} - \hat{\boldsymbol{\eta}}$ are orthogonal for this case.

(d) Develop the relevant analysis of variance table for the suggested value $\beta = \beta_0$.

(e) Find $E(\hat{\beta})$, $E(\text{SS}_e)$, and $E(\text{SS}_r)$.

3.7. Show that the residual sum of squares $\text{SS}_e = \mathbf{e}'\mathbf{e}$ can be expressed as in (3.46) or (3.46a).

3.8. Deduce (3.53). *Hint*: $E[(\mathbf{y} - \boldsymbol{\eta})'A(\mathbf{y} - \boldsymbol{\eta})] = E[\text{tr}\, A(\mathbf{y} - \boldsymbol{\eta})(\mathbf{y} - \boldsymbol{\eta})']$, and is also equal to $E(\mathbf{y}'A\mathbf{y}) - \boldsymbol{\eta}'A\boldsymbol{\eta}$.

3.9. (a) (i) Verify (3.60a). (ii) Show that (3.61) holds.

(b) Suppose $E(\mathbf{y}) = X\boldsymbol{\beta}$, and that it may *not* be assumed that $\boldsymbol{\beta} = \mathbf{0}$. Show that SS_r and SS_e are independent, where, of course, $SS_r = \mathbf{y}'\mathcal{R}\mathbf{y}$, $SS_e = \mathbf{y}'(I - \mathcal{R})\mathbf{y}$, $\mathcal{R} = X(X'X)^{-1}X'$. *Hint*. Show that SS_r can be written as $SS_r = Q + l + c$, where $Q = (\mathbf{y} - X\boldsymbol{\beta})'\mathcal{R}(\mathbf{y} - X\boldsymbol{\beta})$, $l = 2\beta'X'(\mathbf{y} - X\boldsymbol{\beta})$, and $c = \boldsymbol{\beta}'X'X\boldsymbol{\beta}$. Now in view of (i) of part (a) above, $SS_e = (\mathbf{y} - X\boldsymbol{\beta})'(I - \mathcal{R})(\mathbf{y} - X\boldsymbol{\beta})$. Prove that Q and l are each independent of SS_e, using the relevant theorem (and corollary) of Chapter 2, so that $Q + l + c$ (c is the above stated constant) is independent of SS_e.

3.10. Show that the quadratic forms mentioned on the right-hand side of equation (3.62) have χ^2 distributions, and state, giving reasons, their degrees of freedom. State clearly the reasons why these two forms are independent so that their ratio is distributed as a F variable, as given in (3.63). *Hint*: Consult Theorem 2.2.2, Corollary 2.2.2.1, Theorem 2.2.3, and/or Theorem 2.2.5, and (3.9b). See also (2.82) and (2.82a) of Chapter 2.

3.11. Verify (3.67) and (3.68a).

3.12. (i) Verify (3.80c). (ii) Verify (3.82a).

3.13. Deduce (3.87).

3.14. Verify (3.89a) and (3.90).

3.15. Deduce (3.92).

3.16. Verify (3.98).

3.17. (a) Show that $\hat{\boldsymbol{\eta}}(X_1) + \hat{\boldsymbol{\eta}}(X_2) \neq \hat{\boldsymbol{\eta}}(X)$ if $X_1'X_2 \neq \mathrm{O}$. *Hints*: Recall that $\hat{\boldsymbol{\eta}}(X_i) = \mathcal{R}(X_i)\mathbf{y}$, and so on, and also find D_{ij}, where

$$(X'X)^{-1} = \left[\begin{array}{c|c} D_{11} & D_{12} \\ \hline D_{21} & D_{22} \end{array}\right], \qquad \text{if} \quad X'X = \left[\begin{array}{c|c} X_1'X_1 & X_1'X_2 \\ \hline X_2'X_1 & X_2'X_2 \end{array}\right].$$

(b) Use the work of part (a) to show that the vector consisting of the last k_2 components of $\hat{\boldsymbol{\beta}} = (X'X)^{-1}X'\mathbf{y}$ is indeed equal to $(X_{2\cdot1}'X_{2\cdot1})^{-1}X_{2\cdot1}'\mathbf{y}$, where $X_{2\cdot1} = [I - \mathcal{R}(X_1)]X_2$. Show that $E[(X_{2\cdot1}'X_{2\cdot1})^{-1}X_{2\cdot1}'\mathbf{y}] = \boldsymbol{\beta}_2$, where $\boldsymbol{\beta} = (\boldsymbol{\beta}_1' \mid \boldsymbol{\beta}_2')'$.

3.18. Substantiate clearly why the statement made after (3.103a) is correct.

3.19. In terms of A and $X_{2\cdot1}$, find the covariance matrix of $\hat{\boldsymbol{\beta}}_1$ and $\hat{\boldsymbol{\beta}}_2$. *Hint*: $\hat{\boldsymbol{\phi}}$ and $\hat{\boldsymbol{\beta}}_2$ are uncorrelated, and show that $\hat{\boldsymbol{\phi}}$ is unbiased for $\boldsymbol{\phi}$.

3.20. (a) Consider the model $E(\mathbf{y}) = [\mathbf{1} \mid \mathbf{x}](\beta_0 \mid \beta_1)', V(\mathbf{y}) = \sigma^2 I_N$. Show that $\hat{\beta}_1 = \Sigma_{u=1}^N (x_u - \bar{x})(y_u - \bar{y}) / \Sigma_{u=1}^N (x_u - \bar{x})^2$, $\hat{\beta}_0 = \bar{y} - \hat{\beta}_1 \bar{x}$. Orthogonalize when: (i) β_1 is of interest and deduce that the extra regression sum of squares due to β_1, given β_0 is in the model, is $\hat{\beta}_1^2 \Sigma(x_i - \bar{x})^2$; (ii) β_0 is of interest, and show that the extra regression sum of squares due to β_0, given β_1 is in the model, is of the form $K(\bar{y} - \hat{\beta}_1 \bar{x})^2$. What is K?

(b) For the situation given in (a), and using (3.50), show that $E(\text{SS}_r)$ is indeed given by (1.42) of Chapter 1, namely, $E(\text{SS}_r) = 2\sigma^2 + [N\alpha^2 + 2(\Sigma x_i)\alpha\beta + (\Sigma x_i^2)\beta^2]$.

(c) Suppose $E(\mathbf{y}) = X\boldsymbol{\beta}$, with the first column of the full rank matrix X a vector of ones, say $\mathbf{1}_N$. Show that the sum of the components of the residual vector $\mathbf{e} = (\mathbf{y} - \hat{\boldsymbol{\eta}})$ is zero. *Hint*: Differentiating $Q = \Sigma(y_u - \beta_0 - \beta_1 x_{u1} - \cdots - \beta_k x_{uk})^2$ with respect to the β_j, $j = 0, 1, \ldots, k$, gives rise to the normal equations, when we set the $\partial Q / \partial \beta_j$ equal to zero and call the solutions $\hat{\beta}_0, \hat{\beta}_1, \ldots, \hat{\beta}_k$. Consider the equation found when doing this for $j = 0$; or, $X'(\mathbf{y} - \hat{\boldsymbol{\eta}}) = \mathbf{0}$ and the first component of the left-hand side is $\mathbf{1}'(\mathbf{y} - \hat{\boldsymbol{\eta}}) = 0$.

(d) Verify that regressing any vector, for example, $\mathbf{y}$ of order $N \times 1$, on $\mathbf{1}$, leads to residuals $\mathbf{y} - \bar{y}\mathbf{1}$, where $\bar{y} = N^{-1}(\mathbf{1}'\mathbf{y})$.

3.21. (a) Verify that the entries in Table 3.3.5 are correct.

(b) Using (3.114*b*), show that R may be written as in (3.114*e*). *Hints*:

(i) Prove $\mathbf{e}'\mathbf{1} = 0$ so that $\bar{y} = \bar{\hat{\eta}}$.

(ii) Show that the numerator of (3.114*b*) may be written as $\Sigma(\hat{\eta}_i - \bar{\hat{\eta}})^2$ and that $\Sigma(\hat{\eta}_i - \bar{\hat{\eta}})^2 = \Sigma(y_i - \bar{y})(\hat{\eta}_i - \bar{\hat{\eta}})$.

3.22. Show that if $X = (\mathbf{x}_1 \mid \cdots \mid \mathbf{x}_k)$, where $\mathbf{x}_j$ is a $N \times 1$ vector, and if $\mathbf{y} = X\boldsymbol{\beta} + \boldsymbol{\varepsilon}$, $E(\boldsymbol{\varepsilon}) = \mathbf{0}$, $V(\boldsymbol{\varepsilon}) = \sigma^2 I$, then, if $\mathbf{x}_i'\mathbf{x}_j = 0$, for all $i \neq j$, β_j may be estimated separately by regressing $\mathbf{y}$ on $\mathbf{x}_j$.

3.23. Referring to Table 3.4.2, explain clearly how and why a test of $\beta_1 = \cdots = \beta_k = 0$ may be performed, if we assume normality. (*Hint*: Use the discussion of Section 3.7.) Give all details.

3.24. Referring to (3.126), and given that $(\theta_2, \ldots, \theta_k)$ are of interest, eliminate β_1, showing all steps, and verify that the analysis of variance table is as in Table 3.4.3. What is $\phi = \beta_1 + A\boldsymbol{\beta}_2$? (*Hints*: Because of the structure of $\mathbf{x}_j$, $X_{2\cdot 1}$ and $X_{2\cdot 1}'X_{2\cdot 1}$ are easy to find explicitly. The regression sum of squares due to $\boldsymbol{\theta}_2$ is, of course, $\hat{\boldsymbol{\theta}}_2' X_{2\cdot 1}' X_{2\cdot 1} \hat{\boldsymbol{\theta}}_2$.)

3.25. (a) (Continuation of Problem 3.24.) If $\mathbf{y} = N(X\boldsymbol{\beta}, \sigma^2 I)$, what test would be appropriate for the statement $\theta_2 = \cdots = \theta_k$? Cite carefully all results used, giving all details. (*Hint*: use Section 3.7.)

(b) Yields of a certain chemical were recorded using three different catalytic methods, with the results shown below.

Method I	Method II	Method III
47.2	50.1	49.1
49.8	49.3	53.2
48.5	51.5	51.2
48.7	50.9	52.8
		52.3

Test at the 5% level of significance the statement that the catalytic methods are equivalent. [Computational *Hints*: From (3.133), we have that

$$\mathrm{SS}_B = \sum_{j=1}^{k} n_j \bar{y}_j^2 - N\bar{y}^2 = \sum_{j=1}^{k} \frac{T_{\cdot j}^2}{n_j} - \frac{T^2}{N},$$

$$\text{where} \quad T = \sum_{j=1}^{k} \sum_{i=1}^{n_j} y_{ij} \quad \text{and} \quad T_{\cdot j} = \sum_{i=1}^{n_j} y_{ij}$$

is the sum of the observations generated from the jth population or process (in this example the jth process refers to the use of method j). Now using (3.134), it is easy to see that $\mathrm{SS}_T = \Sigma\Sigma y_{ij}^2 - T^2/N$, so that $\mathrm{SS}_W = \mathrm{SS}_T - \mathrm{SS}_B = \Sigma y_{ij}^2 - \Sigma_{j=1}^{B} T_{\cdot j}^2/n_j$.]

3.26. In the notation of this chapter, show that $\mathbf{e}_{12} = \mathbf{e}_1 - \mathcal{R}(X_{2\cdot 1})\,\mathbf{e}_1 = \mathbf{e}_1 - \mathcal{R}(X_{2\cdot 1})\,\mathbf{y} = [I - \mathcal{R}(X_1) - \mathcal{R}(X_{2\cdot 1})]\,\mathbf{y}$.

3.27. Verify that if $\mathbf{e}_1$, given by (3.140), is regressed on $X_{0\cdot 1}$, given by (3.142), then the sum of squares of residuals is indeed given by (3.143). If $X_0 = \mathbf{x}_0$, show that S_{01e} of (3.143) takes the form in (3.143a). Also, verify (3.143d).

3.28. (a) Equation (3.150) could be obtained using the following procedure. Suppose we pretend that we are working with the model $E(\mathbf{y}) = \mathbf{x}_0\beta_0 + \mathbf{1}\beta_1$, where interest is in β_0. Use the two-step procedure and show that the residual sum of squares S_{01e} obtained is as given in (3.152).

(b) Verify that $\mathbf{e}_{12}$ is as stated just after (3.153).

3.29. Referring to Table 3.4.8, show directly that (a) $q_y = q_{y1} + q_{y2}$; (b) $q_x = q_{x1} + q_{x2}$; (c) $q_{xy} = q_{xy1} + q_{xy2}$.

3.30. Verify the results given in (3.160)–(3.160e), and give the form of $H_1 H_2$.

3.31. (a) Derive the entries for the EMS columns of Tables 3.5.1 and 3.5.2. If fitting a polynomial, state the form of these entries, which are the entries of Table 3.5.2. What test would you use, and why, for $\beta_m = \theta_m = 0$?

(b) Verify (3.172b), (3.172c), and (3.174).

3.32. Suppose we are fitting N equispaced points to a third-degree polynomial. Show that the orthogonal polynomials involved are as stated in (3.175). *Hint*: Apply the Gram–Schmidt procedure to the X matrix. You will need to know $\Sigma_{x=1}^{N} x^t$ for $t = 0, 1, 2, 3, 4$, and 5. One can proceed recursively using the method illustrated here with the following example: We know that $x^2 - (x-1)^2 = 2x - 1$. Sum "both sides" over x from 1 to N obtaining $N^2 = 2\Sigma_1^N x - N$, or $2\Sigma_1^N x = N^2 + N = N(N+1)$ or $\Sigma_1^N x = N(N+1)/2$. Using this, take the difference $x^3 - (x-1)^3$, sum, and so on.

3.33. (a) Show that (3.183a) and (3.184) hold.

(b) Suppose $E(\mathbf{y}) = \beta_0 \mathbf{1} + \beta_1 \mathbf{x}$ is fitted, but it is actually the case that $E(\mathbf{y}) = \delta_0 \mathbf{1} + \delta_1 \mathbf{x} + \delta_2 \mathbf{x}^2$ holds [here $\mathbf{x} = (x_1, \ldots, x_u, \ldots, x_N)'$ and $\mathbf{x}^2 = (x_1^2, \ldots, x_u^2, \ldots, x_N^2)'$]. Give an expression for $\boldsymbol{\gamma}^\circ$ and deduce an algebraic expression for Λ^2.

3.34. Derive the expected sum of squares entries of Table 3.6.1.

3.35. Substantiate the following: Referring to (3.186) and (3.187), a test of (3.187) of level α is:

Reject (3.187) if observed value of $F = s^2/s'^2 > F_{N-k,\, v;\, \alpha}$, and so on.

3.36. Referring to the situation dictated by (3.189a), verify the result in (3.193).

3.37. Again referring to (3.189a), show that if $g = k$, $\hat{\eta}_{tr} = \bar{y}_{\cdot r}$. *Hint*: X has a specific structure.

3.38. Given that $E(y) = \beta_0 + \beta_1 x$, and that n_1 observations on y are taken at x_1 and n_2 observations are taken at x_2, show that, in obvious notation, the least-squares regression line is such that $\hat{\eta}_{x_j} = \bar{y}_{\cdot j}$, $j = 1, 2$, supporting the result (3.194).

3.39. Deduce (3.195*b*).

3.40. (a) Show that (3.195*g*) holds when X has first n_1 rows $\mathbf{x}'_{1\cdot}$, and next n_2 rows $\mathbf{x}'_{2\cdot}, \ldots,$ and last n_g rows $\mathbf{x}'_{g\cdot}$, $g > k$. *Hint*: X has a specific structure.

(b) Show that test procedure (3.196) is of significance level α.

3.41. Given the model $\eta = \beta_0 + \beta_1 x_1 + \beta_2 x_2 + \beta_3 x_3$, and the data $\mathbf{y}' = (10, 6, 9, 4, 8)$ generated using

$$X = \begin{bmatrix} 1 & -1 & -1 & 1 \\ 1 & -1 & 1 & -1 \\ 1 & 1 & 1 & 1 \\ 1 & 1 & -1 & 1 \\ 1 & 1 & -1 & 1 \end{bmatrix},$$

calculate the following:

(a) $X'X$;

(b) $(X'X)^{-1}$;

(c) $\hat{\boldsymbol{\beta}}$;

(d) $\hat{\boldsymbol{\eta}}$, the fitted value;

(e) $\mathbf{e} = \mathbf{y} - \hat{\boldsymbol{\eta}}$.

Assume that $V(\mathbf{y}) = \sigma^2 I$ and find

(f) $V(\hat{\beta}_1)$; (g) $V(\hat{\beta}_2)$; (h) $\text{cov}(\hat{\beta}_1, \hat{\beta}_3)$;

(i) Display the variance–covariance matrix of $\mathbf{e}$.

(j) Display the variance–covariance matrix of $\hat{\boldsymbol{\beta}}$.

(k) Give the relevant analysis of variance table for the situation where there is no interest in β_0.

3.42. Assume $E(\mathbf{y}) = X_1\boldsymbol{\beta}_1 + X_2\boldsymbol{\beta}_2$. Regress $\mathbf{y}$ on X_2, yielding residuals of the form $\mathbf{y} - X_2\hat{\boldsymbol{\beta}}_2$. Show that the sum of squares of these residuals are not adequate for estimating σ^2, when $V(\mathbf{y}) = \sigma^2 I$. However, show that, if $X'_1 X_2 = O$, regressing $\mathbf{y}$ on X_2 gives the correct least-squares estimator of $\boldsymbol{\beta}_2$.

3.43. Let $E(\mathbf{y}) = X\boldsymbol{\beta}$, X of full rank, and suppose $V(\mathbf{y}) = \sigma^2 C$, where C is a $N \times N$ positive definite matrix. Let $\tau = \mathbf{a}'\boldsymbol{\beta}$. Show that

(a) $\hat{\tau} = \mathbf{a}'\hat{\boldsymbol{\beta}}$, where $\hat{\boldsymbol{\beta}} = (X'X)^{-1}X'\mathbf{y}$, is unbiased for τ.

(b) The unique linear uniformly unbiased estimator of τ is $\tilde{\tau} = \mathbf{a}'\tilde{\boldsymbol{\beta}}$, where $\tilde{\boldsymbol{\beta}} = (X'C^{-1}X)^{-1}X'C^{-1}\mathbf{y}$ ($\tilde{\boldsymbol{\beta}}$ is known in the literature as Aitken's *generalized least-squares estimator* and sometimes as a weighted least-squares estimator). [*Hint*: since C^{-1} exists and is positive definite, we can write $C^{-1} = P'P$, with P of order $N \times N$ and positive definite. Hence $P\mathbf{y} = \mathbf{z} = PX\boldsymbol{\beta} + P\boldsymbol{\varepsilon}$, or $\mathbf{z} = W\boldsymbol{\beta} + \boldsymbol{\alpha}$, where $W = PX$, $\boldsymbol{\alpha} = P\boldsymbol{\varepsilon}$, and $P'P = C^{-1}$. Note that $E(\boldsymbol{\alpha}) = \mathbf{0}$ and $V(\boldsymbol{\alpha}) = \sigma^2 P[P'P]^{-1}P' = \sigma^2 I$. If we regress $\mathbf{z}$ on W, we have the uniformly minimum variance estimator of $\boldsymbol{\beta}$, which is linear in the y's since $(W'W)^{-1}W'\mathbf{z} = \tilde{\boldsymbol{\beta}}$. Verify all the statements made in this hint.]

3.44. If $E(y_u) = \beta_1 x_{u1} + \beta_2 x_{u2}$, where $y_1,\ldots,y_u,\ldots,y_N$ are N independent random variables, and $V(y_u) = \sigma^2 x_{u1}x_{u2}$ (the x_{uj}'s are assumed to be all positive and such that the associated X matrix is of full rank), obtain

(a) minimum variance unbiased estimators of β_1 and β_2, $\tilde{\beta}_1$ and $\tilde{\beta}_2$, respectively;

(b) the variance–covariance matrix for these estimators;

(c) if $\hat{\beta}_1$ is the usual least-squares estimator for β_1, what is the efficiency of $\hat{\beta}_1$ relative to $\tilde{\beta}_1$ [that is, $V(\tilde{\beta}_1)/V(\hat{\beta}_1)$]?

3.45. Suppose $\mathbf{y} = X\boldsymbol{\beta} + \boldsymbol{\varepsilon}$, X is of order $N \times k$ and full rank, and $\boldsymbol{\varepsilon} = N(\mathbf{0}, \sigma^2 I)$.

(a) Find the joint characteristic function $E\big(\exp\big(i\mathbf{t}_1'\hat{\boldsymbol{\beta}} + it_2 Q\big)\big)$, where $Q = (\mathbf{y} - \hat{\boldsymbol{\eta}})'(\mathbf{y} - \hat{\boldsymbol{\eta}})$, $\hat{\boldsymbol{\eta}} = \mathcal{R}(X)\mathbf{y}$, and $\hat{\boldsymbol{\beta}} = (X'X)^{-1}X'\mathbf{y}$.

(b) Find the distribution of $(N-k)Q_1/kQ_2$, where $Q_1 = (\boldsymbol{\beta} - \hat{\boldsymbol{\beta}})'X'X(\boldsymbol{\beta} - \hat{\boldsymbol{\beta}})$ and $Q_2 = (\mathbf{y} - X\hat{\boldsymbol{\beta}})'(\mathbf{y} - X\hat{\boldsymbol{\beta}})$, where $\hat{\boldsymbol{\beta}}$ is the maximum likelihood estimator of $\boldsymbol{\beta}$.

(c) Find the joint distribution of $t_1 = (\hat{\beta}_1 - \beta_1)/S\sqrt{d^{11}}$ and $t_2 = (\hat{\beta}_2 - \beta_2)/S\sqrt{d^{22}}$, where $S^2 = Q_2/(n-k)$ and $D = (X'X)^{-1} = (d^{ij})$.

3.46. If $E(\mathbf{y}) = X_1\boldsymbol{\beta}_1 + X_2\boldsymbol{\beta}_2$, show, starting from $\hat{\boldsymbol{\beta}} = (X'X)^{-1}X'\mathbf{y}$, that $\hat{\boldsymbol{\beta}}$ has the form $(X_{2\cdot 1}'X_{2\cdot 1})^{-1}X_{2\cdot 1}'\mathbf{y}$ and that

$$E\big[(X_1'X_1)^{-1}X_1\mathbf{y}\big] = \boldsymbol{\phi} = \boldsymbol{\beta}_1 + A\boldsymbol{\beta}_2$$

$$E\big[(X_{2\cdot 1}'X_{2\cdot 1})^{-1}X_{2\cdot 1}'\mathbf{y}\big] = \boldsymbol{\beta}_2.$$

3.47. (Continuation of Problem 3.46.) Suppose $\mathbf{y}$ is regressed on X_1 and residuals $\mathbf{e}_1$ computed, and that $\mathbf{e}_1$ is regressed on X_2. Show that, in general, $\tilde{\boldsymbol{\beta}}_2 = (X_2'X_2)^{-1}X_2'\mathbf{e}_1$ is not unbiased for $\boldsymbol{\beta}_2$. What is the relationship between $\hat{\boldsymbol{\beta}}_2$ and $\tilde{\boldsymbol{\beta}}_2$?

3.48. Orthogonalize when

(a) $E(\mathbf{y}) = \mathbf{x}_1\beta_1 + \mathbf{x}_2\beta_2$, with $\mathbf{x}_1 = \mathbf{1}$, and interest is in β_2.

(b) $E(\mathbf{y}) = \mathbf{x}_1\beta_1 + \mathbf{x}_2\beta_2 + \mathbf{x}_3\beta_3$, with $\mathbf{x}_1 = \mathbf{1}$, and interest is in (i) β_2, β_3, and (ii) β_3 alone. Draw up the analysis of variance tables for each case. What is $\hat{\eta}_{\mathbf{x}}$ for each case?

3.49. (a) Carry out an analysis of covariance for the following, using (i) Table 3.4.7 and (ii) Tables 3.4.8 and 3.4.8a.

Diet 1		Diet 2		Diet 3	
y	x_0	y	x_0	y	x_0
62	12	36	9	98	17
51	9	68	16	52	7
73	13	47	11	86	14
81	15	41	8	62	10

y = gain in weight

x = initial weight

(b) If interest is in $\phi_2, \ldots, \phi_k$, where

$$E(y_u) = \beta_0 x_{u0} + \beta_1 x_{u1} + \phi_2 x_{u2} + \cdots + \phi_k x_{uk},$$

where $x_{u1} \equiv 1$, then using this model as a starting point, eliminate β_0 and β_1 by the usual orthogonalization procedure. Give the analysis of variance table so obtained and compare with the analysis of covariance table. Illustrate this general comparison with the example of part (a).

(c) Find $\mathbf{e} = \mathbf{y} - \hat{\boldsymbol{\eta}}$ and plot e_u against (i) u and (ii) x_{0u}, and comment.

(d) In answering part (a) of this question, Table 3.4.8a must be constructed for the data of the problem. In particular, the various bits and pieces of Table 3.4.8a enable one to examine the hypothesis $\beta_0 = 0$ where, in this problem, β_0 is the effect of including initial weights on diets.

(i) In general, explain how this can be done, substantiating all results used under the assumption that the $\mathbf{y}$ are normally distributed, $V(\mathbf{y}) = \sigma^2 I$ and $E(\mathbf{y}) = \beta_0\mathbf{x}_0 + \beta_1\mathbf{1} + \theta_2\mathbf{x}_2 + \cdots + \theta_k\mathbf{x}_k$.

(ii) Apply (i) to the data of the experiment of (a) stating clearly the value of the (extra) regression sum of squares due to β_0, given $\beta_1, \theta_2, \ldots, \theta_k$ in the model and the value of the error term used, and so on.

(iii) Would different conclusions have been reached for the data of part (a) if x_0 had been ignored and only an analysis of variance been performed on the y's? (The left-hand portion of the counterpart of Table 3.4.8 for this data is the required analysis of variance.) Comment.

3.50. (a) "Fit" the model $\eta = \beta_0 + \beta_1 x + \beta_2 x^2 + \beta_3 x^3 + \beta_4 x^4$ to the data using the usual procedure:

x	-4	-3	-2	-1	0	1	2	3	4
y	4.5	11.7	13.1	17.3	8.9	13.1	8.4	6.0	11.5

Fit using orthogonal polynomials. Show (numerically) the equivalence for $\hat{\eta}_{\mathbf{x}}$.

(b) If it is now known that $\eta = \beta_0 + \Sigma_1^5 \beta_i x^i$, fit the model and discuss in detail. Do the fitting using orthogonal polynomials, and so on.

3.51. Suppose we preselect N equispaced x_i's inside the interval $(-1, 1)$ in such a way that

$$\mathbf{1}'\mathbf{x} = 0, \quad \mathbf{x}'\mathbf{x} = \Sigma x_i^2 = \mathbf{1}'\mathbf{x}^2 = 1, \quad \mathbf{x}^{2'}\mathbf{x} = \Sigma x_i^3 = 0,$$

and that $\mathbf{x}^{2'}\mathbf{x}^2 = \Sigma x_i^4 = a$. If we now observe a y_i at each of the x_i, and fit

$$\eta_{x_i} = \beta_0 + \beta_1 x_i$$

where it is suspected that

$$E(y_i \mid x_i) = \beta_0 + \beta_1 x_i + \beta_2 x_i^2,$$

then show that, using the notation of this chapter,

$$\Lambda^2 = \beta_2^2 \left\{ a - \frac{1}{N} \right\} = (\gamma - \gamma^\circ)'(\gamma - \gamma^\circ), \qquad a = \Sigma x_i^4.$$

3.52. Let $E(\mathbf{y}) = \Sigma_1^3 \theta_i \mathbf{x}_i$, where $X = [\mathbf{1}, \mathbf{x}, \mathbf{x}^2]$ is such that

$$X' = \begin{bmatrix} 1 & 1 & 1 & 1 & 1 & 1 & 1 \\ -3 & -2 & -1 & 0 & 1 & 2 & 3 \\ 9 & 4 & 1 & 0 & 1 & 4 & 9 \end{bmatrix}.$$

(a) Find $\hat{\boldsymbol{\theta}}$. Construct the usual analysis of variance table when the source column is labeled: due to x_1 and x_2; due to x_3, given x_1 and x_2 in model, and so on, and include an expected sum of squares column.

(b) If $y' = [13, 10, 16, 14, 17, 10, 12]$, calculate the actual entries and make the usual tests. Also calculate $\mathbf{y} - \hat{\boldsymbol{\eta}}$ and discuss.

3.53. (This problem is a generalization of the results of Section 1.8 of Chapter 1.)

(a) Suppose $E(y_u) = \Sigma_{j=1}^k x_{uj}\beta_j$, $u = 1, \ldots, N$, and that the associated $N \times k$ matrix X is of full rank, and usual assumptions apply. Suppose it is desired to *predict* the $(N + 1)$st observation y_{N+1} at the "place" $\mathbf{x}'_{N+1} = (x_{N+1,1}, \ldots, x_{N+1,k})$, where y_{N+1} is to be observed independently of $\mathbf{y} = (y_1, \ldots, y_N)'$. Show whether or not there exists a *predictor*, $\tilde{y}_{N+1}$, having the properties

(i) $\tilde{y}_{N+1}$ is a linear function of y;

(ii) $E(\tilde{y}_{N+1} - y_{N+1}) = 0$;

(iii) $E(\tilde{y}_{N+1} - y_{N+1})^2$ is as small as possible.

(b) Using part (a), show that the minimum value of $E(\tilde{y}_{N+1} - y_{N+1})^2$ is given by

$$\min E(\tilde{y}_{N+1} - y_{N+1})^2 = \sigma^2\left(1 + \mathbf{x}'_{N+1}(X'X)^{-1}\mathbf{x}_{N+1}\right) = \sigma^2 V.$$

(c) Suppose that in addition to the assumptions made in part (a), it is also the case that all observations are normally distributed. Using this assumption, show that a $100(1 - \alpha)\%$ confidence prediction interval for y_{N+1} is given by $[y_l, y_u]$, where the end points y_l and y_u are given by

$$\tilde{y}_{N+1} \pm \left(\frac{\mathrm{SS}_e}{N - k}\right)^{1/2}\left[1 + \mathbf{x}'_{N+1}(X'X)^{-1}\mathbf{x}_{N+1}\right]^{1/2} t_{N-k;\,\alpha/2}.$$

Hint: Start with the statement $P(y_l < y_{N+1} < y_u) = 1 - \alpha$ and use the fact that $y_{N+1} - \tilde{y}_{N+1} = N(0, \sigma^2 V)$, with V given in part (b), and prove that $y_{N+1} - \tilde{y}_{N+1}$ is independent of $SS_e = \mathbf{y}'[I - \mathcal{R}(X)]\mathbf{y}$, so that

$$\frac{y_{N+1} - \tilde{y}_{N+1}}{\left([SS_e/(N-k)]V\right)^{1/2}} = t_{N-k}.$$

(d) For the case $E(y) = \beta_0 + \beta_1 x$, give the necessary details to show that the above leads to the prediction belt of Chapter 1, (1.103); that is, for any x, the $(1-\alpha)$ prediction interval for y_{N+1} is $[y_l, y_u]$, with y_l and y_u given by

$$\bar{y} + \hat{\beta}(x - \bar{x}) \pm \left(\frac{SS_e}{N-2}\right)^{1/2} \cdot \left[1 + \frac{1}{N} + \frac{(x-\bar{x})^2}{\Sigma(x_u - \bar{x})^2}\right]^{1/2} t_{N-2;\,\alpha/2},$$

where $$\hat{\beta} = \frac{\sum_{u=1}^{N}(x_u - \bar{x})(y_u - \bar{y})}{\sum_{u=1}^{N}(x_u - \bar{x})^2},$$

$$SS_e = \sum_{u=1}^{N}(y_u - \bar{y})^2 - \hat{\beta}^2 \sum_{u=1}^{N}(x_u - \bar{x})^2.$$

3.54. Suppose y_{ij}, $i = 1, \ldots, n_j$, $j = 1, 2$, are $n_1 + n_2$ independent observations, with $y_{ij} = N(\beta_j, \sigma^2)$. Using the general linear hypothesis framework, derive a test statistic for the hypothesis H_0: $\beta_1 = \beta_2$.

3.55. (a) Suppose $E(y_u) = \beta_0 + \beta_1 x_u = \eta_u$, $u = 1, \ldots, N$ with at least two x_u of different value, and we wish to test H_0: $\beta_1 = b_1$ (b_1 is a known number). If the y_u are independent $N(\eta_u, \sigma^2)$, then derive a test statistic for H_0 using the general linear hypothesis framework.

(b) Do the same as in part (a) for the hypothesis H_0': $\beta_0 = b_0$.

3.56. (a) In Example 3.7.1 of Section 3.7, we may determine $SS_e(H_0')$ by finding the minimum of $\Sigma_{j=1}^{3}(y_j - z_{j1}\theta_1)^2$ with respect to θ_1; that is, by finding the minimum of [see (3.231)]

$$(y_1 + 2\theta_1)^2 + (y_2 + \theta_1)^2 + (y_3 + 3\theta_1)^2.$$

Do so directly by differentiating with respect to θ_1, and so on, and verify that this method produces (3.236a).

(b) Express the numerator of the F statistic given in (3.234) in terms of (y_1, y_2, y_3) and verify that your answer is the same as that obtained from $SS_e(H_0') - SS_e$, where $SS_e(H_0')$ is as in your answer to (a) or (3.236a), and SS_e is as given in (3.235).

3.57. Justify the use of the test procedure for the one-way analysis of variance situation of Table 3.4.3, given by (for level α)

$$\text{Reject } \beta_1 = \cdots = \beta_k, \quad \text{or} \quad \theta_2 = \theta_3 = \cdots = \theta_k = 0, \quad \theta_j = \beta_j - \beta_1,$$

if the observed value of $F = MS_r(\theta_2, \ldots, \theta_k)/MS_e(\beta_1, \theta_2, \ldots, \theta_k) > F_{k-1,\, N-k;\, \alpha}$. Do this by using the method of Section 3.7.

3.58. (a) Using orthogonal polynomials, fit a quadratic to the data of Problem 1.16. (Note that there are $g = 6$ groups.) Test for adequacy of a quadratic fit. (Use the orthogonalization procedure on the columns of the X matrix, or use the tables of orthogonal polynomials for $N = 6$, and then take into account the fact that there are replications at each value of X.)

(b) The results obtained in doing Problem 1.16 could be of use here. Assuming that a linear function of x has been fitted to the data of Problem 1.16, and that $SS_r(\mathbf{1}, \mathbf{x})$ is available, one need only fit a quadratic function in any way possible and obtain $SS_r(\mathbf{1}, \mathbf{x}, \mathbf{x}^2)$ and SS_W for a valid error term, since replications have been carried out. Do this, that is, obtain $SS_r(\mathbf{1}, \mathbf{x}, \mathbf{x}^2)$ and SS_W, and find the extra regression sum of squares due to $\mathbf{x}^2$ (given $\mathbf{1}$ and $\mathbf{x}$ in the model) and test for its significance. (*Hint*: This problem is made very much easier if orthogonal polynomials are used.)

3.59. (a) Consider $E(\mathbf{y}) = \sum_{j=1}^{k} \mathbf{x}_j \beta_j$, $V(\mathbf{y}) = \sigma^2 I_N$, and so on, and such that $X = [\mathbf{x}_1 \cdots \mathbf{x}_k]$ is of full rank. Derive a test statistic for $H_0^{(k)}$: $\beta_k = b_k$ (b_k is a known number) using the general linear hypothesis framework, if the y_u are normal.

(b) Derive a test for the statement $H_0^{(t+1, \ldots, k)}$: $\beta_j = b_j$, $j = t + 1, \ldots, k$ (t is a number such that $1 \le t \le k - 1$).

3.60. Recall from Chapter 1 and Problem 3.20 that if $E(\mathbf{y}) = [\mathbf{1} \mid \mathbf{x}](\beta_0 \mid \beta_1)'$, SS_e may be written

$$SS_e = \sum_{u=1}^{N} (y_u - \bar{y})^2 - \hat{\beta}_1^2 \sum_{u=1}^{N} (x_u - \bar{x})^2$$

(a) Suppose now that we wish to compare two processes and that data are available of the form $(\mathbf{x}_j, \mathbf{y}_j)$ with $\mathbf{y}_j = (y_{1j}, y_{2j}, \ldots, y_{n_j j})'$, $\mathbf{x}_j = (x_{1j}, \ldots, x_{n_j j})'$, where $E(\mathbf{y}_j) = (\mathbf{1}_{n_j} \mid \mathbf{x}_j)(\beta_{0j} \mid \beta_{1j})'$, $j = 1, 2$. Assuming normality,

(i) Use the methods of Section 3.7 to derive a test for $\beta_{11} = \beta_{12}$; that is, the two processes are such that their true regression lines have the same slope. [To do this, you will need to write a model for $\mathbf{y} = (\mathbf{y}_1', \mathbf{y}_2')'$ and to determine the matrix B of (3.205), with $\boldsymbol{\beta}$ of (3.205) given by $\boldsymbol{\beta} = (\beta_{01}, \beta_{02}, \beta_{11}, \beta_{12})'$, and so on.]

(ii) Use the methods of Section 3.7 to derive a test for $\beta_{01} = \beta_{02}$; that is, the two regression lines have common intercept.

(iii) Use the methods of Section 3.7 to derive a test for $\beta_{01} = \beta_{02}$ and $\beta_{11} = \beta_{12}$; that is, the two regression lines are the same (coincident).

(b) The above may be generalized to the case where we wish to compare $p \geq 2$ regression lines. Derive a test for

(i) parallelism of the p lines; (ii) concurrence of the p lines at $x = 0$; and (iii) coincidence of the p lines.

(c) The above may also be generalized to the case where

$$E(\mathbf{y}_j) = \left(\mathbf{1}_{n_j} \mid \mathbf{x}_j \mid \mathbf{x}_j^2 \mid \cdots \mid \mathbf{x}_j^d\right)\left(\beta_{0j} \mid \beta_{1j} \mid \cdots \mid \beta_{dj}\right)'$$

that is, when $E(y_{uj}) = \beta_{0j} + \beta_{1j}x_{uj} + \cdots + \beta_{dj}x_{uj}^d$, a polynomial of degree d ($j = 1, \ldots, p$). Derive tests for the hypotheses (i) $\beta_{dj} = \beta_{dj'}$ for all $j \neq j'$; (ii) $\beta_{0j} = \beta_{0j'}$ for all $j \neq j'$; (iii) $\beta_{rj} = \beta_{rj'}$, $r = 0, \ldots, d$.

3.61. (a) Verify that $X'X$ can be written as in (3.240a), where X is given in (3.240).

(b) Show that maximizing $\prod_1^{d+1} n_j$ subject to $\Sigma n_j = N$ does yield the result stated, namely, $n_j = N/(d+1)$, for all j.

(c) Find the optimal design for the cases $d = 4$ and $d = 5$ if $x \in [-1, 1]$, and so on.

4 Non-Full-Rank Case

We have been assuming until now that, when regressing $\mathbf{y}$ on X, the matrix X is of full rank. In many instances, however, it turns out that X is not of full rank, and in this chapter, and Chapter 5 as well, we explore the intendent analysis for the case X not of full rank. We shall see that the principles implied by Chapter 3 hold fast and will serve us well.

4.1. X NOT OF FULL RANK

In many experimental situations where the model

$$\mathbf{y} = X\boldsymbol{\beta} + \boldsymbol{\varepsilon}, \quad E(\boldsymbol{\varepsilon}) = \mathbf{0}, \quad V(\boldsymbol{\varepsilon}) = \sigma^2 I \tag{4.1}$$

is appropriate, the $N \times k$ matrix X $(N > k)$ is not of full rank, that is, the rank $r(X)$ of X is such that

$$r(X) = p < k. \tag{4.2}$$

This of course means that the space spanned by the columns of X is of dimension $p < k$. A simple example is

$$X = \begin{pmatrix} 1 & 1 & 0 \\ 1 & 1 & 0 \\ 1 & 0 & 1 \\ 1 & 0 & 1 \end{pmatrix}, \tag{4.3}$$

so that

$$(4.3a) \qquad \mathbf{x}_1 = \mathbf{1} = \begin{pmatrix} 1 \\ 1 \\ 0 \\ 0 \end{pmatrix} + \begin{pmatrix} 0 \\ 0 \\ 1 \\ 1 \end{pmatrix} = \mathbf{x}_2 + \mathbf{x}_3.$$

Hence, a linear combination of the $\mathbf{x}_j$, $j = 1$, 2, and 3, is more simply a linear combination of $\mathbf{x}_2$ and $\mathbf{x}_3$, since

$$(4.3b) \qquad \begin{aligned} a_1\mathbf{x}_1 + a_2\mathbf{x}_2 + a_3\mathbf{x}_3 &= \mathbf{x}_2(a_1 + a_2) + \mathbf{x}_3(a_1 + a_3) \\ &= b_2\mathbf{x}_2 + b_3\mathbf{x}_3 \end{aligned}$$

with $b_2 = a_1 + a_2$ and $b_3 = a_1 + a_3$. The space spanned by $\mathbf{x}_1$, $\mathbf{x}_2$, and $\mathbf{x}_3$ is a two-dimensional subspace of R^4, and we would label it as S_2.

We note the fact that the 4×3 matrix X being of rank 2 (and not 3) could pose some interesting complications. For example, suppose that $\mathbf{y}$, a 4×1 random vector, is such that $E(\mathbf{y}) = X\boldsymbol{\beta} = \boldsymbol{\eta}$, with X as given in (4.3), and suppose for the moment we know $\boldsymbol{\eta}$, and in fact

$$(4.3c) \qquad \boldsymbol{\eta}' = (4, 4, 1, 1).$$

Then we note that there is an infinity of vectors $\boldsymbol{\beta}$ that weight $\mathbf{x}_1$, $\mathbf{x}_2$, and $\mathbf{x}_3$ to produce the above $\boldsymbol{\eta}$, for example,

$$(4.3d) \qquad \boldsymbol{\beta}^{(1)} = (3, 1, -2),$$

or

$$(4.3e) \qquad \boldsymbol{\beta}^{(2)} = (2, 2, -1),$$

and so on. Because of this, it is hardly surprising that solutions to the normal equations, derived for the case where X is not of full rank, are not unique (see below). Indeed, the purpose of the experiment that leads to such X's must be carefully borne in mind. For example, the interest of the experimenter might not be in estimating $\boldsymbol{\beta}$, but in making inference about some function of the β_j's, such as $\beta_2 - \beta_1, \beta_3 - \beta_1, \ldots,$ and so on. This and related points will be discussed subsequently in this chapter and Chapter 5.

Now given the space S_p, we are in exactly the same position as in the full-rank situation of the previous chapter, since we seek a point in this space that best estimates the point $\boldsymbol{\eta}$, where $\boldsymbol{\eta} \in S_p$ and is defined by

$$\boldsymbol{\eta} = E(\mathbf{y}) = X\boldsymbol{\beta}. \tag{4.4}$$

Once again, we choose the orthogonal projection of $\mathbf{y}$ into S_p, the subspace generated by the span of the column vectors of X, $\mathbf{x}_j$, $j = 1,\ldots,k$.

As in Chapter 3 then, we are seeking the point $\hat{\boldsymbol{\eta}} \in S_p$ (see Figure 4.1.1) which is such that $(\mathbf{y} - \hat{\boldsymbol{\eta}})$ is orthogonal to the vectors $\mathbf{x}_j$, that is,

$$X'(\mathbf{y} - \hat{\boldsymbol{\eta}}) = \mathbf{0}. \tag{4.5}$$

[Compare with Section 3.1 and Eq. (3.18) in particular.] Since $\hat{\boldsymbol{\eta}}$ is a point in S_p, it is a linear combination of the $\mathbf{x}_j$, and so

$$\hat{\boldsymbol{\eta}} = X\hat{\boldsymbol{\beta}}^0. \tag{4.5a}$$

Equation (4.5) implies that

$$X'X\hat{\boldsymbol{\beta}}^0 = X'\mathbf{y}. \tag{4.5b}$$

That is, $\hat{\boldsymbol{\beta}}^0$ is a solution of the normal equations. However, X and $X'X$ are of rank $p < k$, so that $X'X$ does not have an inverse. Does (4.5b) have a solution, and if so, what are its properties? At this point, we need the following lemma.

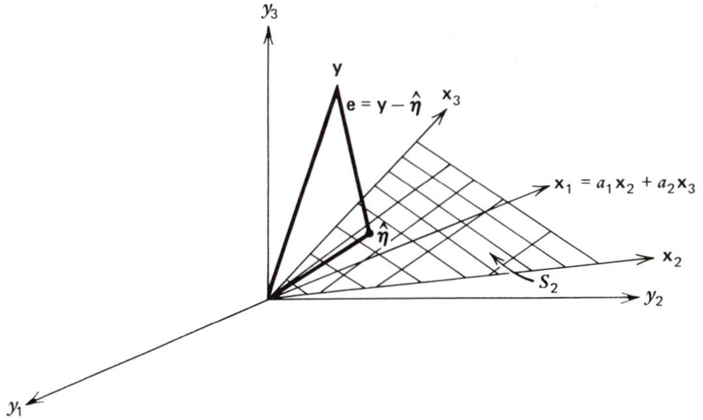

FIGURE 4.1.1. Projection of the vector $\mathbf{y}$ into S_2, where $E(\mathbf{y}) = X\boldsymbol{\beta}$, with $X = (\mathbf{x}_1 \mid \mathbf{x}_2 \mid \mathbf{x}_3)$ of rank 2.

Lemma 4.1.1. *Consider the system of equations*

$$Bz = t. \tag{4.6}$$

The system (4.6) *has a solution if and only if the rank r of B equals the rank of the (column) augmented matrix* $(B \mid t)$, *that is,*

$$r(B) = r(B \mid t). \tag{4.6a}$$

For a proof, see Graybill (1969, pp. 140–141). Now we may use Lemma 4.1.1 to see if the equations (4.5*b*) possess a solution. We have

$$\text{(i)} \qquad r[(X'X \mid X'y)] \geq r(X'X)$$

because we can't diminish the rank by adding a column, and

$$\text{(ii)} \qquad r[(X'X \mid X'y)] = r[X'(X \mid y)] \leq r(X') = r(X'X),$$

since in general, $r(AB) \leq \min[r(A), r(B)]$, and $r(X) = r(X') = r(X'X)$. Using (i) and (ii) above we have

$$r(X'X) \leq r[(X'X \mid X'y)] \leq r(X'X) \tag{4.7}$$

or

$$r(X'X \mid X'y) = r(X'X). \tag{4.7a}$$

Therefore the system (4.5*b*) always has a solution, and we can now say that *the equations* (4.5*b*) *are consistent*. [We are invoking a theorem of linear algebra that says linear equations can be solved if and only if they are consistent—for a proof see Searle (1966, pp. 138–139).] We turn now to the following important theorem.

Theorem 4.1.1. *Consistent equations*

$$Av = u, \tag{4.8}$$

where A is of order $r \times s$, v *is* $s \times 1$, *and* u *is* $r \times 1$, *have a solution* $v = Gu$, *if and only if*

$$AGA = A \tag{4.8a}$$

where G is of order $s \times r$ *and independent of* u.

The proof is given in Appendix A4.1. We note that if indeed $\mathbf{v} = G\mathbf{u}$ is a solution to (4.8) then it must be that

$$AG\mathbf{u} = \mathbf{u}. \tag{4.8b}$$

Furthermore, we remark that the solution $\mathbf{v} = G\mathbf{u}$ is not unique if $GA \neq I_s$, since

$$\mathbf{v}_1 = \mathbf{v} + (GA - I_s)\mathbf{w}, \qquad \mathbf{v} = G\mathbf{u}, \tag{4.8c}$$

satisfies (4.8) for any choice of a $s \times 1$ vector $\mathbf{w}$. This is so because the left-hand side of (4.8) takes the value

$$A\mathbf{v}_1 = A\mathbf{v} + (AGA - A)\mathbf{w} = AG\mathbf{u} + (AGA - A)\mathbf{w} \tag{4.9}$$

and from (4.8a) and (4.8b), we then have from (4.9) that

$$A\mathbf{v}_1 = \mathbf{u}, \tag{4.9a}$$

which is the right-hand side of (4.8).

Thus, referring back again to the normal equations (4.5a), which we know are consistent, we have on applying Theorem 4.1.1 to (4.5a), that a (nonunique) solution of (4.5a) is of the form

$$\hat{\boldsymbol{\beta}}^0 = GX'\mathbf{y} \tag{4.10}$$

if and only if

$$(X'X)G(X'X) = X'X. \tag{4.11}$$

A matrix G satisfying (4.11) is said to be a *generalized inverse* of the matrix $X'X$, and we digress at this point to give a short discussion about generalized inverses in general, and include a discussion of the nature of the solutions (4.10).

4.2. GENERALIZED INVERSES AND USES

We begin with a definition of the *Moore–Penrose inverse*.

Definition 4.2.1. *If A is any $N \times k$ matrix, then the $k \times N$ matrix A^+ is said to be the Moore–Penrose inverse of A if it satisfies the following four* *conditions*:

$$(4.12) \qquad \begin{array}{ll} \text{(i)} \quad AA^+A = A & \text{(ii)} \quad A^+AA^+ = A^+ \\ \text{(iii)} \quad (A^+A)' = A^+A & \text{(iv)} \quad (AA^+)' = AA^+ \end{array}$$

that is, A^+A and AA^+ are symmetric.

We have the following theorems:

Theorem 4.2.1. *A^+ is unique.*

Theorem 4.2.2. *If A is of full rank, then $A^+ = (A'A)^{-1}A'$.*

Theorem 4.2.3. *(Singular Value Decomposition). The $N \times k$ matrix A, where $N > k$, may be written*

$$(4.13) \qquad A = UDV'$$

where the $N \times N$ orthogonal matrix U has columns that are eigenvectors of AA', the $k \times k$ orthogonal matrix V has columns that are eigenvectors of $A'A$, and the $N \times k$ matrix D is of the form:

$$(4.13a) \qquad D = \begin{bmatrix} \sqrt{\lambda_1} & & & & & \\ & \ddots & & & \bigcirc & \\ & & \sqrt{\lambda_r} & & & \\ & & & 0 & & \\ & \bigcirc & & & \ddots & \\ & & & & & 0 \\ \hline & & & \bigcirc & & \end{bmatrix};$$

that is, the upper part of D is a $k \times k$ diagonal matrix, with the ith diagonal element equal to $\sqrt{\lambda_i}$, $i = 1,\ldots,r$, with $r = r(A)$, and λ_j the eigenvalues of $A'A$, and the last $k - r$ diagonal elements zero, while the lower part of D is a $(N - k) \times k$ matrix of zeros.

Theorem 4.2.4. *The Moore–Penrose inverse A may be written*

$$(4.14) \qquad A^+ = VD^*U'$$

where V and U are as in Theorem 4.2.3, and D^* is of the form

$$(4.14a) \qquad D^* = \left[\begin{array}{cccccc|c} 1/\sqrt{\lambda_1} & & & & & & \\ & \ddots & & & & & \\ & & 1/\sqrt{\lambda_r} & & & & O' \\ & & & 0 & & & \\ & & & & \ddots & & \\ & & & & & 0 & \end{array}\right];$$

that is, the left-hand part of D^* is a $k \times k$ diagonal matrix with the ith diagonal element equal to $1/\sqrt{\lambda_i}$, $i = 1,\ldots,r$, with r and λ_j as in Theorem 4.2.3, *and the last $k - r$ diagonal elements zero, while the right-hand part of* (4.14*a*) *is a $k \times (N - k)$ matrix of zeros.*

A proof of Theorem 4.2.1 may be found in Searle (1971, pp. 16–17) while that of Theorem 4.2.2 is left as an exercise (Problem 4.1). Theorems 4.2.3 and 4.2.4 may be found in a paper by Golub (1969). (Given Theorem 4.2.3, we ask the reader to prove Theorem 4.2.4 in Problem 4.1.) We remark that programs based on (4.14) exist and are frequently used [see, e.g., Lapczak (1978)]. The λ_j, the eigenvalues of $A'A$, are often called the *singular values* of A. We also have the following:

Definition 4.2.2. *A matrix G that satisfies only condition (i) of* (4.12), *that is,*

$$(4.15) \qquad AGA = A,$$

is said to be a generalized inverse of A, or a g_1 inverse of A, and denoted by A^- or A_1^-.

Given any matrix A, a quick way to construct a g_1 inverse of A is given in Problem 4.18. However, the g_1 inverse of A as constructed using the method of this problem is not unique. For example, the matrix given in (4.14) may be used; it is a Moore–Penrose inverse of A and automatically is a g_1 inverse of A.

We summarize by stating the following important theorem:

Theorem 4.2.5. *For any matrix A, there exists a matrix G that satisfies* (4.15), *but is not necessarily unique.*

A proof of this theorem is given in Rao and Mitra (1971) and Searle (1971).

We introduce some new notation here by example: B is a g_{14} inverse of A if it satisfies (4.12), (i) and (iv), that is,

$$ABA = A \quad and \quad (AB)' = AB \tag{4.16}$$

and we sometimes denote B by A_{14}^{-}. We note that any B satisfying (4.16) is automatically a g_1 inverse of A. Note also that $A^{+} = A_{1234}^{-}$, that is, the Moore–Penrose inverse is the g_{1234} inverse. We have the following important theorem.

Theorem 4.2.6. *Let $H = AA_1^-$ where A is of order $p \times q$, with $r(A) = r$, and the $q \times p$ matrix A_1^- is a g_1 inverse of A. (H is of course $p \times p$.)*

Then, H is idempotent of rank r, and $I_p - H$ is idempotent of rank $p - r$.

Proof. $H^2 = AA_1^- AA_1^- = (AA_1^- A)A_1^- = AA_1^- = H$ so that H is idempotent. We now show that

$$r(H) = r(AA_1^-) = r(A) = r. \tag{4.17}$$

Since $HA = AA_1^- A = A$, we have that

$$r(A) = r(HA) \le \min(r(H), r(A)) \le r(H). \tag{4.17a}$$

And, of course, $H = AA_1^-$, so that

$$r(H) \le \min(r(A), r(A_1^-)) \le r(A), \tag{4.17b}$$

that is, from (4.17*a*) and (4.17*b*) we have

$$r(A) \le r(H) \le r(A) \tag{4.17c}$$

or

$$r(H) = r(A) = r. \tag{4.17d}$$

Now $I_p - H$ is idempotent, since

$$(I_p - H)(I_p - H) = I_p - H - H + HH = I_p - H \tag{4.18}$$

and this also implies that

$$r(I_p - H) = p - r(H) = p - r. \tag{4.19}$$

It can be shown in a similar manner that $\tilde{H} = A_1^- A$ is idempotent, $r(\tilde{H}) = r(A) = r$, and that $I_q - \tilde{H}$ is idempotent of rank $q - r$. [See Problem 4.1(c).]

We connect the above theorems with the normal equations by the following useful theorem.

Theorem 4.2.7. *A necessary and sufficient condition for*

$$\hat{\boldsymbol{\beta}}^0 = Z\mathbf{y} \tag{4.20}$$

to be a solution of the normal equations (4.5*b*) *is that*

$$Z = X_{14}^-, \tag{4.21}$$

where X_{14}^- *satisfies conditions* (i) *and* (iv) *of Definition* 4.2.1.

Proof. (i) *Necessity.* Assume $\hat{\boldsymbol{\beta}}^0 = Z\mathbf{y}$ is a solution to equations (4.5*b*). We wish to prove that $Z = X_{14}^-$. We have that

$$X'XZ\mathbf{y} = X'\mathbf{y} \tag{4.22}$$

for all $\mathbf{y}$, so that

$$X'XZ = X'. \tag{4.23}$$

Transposing and then pre-multiplying by X',

$$X'Z'X'X = X'X. \tag{4.24}$$

Using the matrix Lemma A4.2.2 of Appendix A4.2, which states that if

$$PX'X = QX'X, \quad \text{then} \quad PX' = QX', \tag{4.25}$$

we have from (4.24) that

$$X'Z'X' = X' \quad \text{or} \quad XZX = X, \tag{4.26}$$

which is to say that Z is a g_1 inverse of X. Again using (4.23), we find on pre-multiplying by Z' that

$$Z'X'XZ = Z'X' = (XZ)' \tag{4.27}$$

or, transposing,

$$XZ = Z'X'XZ \tag{4.28}$$

so that

$$XZ = (XZ)'. \tag{4.29}$$

From (4.26) and (4.29), Z is a g_{14} inverse of X, that is, $Z = X_{14}^{-}$, and the necessity part is proved. [Note from (4.28) and (4.29) that XZ *is symmetric and idempotent*, that is, if $\hat{\boldsymbol{\beta}}^0 = Z\mathbf{y}$ is a solution to $X'X\hat{\boldsymbol{\beta}}^0 = X'\mathbf{y}$, then XZ is the matrix of an orthogonal projection.]

(ii) *Sufficiency*. Here we assume that $Z = X_{14}^{-}$. We wish to prove that

$$\hat{\boldsymbol{\beta}}^0 = X_{14}^{-}\mathbf{y} \tag{4.30}$$

is a solution of (4.5b). Now the left-hand side (LHS) of (4.5a) is

$$\text{LHS} = X'X\hat{\boldsymbol{\beta}}^0 = X'XX_{14}^{-}\mathbf{y} = X'(XX_{14}^{-})'\mathbf{y} \tag{4.31}$$

since, in particular, X_{14}^{-} satisfies condition (iv) of (4.12). Hence

$$\text{LHS} = [(XX_{14}^{-})X]'\mathbf{y} = [XX_{14}^{-}X]'\mathbf{y} = X'\mathbf{y} \tag{4.31a}$$

since X_{14}^{-} also satisfies condition (i) of (4.12). Thus,

$$\text{LHS} = X'\mathbf{y} = \text{RHS} \tag{4.31b}$$

and the sufficiency part is proved, and the theorem is now proven.

We remark that, in view of Theorem 4.2.7, many authors label g_{14} inverses of X as X_{ls}^{-}, and X_{ls}^{-} is referred to as a *least-squares (generalized) inverse* of X. We will use this theorem often; see, for example, Theorem 4.6.1.

Now of course when X is not of full rank, the consistent system (4.5b) has many solutions. We know that the set (4.5b) is consistent, from (4.7a), so Theorem 4.1.1 tells us that there is a solution, and the remark after shows that this solution is not unique. Since this solution is not unique, we do not call $\hat{\boldsymbol{\beta}}^0$ an estimate of $\boldsymbol{\beta}$. Indeed, as we will see below, the expectation of $\hat{\boldsymbol{\beta}}^0$ depends on which g_{14} inverse of X is chosen.

Now, although $\hat{\boldsymbol{\beta}}^0$ is not unique, the vector $\hat{\boldsymbol{\eta}}$, where

$$\hat{\boldsymbol{\eta}} = X\hat{\boldsymbol{\beta}}^0 = XX_{14}^{-}\mathbf{y}, \tag{4.32}$$

is unique, because XX_{14}^{-} is a projection matrix—see Figure 4.2.1. Directly, we have by definition that XX_{14}^{-} is symmetric and also

$$XX_{14}^{-}XX_{14}^{-} = (XX_{14}^{-}X)X_{14}^{-} = XX_{14}^{-}, \tag{4.32a}$$

so that XX_{14}^{-} is idempotent and hence a projection matrix; and/or we may also apply Theorem 4.2.6, or consult the necessity part of the proof of Theorem 4.2.7—see the remark made just after the conclusion of the proof of the necessity part of that theorem. Now, of course, $X\hat{\boldsymbol{\beta}}^0 = XX_{14}^{-}\mathbf{y} = X\mathbf{t}$ —that is, a linear combination of the columns of X—so that indeed $X\hat{\boldsymbol{\beta}}^0 = XX_{14}^{-}\mathbf{y}$ is a projection of $\mathbf{y}$ into S_p.

Now since the symmetric matrix XX_{14}^{-} is idempotent and hence a projection matrix, and since the projection of $\mathbf{y}$ into S_p is unique, we have that

$$XX_{14}^{-}\mathbf{y} = X\hat{\boldsymbol{\beta}}^0 = \hat{\boldsymbol{\eta}} \tag{4.32b}$$

is unique, and, as we will see, $\hat{\boldsymbol{\eta}}$ is unbiased for $\boldsymbol{\eta} = X\boldsymbol{\beta}$. [The reader is referred to the discussion in Chapter 3, Section 3.1, where, of course, geometrically, we are performing the equivalent operation of projecting $\mathbf{y}$ into a subspace by means of $\mathcal{R}(X) = X[(X'X)^{-1}X']$, and indeed, in the full-rank case $(X'X)^{-1}X'$ is a g_{14} inverse of X (in fact, it is a Moore–Penrose inverse of X).]

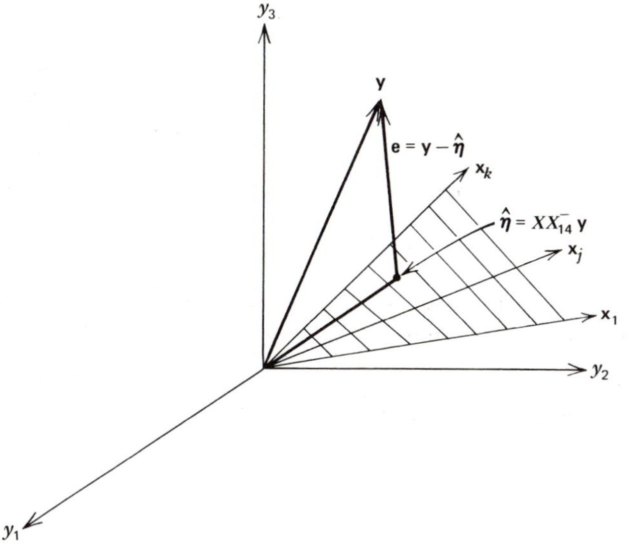

FIGURE 4.2.1. Projection of $\mathbf{y}$ into $S_p(p < k)$ at $\hat{\boldsymbol{\eta}} = X\hat{\boldsymbol{\beta}}^0 = XX_{14}^{-}\mathbf{y}$. We note that, using the Pythagoras theorem, $\mathbf{y}'\mathbf{y} = \hat{\boldsymbol{\eta}}'\hat{\boldsymbol{\eta}} + \mathbf{e}'\mathbf{e}$.

In summary then, we have used *geometrical considerations* to find the normal equations (4.5*b*). Applying Theorem 4.1.1 to these equations, we find that $\hat{\boldsymbol{\beta}}^0 = GX'\mathbf{y}$ is a solution to these equations if and only if G is a g_1 inverse of $X'X$. So from Theorem 4.2.7, we have established that

$$GX' = X_{14}^{-}; \tag{4.33}$$

in words, if G is a g_1 inverse of $X'X$, then GX' is a g_{14} inverse of X. In turn, (4.33) implies, of course, that

$$XGX' = XX_{14}^{-}, \tag{4.34}$$

that is, since XX_{14}^{-} is idempotent, XGX' is a (unique) projection matrix that projects $\mathbf{y}$ into S_p. (See also the remark at the end of the proof of the necessity part of Theorem 4.2.7.) These and other results may be proved without geometry, using only algebraic considerations, and we do so now in the following theorem.

Theorem 4.2.8. *When G is a generalized inverse of $X'X$, then*

(i) *G' is also a generalized inverse of $X'X$,*
(ii) *GX' is a generalized inverse of X,*
(iii) *XGX' is invariant to G,*
(iv) *XGX' is symmetric (whether or not G is).*

We remark, before proving the theorem, that (ii) states that $GX' = X^{-}$ and (iii) and (iv) together state that $XGX' = XX_{14}^{-}$ is unique: We know the latter is true since, in general, XX_{14}^{-} is symmetric and idempotent and hence a projection matrix, and, again from (iv), we have that $XGX' = XX^{-}$ is symmetric, so that the matrix GX' is a g_{14} inverse, and we may then write

$$GX' = X_{14}^{-}. \tag{4.35}$$

This in turn implies, from Theorem 4.2.7, that $\hat{\boldsymbol{\beta}}^0 = GX'\mathbf{y} = X_{14}^{-}\mathbf{y}$ [see (4.10)] is a solution to the normal equations (4.5*b*).

Proof of Theorem 4.2.8. [Due to Searle (1971)]. Since G is a g_1 inverse of $X'X$, we have by definition that

$$X'XGX'X = X'X. \tag{4.36}$$

Transposing yields

$$X'XG'X'X = X'X, \tag{4.37}$$

that is, G' is also a g_1 inverse of X, proving part (i). Applying Lemma A4.2.2 of Appendix A4.2 yields

$$X'XG'X' = X' \tag{4.38}$$

or, transposing,

$$XGX'X = X, \tag{4.39}$$

that is, GX' is a g_1 inverse of X, proving (ii).

Suppose now F is some other generalized inverse of $X'X$, different from G. Then (ii) gives

$$XGX'X = XFX'X, \tag{4.40}$$

and the use again of Lemma A4.2.2 of Appendix A4.2 yields

$$XGX' = XFX', \tag{4.40a}$$

that is, XGX' is the same for all generalized inverses of $X'X$, proving (iii). Now suppose S is a symmetric generalized inverse of $X'X$. [It is easy to see from (i) that an S always exists: Simply use $S = \frac{1}{2}(G + G')$, and so on.] Since S is symmetric, XSX' is symmetric (the transpose of XSX' is clearly XSX'). But from (iii)

$$XSX' = XGX', \tag{4.41}$$

so that XGX' is symmetric, proving (iv).

Corollary 4.2.8.1. *Using the fact that G' is a generalized inverse of $X'X$ if $G = (X'X)_1^-$, we may prove*

(4.42)

(i) $XG'X'X = X$; (ii) $X'XGX' = X'$

(iii) $X'XG'X = X'$; (iv) $XG'X' = XGX'$

(v) $XG'X'$ *is symmetric.*

The proof is straightforward and left to the reader (see Problem 4.2).

We now state a theorem that is very useful in practice (see, e.g., Section 4.5, where an analysis of variance problem is discussed) when constructing g_1 inverses of $X'X$. We refer the reader to Appendix A4.3 for a proof.

Theorem 4.2.9. *If X is of order $N \times k$ and of rank $r = k - q$, $0 < k - q < k < N$, and B is of order $q \times k$, whose q rows are linearly independent of the rows of X, and B is of full row rank, and if the so-called augmented matrix A, where A is of order $(N + q) \times k$, is such that*

$$A = \begin{pmatrix} X \\ -- \\ B \end{pmatrix}, \tag{4.43}$$

then $A'A$ possesses an inverse, and indeed

$$(A'A)^{-1} = (X'X + B'B)^{-1} = (X'X)_1^- = (B'B)_1^- . \tag{4.44}$$

That is, $(A'A)^{-1}$ is a g_1 inverse of $X'X$, and a g_1 inverse of $B'B$. Furthermore,

$$X(X'X + B'B)^{-1}B' = \text{O}_{N \times q}. \tag{4.44a}$$

We remark here that Theorem 4.2.9 gives us at this point a procedure for analyzing the non-full-rank situation. From Theorem 4.2.9, we know that since $G = (X'X + B'B)^{-1}$ is a g_1 inverse of $X'X$, then

$$GX' = (X'X + B'B)^{-1}X' = X_{14}^-, \tag{4.44b}$$

so that $\hat{\boldsymbol{\eta}} = XX_{14}^- \mathbf{y}$, from which we may find the regression sum of squares SS_r $(= \hat{\boldsymbol{\eta}}'\hat{\boldsymbol{\eta}})$ due to $\boldsymbol{\beta}$, and the error sum of squares $SS_e = \mathbf{y}'\mathbf{y} - SS_r$. Thus we need only find a matrix B, of order $q \times k$, whose q rows are linearly independent of the rows of X, and so on. We will use this and other methods for the non-full-rank case in subsequent sections of this chapter and Chapter 5.

4.3. SOME PROPERTIES OF THE SOLUTIONS: REGRESSION AND ERROR SUM OF SQUARES—THE ERROR AND SOLUTION SPACES REVISITED

Since the process we have been describing in this chapter so far is that of projection of a sample point $\mathbf{y}'$ in R^N into a linear subspace of dimension $p < k$, labeled here as S_p, it is to be anticipated that many of the properties and entities that evolve are similar to the properties and entities that we examined in Chapter 3. Of course, this is brought about because the process used in Chapter 3 is also that of projection of $\mathbf{y}'$ into a linear subspace. Indeed we have the following as a first property [compare (3.26)]:

PROPERTY 1. If $\hat{\boldsymbol{\eta}}$ is chosen by the orthogonal projection procedure of this chapter, that is, $\hat{\boldsymbol{\eta}}$ is as given in (4.32*b*), then this choice *minimizes the squared distance between* $\mathbf{y}$ *and points in the solution surface* S_p. Therefore,

$$\min_{\boldsymbol{\tau}\in S_p}(\mathbf{y}-\boldsymbol{\tau})'(\mathbf{y}-\boldsymbol{\tau})=(\mathbf{y}-\hat{\boldsymbol{\eta}})'(\mathbf{y}-\hat{\boldsymbol{\eta}}). \tag{4.45}$$

The algebraic proof parallels that in Chapter 3 and is left as an exercise (Problem 4.3). Note the correspondence of the properties listed here with the properties enumerated in Chapter 3, Section 3.1.

At this point we need some notation, terminology, and algebraic identities.

Terminology

$$1.\quad \hat{\boldsymbol{\beta}}^0 = X_{14}^-\mathbf{y} \tag{4.46a}$$

is referred to as a *least-squares solution* to the normal equations (4.5*a*).

$$2.\quad \hat{\boldsymbol{\eta}} = X\hat{\boldsymbol{\beta}}^0 = XX_{14}^-\mathbf{y} = \mathcal{R}^{(\mathrm{nf})}(X)\mathbf{y}, \qquad \mathcal{R}^{(\mathrm{nf})}(X) = XX_{14}^- \tag{4.46b}$$

is called the vector of fitted values, or *fitted value* for short, and $\hat{\boldsymbol{\eta}}$ is of course the projection of $\mathbf{y}$ into S_p, projected by use of the projection matrix $\mathcal{R}^{(\mathrm{nf})}(X)=\mathcal{R}^{(\mathrm{nf})}$ (nf stands for non-full), that is, $\hat{\boldsymbol{\eta}}\in S_p$.

$$3.\quad \mathbf{e} = (\mathbf{y}-\hat{\boldsymbol{\eta}}) = \left[I_N - \mathcal{R}^{(\mathrm{nf})}\right]\mathbf{y} = \mathcal{M}^{(\mathrm{nf})}\mathbf{y} \tag{4.46c}$$

with

$$\mathcal{M}^{(\mathrm{nf})} = \left(I_N - \mathcal{R}^{(\mathrm{nf})}\right) = \left(I_N - XX_{14}^-\right), \tag{4.46d}$$

is said to be the *residual vector*. The sum of squares

$$\mathbf{e}'\mathbf{e} = (\mathbf{y}-\hat{\boldsymbol{\eta}})'(\mathbf{y}-\hat{\boldsymbol{\eta}}) = \mathrm{SS}_e \tag{4.46e}$$

is called the *residual sum of squares* and sometimes the *error sum of squares*. SS_e is of course the squared length of $\mathbf{e}$. The vector $\mathbf{e}$ is said to belong to the *error space*, which is the set of vectors orthogonal to the columns of X, so that

$$X'\mathbf{a} = \mathbf{0} \tag{4.46f}$$

—see Figure 4.1.1; or **a** belongs to the *error space* if it is orthogonal to vectors lying in the *estimation space* or *solution surface* S_p. The error space is of dimension $N - p$ [see (4.47c)].

$$\text{4.} \quad \hat{\boldsymbol{\eta}}'\hat{\boldsymbol{\eta}} = \hat{\boldsymbol{\beta}}^{0\prime}X'X\hat{\boldsymbol{\beta}}^0 = \mathbf{y}'X_{14}^{-\prime}X'XX_{14}^{-}\mathbf{y} = \text{SS}_r \tag{4.46g}$$

is the squared length of the *fitted value* or *regression* vector and is called the *regression sum of squares*. Problem 4.4 shows that we may write (4.46g) as

$$\hat{\boldsymbol{\eta}}'\hat{\boldsymbol{\eta}} = \mathbf{y}'\mathcal{R}^{(\text{nf})}\mathbf{y} = \text{SS}_r. \tag{4.46h}$$

Algebraic Facts

1. $\mathcal{R}^{(\text{nf})}$ and $\mathcal{M}^{(\text{nf})}$ are symmetric and idempotent, that is,

$$\mathcal{R}^{(\text{nf})\prime} = \mathcal{R}^{(\text{nf})}, \qquad \mathcal{R}^{(\text{nf})}\mathcal{R}^{(\text{nf})} = \mathcal{R}^{(\text{nf})} \tag{4.47a}$$

and similarly

$$\mathcal{M}^{(\text{nf})\prime} = \mathcal{M}^{(\text{nf})}, \qquad \mathcal{M}^{(\text{nf})}\mathcal{M}^{(\text{nf})} = \mathcal{M}^{(\text{nf})}. \tag{4.47b}$$

Using Theorem 4.2.6, this means that

$$r(\mathcal{R}^{(\text{nf})}) = \text{tr}(\mathcal{R}^{(\text{nf})}) = p; \qquad r(\mathcal{M}^{(\text{nf})}) = \text{tr}(\mathcal{M}^{(\text{nf})}) = N - p, \tag{4.47c}$$

where $p = r(X)$. The results (4.47a), (4.47b), and (4.47c) are easily proved (see Problem 4.5).

The next series of algebraic facts are anticipated by recalling that the error space is orthogonal to the solution space.

2. We have

$$\mathcal{R}^{(\text{nf})}X = X, \qquad X'\mathcal{R}^{(\text{nf})} = X' \tag{4.48a}$$

so that

$$X'(I_N - \mathcal{R}^{(\text{nf})}) = \bigcirc, \tag{4.48b}$$

or

$$(I_N - \mathcal{R}^{(\text{nf})})'X = (I_N - \mathcal{R}^{(\text{nf})})X = \bigcirc.$$

The results [(4.48a) and (4.48b)] are also easily proven (see Problem 4.5).

We note that $\mathbf{e}$ belongs to the error space since

$$X'\mathbf{e} = \mathbf{0}_k; \qquad \mathbf{e}'X = \mathbf{0}'_k. \tag{4.48c}$$

Some Assumptions: $\mathbf{y} = X\boldsymbol{\beta} + \boldsymbol{\varepsilon}$

$AE(\mathbf{y})$	1.	$E(\mathbf{y}) = \boldsymbol{\eta} = X\boldsymbol{\beta}$ or $E(\boldsymbol{\varepsilon}) = \mathbf{0}$, $\boldsymbol{\varepsilon} = \mathbf{y} - X\boldsymbol{\beta}$.
$AV(\mathbf{y})$	2.	$V(\mathbf{y}) = \sigma^2 I_N$.
$AD\mathbf{y}$	3.	$\mathbf{y} = N(\boldsymbol{\eta}, \sigma^2 I_N)$.
$Ar(X)$	4.	X is of order $N \times k$ with $r(X) = p < k < N$.

We now are able to list additional properties of our least-squares procedure when X is not of full rank.

PROPERTY 2. For any X_{14}^-, we have, if $AE(\mathbf{y})$ holds, that

$$E(\hat{\boldsymbol{\beta}}^0) = E(X_{14}^- \mathbf{y}) = X_{14}^- X\boldsymbol{\beta} \neq \boldsymbol{\beta}, \tag{4.49}$$

that is, $\hat{\boldsymbol{\beta}}^0$ is *not unbiased* for $\boldsymbol{\beta}$ and indeed its expectation depends on which g_{14} matrix is chosen for the solution $\hat{\boldsymbol{\beta}}^0$.

The above raises the question as to whether, when $AE(\mathbf{y})$ holds, with X of rank $p < k$, we may ever find an estimator which is linear in the y_j's and unbiased for $\boldsymbol{\beta}$. *The answer is that we cannot*: Suppose there were an estimator $L\mathbf{y}$, where L is of order $k \times N$. Then

$$E(L\mathbf{y}) = \boldsymbol{\beta} \tag{4.49a}$$

implies that

$$LX\boldsymbol{\beta} = \boldsymbol{\beta} \tag{4.49b}$$

for any value that $\boldsymbol{\beta}$ may be, so that

$$LX = I_k. \tag{4.49c}$$

Hence it would have to be the case that

$$r(LX) = r(I_k) = k. \tag{4.49d}$$

But the rank of LX is such that

$$r(LX) \leq \min\{r(L), r(X)\} \leq p < k, \tag{4.49e}$$

providing a contradiction to (4.49d), so that there does not exist a linear unbiased estimator for $\boldsymbol{\beta}$ in the non-full-rank case. (However, see Theorem 4.6.1.)

PROPERTY 3. For any X_{14}^{-}, we have, if $AE(\mathbf{y})$ holds, that

$$(4.50) \qquad E(\hat{\boldsymbol{\eta}}) = E(X\hat{\boldsymbol{\beta}}^0) = E(XX_{14}^{-}\mathbf{y}) = XX_{14}^{-}X\boldsymbol{\beta} = X\boldsymbol{\beta} = \boldsymbol{\eta},$$

that is, $\hat{\boldsymbol{\eta}}$ is unbiased for $\boldsymbol{\eta}$.

PROPERTY 4. The residual $\mathbf{e}$ reflects only error, because if $AE(\mathbf{y})$ holds, then

$$(4.51) \qquad E(\mathbf{e}) = E(M^{(\mathrm{nf})}\mathbf{y}) = M^{(\mathrm{nf})}X\boldsymbol{\beta} = \mathbf{0},$$

that is, $\mathbf{e}$ is unbiased for $\mathbf{0}$.

PROPERTY 5. If $AV(\mathbf{y})$ holds, then

$$\text{(i)} \quad V(\hat{\boldsymbol{\beta}}^0) = \sigma^2 X_{14}^{-}(X_{14}^{-})';$$

$$\text{(ii)} \quad V(\hat{\boldsymbol{\eta}}) = \sigma^2 \mathfrak{R}^{(\mathrm{nf})};$$

$$\text{(iii)} \quad V(\mathbf{e}) = \sigma^2 \mathfrak{M}^{(\mathrm{nf})}.$$

4.4. ESTIMABLE FUNCTIONS: THE GAUSS THEOREM FOR THE NON-FULL-RANK CASE

If dealing with the model (4.1) when the condition (4.2) holds, that is, X is not of full rank, then certain functions of $\boldsymbol{\beta}$ that may be of interest to the experimenter may or may not be *estimable* in the sense of the following definition.

Definition 4.4.1. *We suppose that* (4.1) *and* (4.2) *hold. Then, the (linear) function* $\mathbf{q}'\boldsymbol{\beta}$ *is said to be estimable if there exists a vector* $\mathbf{t}'$ *such that* $\mathbf{t}'\mathbf{y}$ *is unbiased for* $\mathbf{q}'\boldsymbol{\beta}$, *that is, if*

$$(4.52) \qquad E(\mathbf{t}'\mathbf{y}) = \mathbf{q}'\boldsymbol{\beta}.$$

We make the following remarks:

(a) If $\mathbf{q}'\boldsymbol{\beta}$ is estimable, we have from (4.52) and (4.1) that

$$(4.53) \qquad \mathbf{t}'X\boldsymbol{\beta} = \mathbf{q}'\boldsymbol{\beta}$$

no matter what the value of $\boldsymbol{\beta}$ may be, so that a consequence of estimability is

$$\mathbf{t}'X = \mathbf{q}'. \tag{4.53a}$$

(b) If $\hat{\boldsymbol{\beta}}^0$ is any solution to the normal equations (4.5a), and if $\mathbf{q}'\boldsymbol{\beta}$ is estimable, then $\mathbf{q}'\hat{\boldsymbol{\beta}}^0$ is invariant to whatever solution $\hat{\boldsymbol{\beta}}^0$ is used and is unbiased for $\mathbf{q}'\boldsymbol{\beta}$.

The invariance follows from the fact that $\hat{\boldsymbol{\beta}}^0 = X_{14}^- \mathbf{y}$ and hence

$$\mathbf{q}'\hat{\boldsymbol{\beta}}^0 = \mathbf{t}'XX_{14}^- \mathbf{y} \tag{4.54}$$

and XX_{14}^- is unique, as discussed before. Also, taking expectations in (4.54), we find

$$E(\mathbf{q}'\hat{\boldsymbol{\beta}}^0) = \mathbf{t}'XX_{14}^- X\boldsymbol{\beta} = \mathbf{t}'X\boldsymbol{\beta} = \mathbf{q}'\boldsymbol{\beta}, \tag{4.54a}$$

that is, $\mathbf{q}'\hat{\boldsymbol{\beta}}^0$ is unbiased for $\mathbf{q}'\boldsymbol{\beta}$, when $\mathbf{q}'\boldsymbol{\beta}$ is estimable. We note from (4.54) that $\mathbf{q}'\hat{\boldsymbol{\beta}}^0$ is linear in the y_j's. We refer to $\mathbf{q}'\hat{\boldsymbol{\beta}}^0$ as the *least-squares estimator* of $\mathbf{q}'\boldsymbol{\beta}$.

(c) The variance of $\mathbf{q}'\hat{\boldsymbol{\beta}}^0$ is

$$V(\mathbf{q}'\hat{\boldsymbol{\beta}}^0) = \sigma^2\mathbf{t}'XX_{14}^-\mathbf{t}. \tag{4.55}$$

This may be seen as follows. We have, using (4.54) and (4.53a), that

$$\begin{aligned} V(\mathbf{q}'\hat{\boldsymbol{\beta}}^0) &= V(\mathbf{q}'X_{14}^- \mathbf{y}) = V(\mathbf{t}'XX_{14}^- \mathbf{y}) \\ &= \sigma^2\mathbf{t}'XX_{14}^-(XX_{14}^-)'\mathbf{t}. \end{aligned} \tag{4.55a}$$

But X_{14}^- is a g_{14} inverse, so that XX_{14}^- is symmetric and $XX_{14}^- X = X$, which yields

$$V(\mathbf{q}'\hat{\boldsymbol{\beta}}^0) = \sigma^2\mathbf{t}'XX_{14}^-\mathbf{t}. \tag{4.55b}$$

(d) Suppose $\mathbf{k}'\mathbf{y}$ is another unbiased estimator of $\mathbf{q}'\boldsymbol{\beta}$, different from $\mathbf{q}'\hat{\boldsymbol{\beta}}^0$, that is, $\mathbf{k}' \neq \mathbf{q}'X_{14}^- = \mathbf{t}'XX_{14}^-$; then

$$\operatorname{cov}(\mathbf{q}'\hat{\boldsymbol{\beta}}^0, \mathbf{k}'\mathbf{y}) = V(\mathbf{q}'\hat{\boldsymbol{\beta}}^0). \tag{4.56}$$

To see this, we have first that, from the property of unbiasedness,

$$E(\mathbf{k}'\mathbf{y}) = \mathbf{k}'X\boldsymbol{\beta} = \mathbf{q}'\boldsymbol{\beta} \tag{4.56a}$$

for any value that $\boldsymbol{\beta}$ may be, so that

$$\mathbf{k}'X = \mathbf{q}'. \tag{4.56b}$$

Furthermore, we have that

$$\begin{aligned} \operatorname{cov}(\mathbf{q}'\hat{\boldsymbol{\beta}}^0, \mathbf{k}'\mathbf{y}) &= \operatorname{cov}(\mathbf{q}'X_{14}^- \mathbf{y}, \mathbf{k}'\mathbf{y}) \\ &= \sigma^2 \mathbf{q}'X_{14}^- \mathbf{k}. \end{aligned} \tag{4.56c}$$

Using (4.53a) and the symmetry of XX_{14}^-, we find

$$\begin{aligned} \operatorname{cov}(\mathbf{q}'\hat{\boldsymbol{\beta}}^0, \mathbf{k}'\mathbf{y}) &= \sigma^2 \mathbf{t}'(X_{14}^-)'X'\mathbf{k} \\ &= \sigma^2 \mathbf{t}'(X_{14}^-)'\mathbf{q}, \end{aligned} \tag{4.56d}$$

from (4.56b). Using (4.53a) again, and the symmetry of XX_{14}^-, we obtain

$$\operatorname{cov}(\mathbf{q}'\hat{\boldsymbol{\beta}}^0, \mathbf{k}'\mathbf{y}) = \sigma^2 \mathbf{t}'XX_{14}^-\mathbf{t} \tag{4.56e}$$

which from (4.55b) establishes (4.56).

(e) We now state the analogue of the Gauss theorem (Theorem 3.1.1).

Theorem 4.4.1. *If* (4.1) *and* (4.2) *hold, and if* $\mathbf{q}'\boldsymbol{\beta}$ *is estimable, then among all linear (in the* y_i*'s) unbiased estimators of* $\mathbf{q}'\boldsymbol{\beta}$*, the least-squares estimator* $\mathbf{q}'\hat{\boldsymbol{\beta}}^0$ *has minimum variance, where* $\hat{\boldsymbol{\beta}}^0$ *is a solution to the normal equations* (4.5b).

Proof. Suppose that $\mathbf{k}'\mathbf{y}$ is another unbiased estimator of $\mathbf{q}'\boldsymbol{\beta}$, different from the (unbiased) estimator $\mathbf{q}'\hat{\boldsymbol{\beta}}^0$. Then

$$0 \le V(\mathbf{k}'\mathbf{y} - \mathbf{q}'\hat{\boldsymbol{\beta}}^0) = V(\mathbf{k}'\mathbf{y}) + V(\mathbf{q}'\hat{\boldsymbol{\beta}}^0) - 2\operatorname{cov}(\mathbf{q}'\hat{\boldsymbol{\beta}}^0, \mathbf{k}'\mathbf{y}), \tag{4.57}$$

and from (4.56) we have

$$V(\mathbf{k}'\mathbf{y}) - V(\mathbf{q}'\hat{\boldsymbol{\beta}}^0) \ge 0 \tag{4.57a}$$

with equality if and only if $\mathbf{k}'\mathbf{y} = \mathbf{q}'X_{14}^- \mathbf{y} = \mathbf{t}'XX_{14}^- \mathbf{y}$, ruled out by assumption. Hence

$$V(\mathbf{q}'\hat{\boldsymbol{\beta}}^0) < V(\mathbf{k}'\mathbf{y}) \tag{4.57b}$$

and the theorem is proved.

As indicated in the previous sections [e.g., see (4.35)], we very often employ for a solution to the normal equations

$$\hat{\boldsymbol{\beta}}^0 = GX'\mathbf{y}, \qquad G = (X'X)_1^- \tag{4.58}$$

since GX' is a g_{14} inverse of X (see Theorem 4.2.8). Hence, when $\mathbf{q}'\boldsymbol{\beta}$ is estimable, the least-squares estimator has the form

$$\mathbf{q}'\hat{\boldsymbol{\beta}}^0 = \mathbf{q}'GX'\mathbf{y} = \mathbf{t}'XGX'\mathbf{y}, \tag{4.58a}$$

and using this, or proceeding from (4.55*b*), we easily find (Problem 4.6) that

$$V(\mathbf{q}'\hat{\boldsymbol{\beta}}^0) = \sigma^2\mathbf{q}'G\mathbf{q} \tag{4.58b}$$

[compare the full-rank case where $G = (X'X)^{-1}$, etc.]

(f) A condition for estimability is contained in the following theorem.

Theorem 4.4.2. *Suppose $E(\mathbf{y}) = X\boldsymbol{\beta}$, $V(\mathbf{y}) = \sigma^2 I$ with the $N \times k$ matrix X of rank $p < k < N$. A necessary and sufficient condition that $\mathbf{q}'\boldsymbol{\beta}$ be estimable is that*

$$\mathbf{q}'GX'X = \mathbf{q}' \tag{4.58c}$$

where G is any g_1 inverse of $X'X$.

Proof. Necessity. We assume for this part that $\mathbf{q}'\boldsymbol{\beta}$ is estimable, so that there exists $\mathbf{t}'$ such that

$$\mathbf{q}' = \mathbf{t}'X \tag{4.58d}$$

or

$$\mathbf{q}' = (\mathbf{t}'X)[GX']X, \tag{4.58e}$$

that is, from (4.58*d*) we have

$$\mathbf{q}' = \mathbf{q}'GX'X \tag{4.58f}$$

and the necessity part is now proved.

Sufficiency. Here we assume that

$$\mathbf{q}'GX'X = \mathbf{q}'. \tag{4.58g}$$

Consider the expectation of $\mathbf{q}'\hat{\boldsymbol{\beta}}^0$. We have

$$E(\mathbf{q}'\hat{\boldsymbol{\beta}}^0) = \mathbf{q}'E(X_{14}^{-}\mathbf{y}) \tag{4.58h}$$

or

$$\begin{aligned} E(\mathbf{q}'\hat{\boldsymbol{\beta}}^0) &= \mathbf{q}'X_{14}^{-}X\boldsymbol{\beta} \\ &= (\mathbf{q}'GX'X)\boldsymbol{\beta} \end{aligned} \tag{4.58i}$$

so that from (4.58g), we have

$$E(\mathbf{q}'\hat{\boldsymbol{\beta}}^0) = \mathbf{q}'\boldsymbol{\beta} \tag{4.58j}$$

and the sufficiency part is proved.

The above theorem then states that in order to establish whether $\mathbf{q}'\boldsymbol{\beta}$ is estimable, we need only determine whether condition (4.58c) holds. An example follows.

EXAMPLE 4.4.1.

Suppose $E(\mathbf{y}) = X\boldsymbol{\beta}$ with X given by

$$X = \begin{bmatrix} 1 & -1 \\ 1 & -1 \\ -1 & 1 \\ -1 & 1 \end{bmatrix}. \tag{4.58k}$$

Here, $r(X) = 1 < 2$, and it is easy to see, using Theorem 4.2.9, for example, that by augmenting X with $B = \mathbf{b} = (1, 1)$

$$G = \frac{1}{16}\begin{pmatrix} 5 & 3 \\ 3 & 5 \end{pmatrix}, \qquad GX'X = \frac{1}{2}\begin{pmatrix} 1 & -1 \\ -1 & 1 \end{pmatrix}. \tag{4.58l}$$

Suppose we wish to determine whether $\beta_1 - \beta_2$ is estimable. We have

$$\mathbf{q}'\boldsymbol{\beta} = (1, -1)\boldsymbol{\beta} = \beta_1 - \beta_2, \tag{4.58m}$$

that is, $\mathbf{q}' = (1, -1)$. Now, the left-hand side of (4.58c) of Theorem 4.4.2 is

$$\begin{aligned} \mathbf{q}'GX'X &= (1, -1)\left[\frac{1}{2}\begin{pmatrix} 1 & -1 \\ -1 & 1 \end{pmatrix}\right] \\ &= (1, -1) = \mathbf{q}' \end{aligned} \tag{4.58n}$$

so that, by Theorem 4.2.2, we have that $\beta_1 - \beta_2$ is estimable.

In this example, of course, we could have used the test for estimability implied by Definition 4.4.1 itself, since it is easy to see that, for Example 4.4.1,

$$\mathbf{t}'X = \mathbf{q}' = (1, -1) \tag{4.58o}$$

with $\mathbf{t}' = \frac{1}{4}(1, 1, -1, -1)$. However, when N is large, it is sometimes difficult to ascertain whether there does or does not exist a $\mathbf{t}'$ such that $\mathbf{t}'X = \mathbf{q}'$, and the condition for estimability given in Theorem 4.4.2 is easier to use.

In general, we say that the functions $Q\boldsymbol{\beta}$ are estimable (Q is a $s \times k$ matrix, $\boldsymbol{\beta}$ a $k \times 1$ vector, and so on) if each of the functions $\mathbf{q}'_j\boldsymbol{\beta}$ are estimable, where $\mathbf{q}'_j$ is the jth row of Q. This then implies that there exists a $s \times N$ matrix T such that $E(T\mathbf{y}) = Q\boldsymbol{\beta}$, or $TX = Q$ (see Problem 4.21).

We are interested in the estimability of a linear function, or linear functions of $\boldsymbol{\beta}$, because we often wish to examine the statement

$$H_0: B\boldsymbol{\beta} = \mathbf{b} \tag{4.58p}$$

where B is a $s \times k$ matrix, $s \leq k$, with $r(B) = s$. The statement (4.58p) is said to be a *testable hypothesis* if $B\boldsymbol{\beta}$ is estimable. [The consequences of $B\boldsymbol{\beta}$ not being estimable when we wish to test (4.58p) is explored in Section 5.4.] Hence, when dealing with a linear model of non-full rank, and if a statement of the form (4.58p) is of concern, we must first establish whether H_0 is testable or not, that is, whether $B\boldsymbol{\beta}$ is estimable, and Theorem 4.4.2 [and its generalization given in Problem 4.21(a)] is helpful in this regard.

4.5. DETAILED LOOK AT THE REGRESSION AND RESIDUAL SUM OF SQUARES

This section parallels Section 3.2 of Chapter 3, and the reader should expect and suspect that the results obtained in this section will be very similar to those of Section 3.2. This is so because this chapter utilizes the least-squares procedure of orthogonally projecting $\mathbf{y}$ into a linear subspace defined by the columns of X, just as was done in Chapter 3. The projection here is obtained by using the (symmetric and idempotent) projection matrix $\mathcal{R}^{(\mathrm{nf})} = XX_{14}^{-}$: The projection in Chapter 3 is obtained using $\mathcal{R} = X(X'X)^{-1}X'$, and indeed $(X'X)^{-1}X'$ is a g_{14} inverse of X when the rank of X is of full rank k.

To begin with, we have that the residual sum of squares, which is the squared length of the residual vector $\mathbf{e}$ defined at (4.46c), is

$$\mathrm{SS}_e = \mathbf{e}'\mathbf{e} = (\mathbf{y} - \hat{\boldsymbol{\eta}})'(\mathbf{y} - \hat{\boldsymbol{\eta}}) = (\mathbf{y} - X\hat{\boldsymbol{\beta}}^0)'(\mathbf{y} - X\hat{\boldsymbol{\beta}}^0) \tag{4.59}$$

and it is very easy to see that

$$\text{(4.59}a\text{)} \qquad \mathrm{SS}_e = \mathbf{e}'\mathbf{e} = \mathbf{y}'\mathbf{y} - \hat{\boldsymbol{\beta}}^0 X'\mathbf{y} = \mathbf{y}'\mathbf{y} - \hat{\boldsymbol{\beta}}^{0\prime} X'X\hat{\boldsymbol{\beta}}^0$$

[compare (3.46)]. Furthermore, the regression sum of squares SS_r, the squared length of $\hat{\boldsymbol{\eta}} = X\hat{\boldsymbol{\beta}}^0$, is

$$\text{(4.59}b\text{)} \qquad \mathrm{SS}_r = \hat{\boldsymbol{\eta}}'\hat{\boldsymbol{\eta}} = \hat{\boldsymbol{\beta}}^{0\prime} X'\mathbf{y} = \hat{\boldsymbol{\beta}}^{0\prime} X'X\hat{\boldsymbol{\beta}}^0,$$

and we may write (see Problem 4.7)

$$\text{(4.59}c\text{)} \qquad \mathrm{SS}_r = \mathbf{y}'XX_{14}^- \mathbf{y} = \mathbf{y}'\mathcal{R}^{(\mathrm{nf})}\mathbf{y},$$

where $\mathcal{R}^{(\mathrm{nf})} = XX_{14}^-$.

In fact, the results of Section 4.3 establish that when $AE(\mathbf{y})$ holds, that is, when $E(\mathbf{y}) = X\boldsymbol{\beta}$, $r(X) = p < k < N$, then $\hat{\boldsymbol{\eta}}$ is orthogonal to $\mathbf{e}$ (Problem 4.8). Hence the Pythagoras theorem can be invoked and we have (see Figure 4.5.1)

$$\text{(4.60)} \qquad \mathbf{y}'\mathbf{y} = \mathbf{e}'\mathbf{e} + \hat{\boldsymbol{\eta}}'\hat{\boldsymbol{\eta}}.$$

We recall that $\mathbf{e} = (I - \mathcal{R}^{(\mathrm{nf})})\mathbf{y} = \mathcal{M}^{(\mathrm{nf})}\mathbf{y}$, where $\mathcal{M}^{(\mathrm{nf})}$ is symmetric and idempotent, so that we may now write (and see Figure 4.5.1)

$$\text{(4.60}a\text{)} \qquad \begin{aligned} \mathbf{y}'\mathbf{y} &= \mathbf{y}'(I - \mathcal{R}^{(\mathrm{nf})})\mathbf{y} + \mathbf{y}'\mathcal{R}^{(\mathrm{nf})}\mathbf{y} \\ &= \qquad \mathrm{SS}_e \qquad + \quad \mathrm{SS}_r, \end{aligned}$$

that is, we have the usual analysis of variance breakdown of the total sum of squares into a part SS_r, due to regression ($X\boldsymbol{\beta}$ is the model), and a part due to error, SS_e. It is easy to see (Problem 4.9) that, if $AE(\mathbf{y})$ and $AV(\mathbf{y})$ hold,

(4.60b)

(i) $E(\mathrm{SS}_r) = p\sigma^2 + \boldsymbol{\beta}'X'X\boldsymbol{\beta} = p\sigma^2 + \boldsymbol{\eta}'\boldsymbol{\eta}$

(ii) $E(\mathrm{SS}_e) = (N - p)\sigma^2$

(iii) SS_r and SS_e are independent if, in addition, $AD(\mathbf{y})$ holds.

We may indeed at this point construct an analysis of variance table similar to that of Table 3.2.1, with the only change being, of course, that the regression degrees of freedom are p, since $\hat{\boldsymbol{\eta}}$ lies in S_p (and not a space S_k); or put another way, we are projecting $\mathbf{y}$ into a subspace S_p since X is not of

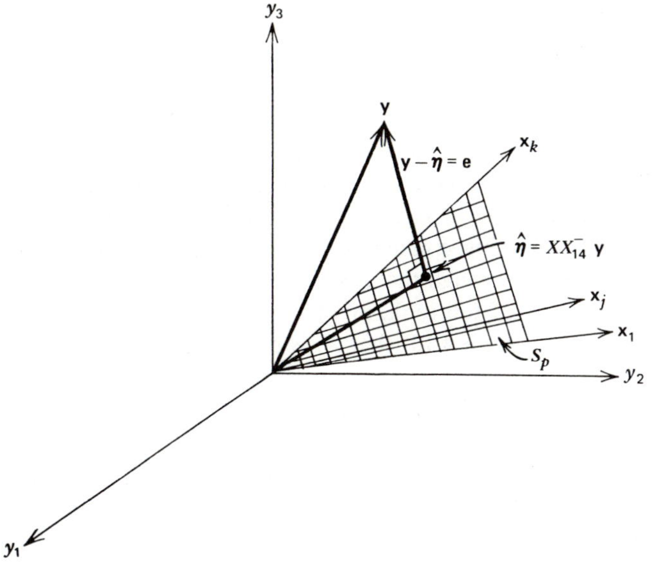

FIGURE 4.5.1. Pythagoras and the analysis of variance breakdown (4.60) and (4.60*a*); $|\mathbf{y}|^2 = |\hat{\boldsymbol{\eta}}|^2 + |\mathbf{e}|^2$ or $\mathbf{y}'\mathbf{y} = \hat{\boldsymbol{\eta}}'\hat{\boldsymbol{\eta}} + \mathbf{e}'\mathbf{e}$.

full rank—indeed $r(X) = p < k$. This of course means there are $N - p$ degrees of freedom for error. We note that the statement "H_0: $\boldsymbol{\eta} = \mathbf{0}$" is a testable hypothesis, because $X\boldsymbol{\beta}$ is estimable. Hence in view of the three results given in (4.60*b*), it is evident that a test of H_0 of significance level α is

(4.60*c*)
$$\text{Reject "}\boldsymbol{\eta} = \mathbf{0}\text{" if } F = \frac{\mathrm{SS}_r/p}{\mathrm{SS}_e/(N-p)} > F_{p,\,N-p;\,\alpha}$$

Accept otherwise.

Note too that $\boldsymbol{\eta} = \mathbf{0}$ does *not* imply that $\boldsymbol{\beta} = \mathbf{0}$, since X is not of full rank [see the discussion of Property 2 in Section 4.3, and also Problem 4.19(c) for an example].

Since we are basically involved with the same geometric operations, all the methods of Chapter 3 carry over. For example, suppose interest lies in the estimable function $\boldsymbol{\tau} = B_2\boldsymbol{\beta}_2$, where

$$E(\mathbf{y}) = X_1\boldsymbol{\beta}_1 + X_2\boldsymbol{\beta}_2 \tag{4.61}$$

and where $X = (X_1 \mid X_2)$ is not of full rank. Here $\boldsymbol{\beta}_1$ is of nuisance status.

We may orthogonalize and eliminate $X_1\boldsymbol{\beta}_1$ in the same manner as in Chapter 3. Defining

$$(4.62) \qquad X_{2\cdot 1} = [I - X_1(X_1)_{14}^{-}]X_2 = X_2 - X_1 A^{(\mathrm{nf})},$$

where $A^{(\mathrm{nf})} = (X_1)_{14}^{-} X_2$ may be called an alias matrix; we thus have the analogue of the *orthogonalization* process of Chapter 3. That is to say, $X_{2\cdot 1}$ is obtained by regressing columns of X_2 on X_1, and taking residuals, and this is accomplished through the use of the projection matrix $X_1(X_1)_{14}^{-}$. [X_1 is often, in practice, of full rank, in which case $X_1(X_1)_{14}^{-} = X_1(X_1'X_1)^{-1}X_1'$.] It is easy to see (Problem 4.10) that X_1 is orthogonal to $X_{2\cdot 1}$, that is,

$$(4.62a) \qquad X_1'X_{2\cdot 1} = \mathrm{O} \quad \text{or} \quad X_{2\cdot 1}'X_1 = \mathrm{O}'.$$

It is also straightforward to see that we may isolate and examine the regression sum of squares due to $\boldsymbol{\beta}_2$, given that X_1 is in the model: One has to be careful with the ranks of X_1 and $X_{2\cdot 1}$ because it may be that X_1 is of full rank, $X_{2\cdot 1}$ of non-full rank, or vice versa, or X_1 and $X_{2\cdot 1}$ both of non-full rank, and so on. (See Example 4.5.1 for a case where X_1 is of full rank, but $X_{2\cdot 1}$ is of non-full rank.) Problem 4.11 asks the reader to construct the analysis of variance table for this procedure for the general case. (The reader is again cautioned that the statement of interest, $B_2\boldsymbol{\beta}_2 = \mathbf{b}_2$, must be a testable hypothesis; see Section 5.4 for discussion.)

The key to the above again lies in the fact that nonsingular transformations on the X matrix leave $\hat{\boldsymbol{\eta}}$ invariant, which means that $\mathrm{SS}_r = \hat{\boldsymbol{\eta}}'\hat{\boldsymbol{\eta}}$ and $\mathrm{SS}_e = \mathbf{y}'\mathbf{y} - \mathrm{SS}_r$ are left invariant under nonsingular transformations on X. *This is obvious geometrically*, since a transformation of the form XH, where H is of order $k \times k$ and nonsingular, is merely substituting one set of basis vectors for S_p with another set of basis vectors, so that this type of transformation means that we are still projecting into the same subspace S_p (compare with Section 3.3 of Chapter 3).

Algebraically, we may demonstrate this by rewriting $AE(\mathbf{y})$ as

$$(4.63) \qquad E(\mathbf{y}) = XHH^{-1}\boldsymbol{\beta} = Z\boldsymbol{\phi}, \quad Z = XH, \quad \boldsymbol{\phi} = H^{-1}\boldsymbol{\beta},$$

where H is of order $k \times k$ and nonsingular. Now Z is of order $N \times k$, but its rank is non-full, since clearly $r(Z) = r(XH) = r(X) = p < k$.

Suppose now $\hat{\boldsymbol{\eta}}_x$ is the projection of $\mathbf{y}$ found by regressing $\mathbf{y}$ on X; that is, we use the model $E(\mathbf{y}) = X\boldsymbol{\beta}$. The normal equations are

$$(4.64) \qquad X'X\hat{\boldsymbol{\beta}}^0 = X'\mathbf{y}$$

and we then find

(4.64a) $$\hat{\boldsymbol{\eta}}_x = X\hat{\boldsymbol{\beta}}^0.$$

Note that pre-multiplying by X' in (4.64a), yields, from (4.64),

(4.64b) $$X'\hat{\boldsymbol{\eta}}_x = X'X\hat{\boldsymbol{\beta}}^0 = X'\mathbf{y}.$$

Suppose we now "use" the model (4.63), that is, $E(\mathbf{y}) = Z\boldsymbol{\phi}$ where Z and $\boldsymbol{\phi}$ are as specified in (4.63). Let $\hat{\boldsymbol{\eta}}_z$ be the projection of $\mathbf{y}$ found by regressing $\mathbf{y}$ on Z: The normal equations are

(4.65) $$Z'Z\hat{\boldsymbol{\phi}}^0 = Z'\mathbf{y}.$$

But $Z = XH$, so (4.65) may be written

(4.65a) $$H'X'XH\hat{\boldsymbol{\phi}}^0 = H'X'\mathbf{y}$$

with H, or H', nonsingular, so that

(4.65b) $$X'XH\hat{\boldsymbol{\phi}}^0 = X'\mathbf{y}.$$

Furthermore,

(4.65c) $$\hat{\boldsymbol{\eta}}_z = Z\hat{\boldsymbol{\phi}}^0 = XH\hat{\boldsymbol{\phi}}^0$$

so that

(4.65d) $$Z'\hat{\boldsymbol{\eta}}_z = Z'Z\hat{\boldsymbol{\phi}}^0 = Z'\mathbf{y}$$

from (4.65). We note that, using (4.64a) and (4.65c),

(4.66) $$\hat{\boldsymbol{\eta}}_x - \hat{\boldsymbol{\eta}}_z = X\hat{\boldsymbol{\beta}}^0 - XH\hat{\boldsymbol{\phi}}^0 = X(\hat{\boldsymbol{\beta}}^0 - H\hat{\boldsymbol{\phi}}^0),$$

and we know that

(4.66a) $$(\hat{\boldsymbol{\eta}}_x - \hat{\boldsymbol{\eta}}_z)'(\hat{\boldsymbol{\eta}}_x - \hat{\boldsymbol{\eta}}_z) \geq 0$$

with equality if and only if $\hat{\boldsymbol{\eta}}_x = \hat{\boldsymbol{\eta}}_z$. But the left-hand side of (4.66a) is, using (4.66),

(4.66b) $$\begin{aligned}(\hat{\boldsymbol{\eta}}_x - \hat{\boldsymbol{\eta}}_z)'(\hat{\boldsymbol{\eta}}_x - \hat{\boldsymbol{\eta}}_z) &= (\hat{\boldsymbol{\beta}}^0 - H\hat{\boldsymbol{\phi}}^0)'X'(X\hat{\boldsymbol{\beta}}^0 - XH\hat{\boldsymbol{\phi}}^0)\\ &= (\hat{\boldsymbol{\beta}}^0 - H\hat{\boldsymbol{\phi}}^0)'(X'X\hat{\boldsymbol{\beta}}^0 - X'XH\hat{\boldsymbol{\phi}}^0)\end{aligned}$$

and from (4.64) and (4.65*b*) we have

$$(4.66c)\quad (\hat{\boldsymbol{\eta}}_x - \hat{\boldsymbol{\eta}}_z)'(\hat{\boldsymbol{\eta}}_x - \hat{\boldsymbol{\eta}}_z) = (\hat{\boldsymbol{\beta}}^0 - H\boldsymbol{\phi}^0)'(X'\mathbf{y} - X'\mathbf{y}) = 0.$$

Hence $\hat{\boldsymbol{\eta}}_x - \hat{\boldsymbol{\eta}}_z = \mathbf{0}$; that is, *the projection is left invariant under nonsingular transformations* $Z = XH$.

It is interesting to note that to effect the utilization of $X_{2\cdot1}$, that is, to consider (4.61) rewritten as

$$(4.67)\qquad E(\mathbf{y}) = X_1\boldsymbol{\phi} + X_{2\cdot1}\boldsymbol{\beta}_2,$$

we have

$$(4.67a)\qquad E(\mathbf{y}) = (XH)(H^{-1}\boldsymbol{\beta}), \qquad X = [X_1 \mid X_2]$$

and, recalling (4.62),

$$(4.67b)\qquad H = \left[\begin{array}{c|c} I_{k_1} & -A^{(\mathrm{nf})} \\ \hline O & I_{k_2} \end{array}\right], \qquad H^{-1} = \left[\begin{array}{c|c} I_{k_1} & A^{(\mathrm{nf})} \\ \hline O & I_{k_2} \end{array}\right].$$

We would now eliminate X_1 by regressing $\mathbf{y}$ on X_1, taking residuals $\mathbf{e}_1$, regressing these on $X_{2\cdot1}$, and so on (compare Section 3.3). We illustrate all the above with the following example.

Example 4.5.1.

For this example, we suppose that two observations are generated independently using each of two different treatments and the response (y) measured. We postulate the model

$$(4.68)\qquad \begin{aligned} &E(y_{ij}) = \mu + \beta_j, \qquad i = 1,2; \quad j = 1,2 \\ &V(y_{ij}) \equiv \sigma^2, \qquad \operatorname{cov}(y_{ij}, y_{rs}) = 0. \end{aligned}$$

Let $\mathbf{y} = (y_{11}, y_{21}, y_{12}, y_{22})'$. Then denote sample means as follows:

$$(4.69)\qquad \bar{y} = \frac{(\mathbf{y}'\mathbf{1})}{4} = \frac{(\bar{y}_{\cdot1} + \bar{y}_{\cdot2})}{2},$$

where

$$(4.69a)\qquad \bar{y}_{\cdot j} = \frac{y_{1j} + y_{2j}}{2}, \qquad j = 1,2.$$

Note that

$$(4.69b)\qquad (\bar{y}_{\cdot 1} - \bar{y}) + (\bar{y}_{\cdot 2} - \bar{y}) = 0.$$

In matrix form, our model may be written

$$(4.70)\qquad E(\mathbf{y}) = X\boldsymbol{\beta} = \begin{bmatrix} 1 & 1 & 0 \\ 1 & 1 & 0 \\ 1 & 0 & 1 \\ 1 & 0 & 1 \end{bmatrix} \begin{bmatrix} \mu \\ \hline \beta_1 \\ \beta_2 \end{bmatrix},$$

and it is usually the case in this simple one-way analysis of variance situation (two groups, two observations each) that μ is of nuisance value, and the real interest lies in various statements concerning the effects of the two treatments, such as $\beta_1 = \beta_2 = 0$ (no effect), and so on.

Indeed, we note that if $\beta_1 = \beta_2 = 0$, μ represents an overall mean, and so we are in the classical situation of wishing to eliminate the mean and finding the (extra) regression sum of squares due to the two treatments (i.e., due to β_1 and β_2) and the relevant error term. Using the preceding prescriptions, we have ($\mathbf{1}_4$ denotes a 4×1 vector of ones)

$$(4.71)\qquad E(\mathbf{y}) = (X_1 \mid X_2)\begin{bmatrix} \mu \\ \hline \boldsymbol{\beta} \end{bmatrix} = \left[\begin{array}{c|cc} & 1 & 0 \\ \mathbf{1}_4 & 1 & 0 \\ & 0 & 1 \\ & 0 & 1 \end{array}\right]\begin{bmatrix} \mu \\ \hline \boldsymbol{\beta} \end{bmatrix}$$

$$= \mathbf{1}_4\phi + X_{2\cdot 1}\boldsymbol{\beta},$$

where

$$(4.71a)\qquad \phi = \mu + \tfrac{1}{4}\mathbf{1}_4' X_2\boldsymbol{\beta}, \qquad X_{2\cdot 1} = \left(I_4 - \tfrac{1}{4}\mathbf{1}_4\mathbf{1}_4'\right)X_2.$$

We recall (see previous work of this section and Section 3.3) that the mean is eliminated by regressing $\mathbf{y}$ on $X_1 = \mathbf{1}_4$ and taking residuals—$\mathbf{e}_1$. That is, we pretend we are dealing with the model

$$(4.72)\qquad E(\mathbf{y}) = \mathbf{1}_4\phi$$

and quickly find that

$$(4.73)\qquad \mathbf{e}_1 = \mathbf{y} - \bar{y}\mathbf{1}_4.$$

To get at the regression sum of squares due to $\boldsymbol{\beta} = (\beta_1, \beta_2)'$, given that μ is in the model, we now pretend that our model is

$$E(\mathbf{e}_1) = X_{2\cdot 1}\boldsymbol{\beta} \tag{4.74}$$

which is to say that we manipulate as if $\mathbf{e}_1$ is the dependent variable, $X_{2\cdot 1}$ the "X matrix", and so on. From (4.71a), we have that

$$X_{2\cdot 1} = \frac{1}{2}\begin{bmatrix} 1 & -1 \\ 1 & -1 \\ -1 & 1 \\ -1 & 1 \end{bmatrix}. \tag{4.75}$$

We note that (4.74) and (4.75) imply that

$$\begin{aligned} E(\mathbf{e}_1) &= \tfrac{1}{2}(\beta_1 - \beta_2, \beta_1 - \beta_2, -(\beta_1 - \beta_2), -(\beta_1 - \beta_2))' \\ &= \tfrac{1}{2}(\beta_1 - \beta_2)(1, 1, -1, -1)'. \end{aligned} \tag{4.75a}$$

Furthermore, it is clear that $r(X_{2\cdot 1}) = 1$, that is, $X_{2\cdot 1}$ is not of full rank. Hence, we will need a g_{14} inverse of $X_{2\cdot 1}$. Such a matrix is provided by

$$(X_{2\cdot 1})_{14}^{-} = \frac{1}{4}\begin{bmatrix} 1 & 1 & -1 & -1 \\ -1 & -1 & 1 & 1 \end{bmatrix} = \tfrac{1}{2}X'_{2\cdot 1}. \tag{4.75b}$$

That (4.75b) is indeed a g_{14} inverse of $X_{2\cdot 1}$ is easily verified (Problem 4.12). It was obtained using Theorem 4.2.9: Problem 4.13 gives details, and we use the same method and give all details in, for example, the discussion of the one-way analysis of variance in Chapter 5. Now from the form of the regression sum of squares given in (4.59c), and recalling that we are using (4.74), we have from (4.75b) that the (extra) regression sum of squares $SS_r(\beta_1, \beta_2)$ due to (β_1, β_2), given that μ is in the model, is

$$\begin{aligned} SS_r(\beta_1, \beta_2) &= \mathbf{e}'_1 X_{2\cdot 1}(X_{2\cdot 1})_{14}^{-}\mathbf{e}_1 \\ &= \tfrac{1}{2}\mathbf{e}'_1 X_{2\cdot 1}X'_{2\cdot 1}\mathbf{e}_1. \end{aligned} \tag{4.76}$$

But it is easy to verify that (Problem 4.14)

$$X'_{2\cdot 1}\mathbf{e}_1 = 2\begin{bmatrix} \bar{y}_{\cdot 1} - \bar{y} \\ \bar{y}_{\cdot 2} - \bar{y} \end{bmatrix}. \tag{4.76a}$$

Hence we may write (4.76) as

$$SS_r(\beta_1, \beta_2) = \sum_{j=1}^{2} 2(\bar{y}_{\cdot j} - \bar{y})^2 = SS_B. \tag{4.76b}$$

This quantity, usually denoted by SS_B, is often called in the literature the between sum of squares, and since the rank of $X_{2\cdot 1}$ is 1, the degrees of freedom allotted to it in this example is 1. Referring now to (4.60a) and recalling (4.74), we have that the residual sum of squares is

$$SS_e = \mathbf{e}_1'\mathbf{e}_1 - SS_r(\beta_1, \beta_2), \tag{4.77}$$

and from (4.73) we may write this as

$$\begin{aligned} SS_e &= \sum_{j=1}^{2}\sum_{i=1}^{2}(y_{ij}-\bar{y})^2 - \sum_{j=1}^{2} 2(\bar{y}_{\cdot j}-\bar{y})^2 \\ &= \sum_{j=1}^{2}\sum_{i=1}^{2}(y_{ij}-\bar{y})^2 - \sum_{j=1}^{2}\sum_{i=1}^{2}(\bar{y}_{\cdot j}-\bar{y})^2. \end{aligned} \tag{4.77a}$$

As is well known and easily derived, we may rewrite (4.77a) as (Problem 4.15)

$$SS_e = \sum_{j=1}^{2}\sum_{i=1}^{2}(y_{ij}-\bar{y}_{\cdot j})^2 = SS_W, \tag{4.77b}$$

usually referred to in the literature as the within sum of squares. The degrees of freedom associated with SS_W is easily seen to be 2 and may be arrived at by a variety of routes. Problem 4.16 asks for the verification using the trace of the idempotent matrix involved in the associated quadratic form for SS_W in the variables y_{ij}. But an easier argument is that we are working with (4.73) and (4.74), and the 4×1 vector $\mathbf{e}_1$ clearly lies in $(4-1) = 3$-dimensional space, since $\mathbf{1}'\mathbf{e}_1 = 0$. Hence the dimensionality of the "sample space" of the "dependent" variables e_{1ij} is 3; the estimation space has dimension $r(X_{2\cdot 1}) = 1$, so that the degrees of freedom for $SS_e = SS_W$ is $3 - 1$ or 2.

We may now enter the ingredients in an analysis of variance table given in Table 4.5.1. [Recall that at this stage we are working with the model (4.74) or (4.75a), $E(\mathbf{e}_1) = X_{2\cdot 1}\boldsymbol{\beta}$, where the 4×2 matrix $X_{2\cdot 1}$ is given by (4.75) and is of rank 1.]

If $AD(\mathbf{y})$ holds, that is,

$$y = X\begin{bmatrix} \mu \\ \hline \boldsymbol{\beta} \end{bmatrix} + \boldsymbol{\varepsilon}, \qquad \boldsymbol{\varepsilon} = N(\mathbf{0}, \sigma^2 I) \tag{4.78}$$

with X as before, then, under the condition $\beta_1 - \beta_2 = (1, -1)\boldsymbol{\beta} = 0$, it is

TABLE 4.5.1. Analysis of Variance for the Data Generated as in (4.68)

Source	Degrees of Freedom	Sum of Squares	Expected Sum of Squares
Due to effects β_1, β_2	1	SS_B	$\sigma^2 + \boldsymbol{\beta}' X'_{2\cdot 1} X_{2\cdot 1} \boldsymbol{\beta} = \sigma^2 + (\beta_1 - \beta_2)^2$
Residual (error)	2	SS_W	$2\sigma^2$
Total	3	$\Sigma\Sigma(y_{ij} - \bar{y})^2$	

easy to see that (Problem 4.15)

$$\frac{SS_B/1}{SS_W/2} = F_{1,2} \tag{4.79}$$

and this may be employed in the usual way to test "$\beta_1 - \beta_2 = 0$," and so on.

It is interesting to note that the total degrees of freedom is 4, yet Table 4.5.1 shows 3 as the total. This, of course, is due to the fact that in Table 4.5.1 a line corresponding to information about μ through ϕ [see (4.71a)] is omitted. Another explanation, as we have seen, is the fact that Table 4.5.1 tabulates the results of regressing $\mathbf{e}_1$ on $X_{2\cdot 1}$, and the dimension of the sample space for $\mathbf{e}_1$ is 3, and so on.

A word of caution: The above example works because it turns out that $\beta_1 - \beta_2$ is estimable, so that the hypothesis $\beta_1 - \beta_2 = 0$ is a testable hypothesis. The fact that $\beta_1 - \beta_2$ is estimable may be easily verified by applying Theorem 4.4.2 [see Problem 4.21(c)].

4.6. SOME REMARKS ABOUT CONSTRAINTS ON PARAMETERS

Frequently it happens that we deal with a general model that may be stated as

$$E(\mathbf{y}) = X\boldsymbol{\beta}, \quad r(X) = p < k \tag{4.80}$$

and where the β's are *subject to the q constraints*, $q + p = k$, given by

$$B\boldsymbol{\beta} = \boldsymbol{\delta} \tag{4.80a}$$

(B and $\boldsymbol{\delta}$ are of course given). That is, B is of order $q \times k$ and assumed to have rows that are linearly independent and linearly independent of the rows of X, so that, in particular, $r(B) = q$. It turns out that when (4.80) and (4.80a) apply, *an unbiased estimate of* $\boldsymbol{\beta}$ *exists and obeys the constraint* (4.80a). We first have the following theorem:

Theorem 4.6.1. *Referring to the normal equations* (4.5b), *and assuming that* (4.80) *and* (4.80a) *apply, then* $\hat{\boldsymbol{\beta}}_B^{(0)}$, $\hat{\boldsymbol{\beta}}_B^{(1)}$ *satisfy* (4.5b), *where*

$$\hat{\boldsymbol{\beta}}_B^{(1)} = \hat{\boldsymbol{\beta}}_B^{(0)} + \boldsymbol{\gamma} \tag{4.81}$$

with

$$\hat{\boldsymbol{\beta}}_B^{(0)} = (X'X + B'B)^{-1}X'\mathbf{y}, \qquad \boldsymbol{\gamma} = (X'X + B'B)^{-1}B'\boldsymbol{\delta}. \tag{4.81a}$$

Furthermore, $\hat{\boldsymbol{\beta}}_B^{(1)}$ *is unbiased for* $\boldsymbol{\beta}$ *if and only if*

$$B\boldsymbol{\beta} = \boldsymbol{\delta}. \tag{4.82}$$

Proof. We demonstrate below—see (4.91) and the following— that $\hat{\boldsymbol{\beta}}_B^{(0)}$ and $\hat{\boldsymbol{\beta}}_B^{(1)}$, as given by (4.81$a$) and (4.81), respectively, satisfy the normal equations (4.5b), and we discuss now the unbiasedness of $\hat{\boldsymbol{\beta}}_B^{(1)}$, and so on.

(i) *Sufficiency.* Assume that $B\boldsymbol{\beta} = \boldsymbol{\delta}$. Then using (4.81$a$) and (4.80), we have

$$\begin{aligned} E(\hat{\boldsymbol{\beta}}_B^{(1)}) &= (X'X + B'B)^{-1}X'X\boldsymbol{\beta} + \boldsymbol{\gamma} \\ &= (X'X + B'B)^{-1}[(X'X + B'B) - B'B]\boldsymbol{\beta} + \boldsymbol{\gamma} \\ &= \left[I - (X'X + B'B)^{-1}B'B\right]\boldsymbol{\beta} + \boldsymbol{\gamma} \\ &= \boldsymbol{\beta} - (X'X + B'B)^{-1}B'(B\boldsymbol{\beta}) + \boldsymbol{\gamma}. \end{aligned} \tag{4.83}$$

But from the sufficiency assumption, $B\boldsymbol{\beta} = \boldsymbol{\delta}$ so that

$$E\left[\hat{\boldsymbol{\beta}}_B^{(1)}\right] = \boldsymbol{\beta} - \boldsymbol{\gamma} + \boldsymbol{\gamma} = \boldsymbol{\beta} \tag{4.83a}$$

and the sufficiency is proved.

(ii) *Necessity.* Assume that $E(\hat{\boldsymbol{\beta}}_B^{(1)}) = \boldsymbol{\beta}$.This implies that

$$(X'X + B'B)^{-1}X'X\boldsymbol{\beta} + \boldsymbol{\gamma} = \boldsymbol{\beta} \tag{4.84}$$

or, pre-multiplying by $(X'X + B'B)$,

$$X'X\boldsymbol{\beta} + B'\boldsymbol{\delta} = (X'X + B'B)\boldsymbol{\beta} \tag{4.84a}$$

or

$$B'\boldsymbol{\delta} = B'B\boldsymbol{\beta}. \tag{4.84b}$$

Pre-multiplying by B and using the easily verified fact that B described below (4.80a) is such that BB' is of full rank, we find

$$BB'B\boldsymbol{\beta} = BB'\boldsymbol{\delta} \tag{4.85}$$

or

$$B\boldsymbol{\beta} = \boldsymbol{\delta}, \tag{4.85a}$$

and the necessity part is proved.

Corollary 4.6.1.1. *If* $E(\mathbf{y}) = X\boldsymbol{\beta}$, *and* $\boldsymbol{\delta} = \mathbf{0}$, *then*

$$\hat{\boldsymbol{\beta}}_B^{(0)} = (X'X + B'B)^{-1}X'\mathbf{y} \tag{4.86}$$

is unbiased for $\boldsymbol{\beta}$ *if and only if* $B\boldsymbol{\beta} = \mathbf{0}$.

The implication of the above is immediate: *if you have constraints of the form* (4.80a), use them to augment the X matrix, and use Theorem 4.2.9 which implies that

$$(X'X + B'B)^{-1}X' = X_{14}^{-}. \tag{4.87}$$

[Constraints of the form (4.80a) are often called identifiability constraints.] Now we also recall the result (4.44a) of Theorem 4.2.9, which, on transposing, gives

$$B(X'X + B'B)^{-1}X' = O. \tag{4.88}$$

This result is important. It can be used immediately to show that

$$B\hat{\boldsymbol{\beta}}_B^{(1)} = \boldsymbol{\delta}, \tag{4.89}$$

since the left-hand side of (4.89) is

$$(4.89a)\quad B\hat{\boldsymbol{\beta}}_B^{(1)} = B(X'X + B'B)^{-1}X'\mathbf{y} + B(X'X + B'B)^{-1}B'\boldsymbol{\delta},$$

and from (4.88) and (4.82) we have that

$$(4.89b)\qquad B\hat{\boldsymbol{\beta}}_B^{(1)} = B(X'X + B'B)^{-1}B'B\boldsymbol{\beta}.$$

Now from Theorem 4.2.9, we know that $(X'X + B'B)^{-1}$ is a g_1 inverse of $B'B$, so that, using Theorem 4.2.8,

$$(4.89c)\qquad (X'X + B'B)^{-1}B' = B_1^-,$$

that is, $(X'X + B'B)^{-1}B'$ is a g_1 inverse of B, and we have

$$(4.89d)\qquad B\left[(X'X + B'B)^{-1}B'\right]B = B.$$

Hence from (4.89b) and (4.89d) we have

$$(4.90)\qquad B\hat{\boldsymbol{\beta}}_B^{(1)} = B\boldsymbol{\beta}.$$

This in turn implies, using (4.82),

$$(4.90a)\qquad B\hat{\boldsymbol{\beta}}_B^{(1)} = \boldsymbol{\delta},$$

that is, the estimator (4.81) obeys the constraints (4.82). Note that if $\boldsymbol{\delta} = \mathbf{0}$, $\hat{\boldsymbol{\beta}}_B^{(1)} = \hat{\boldsymbol{\beta}}_B^{(0)}$ so that $B\hat{\boldsymbol{\beta}}_B^{(0)} = \mathbf{0}$, that is, $\hat{\boldsymbol{\beta}}_B^{(0)}$ obeys the constraints $B\boldsymbol{\beta} = \mathbf{0}$.

Another important point, as stated in Theorem 4.6.1, is that both $\hat{\boldsymbol{\beta}}_B^{(0)}$ and $\hat{\boldsymbol{\beta}}_B^{(1)}$ satisfy the normal equations.

Proof. (For the first part of Theorem 4.6.1.) We note from (4.87) that $\hat{\boldsymbol{\beta}}_B^{(0)}$, as given by (4.81$a$), is of the form

$$(4.91)\qquad \hat{\boldsymbol{\beta}}_B^{(0)} = X_{14}^-\mathbf{y}, \qquad X_{14}^- = (X'X + B'B)^{-1}X'.$$

By Theorem 4.2.7, we then have that $\hat{\boldsymbol{\beta}}_B^0$ satisfies the normal equations, that is,

$$(4.92)\qquad X'X\hat{\boldsymbol{\beta}}_B^{(0)} = X'\mathbf{y}.$$

Now to see that $\hat{\boldsymbol{\beta}}_B^{(1)}$ is also a solution, note that

$$X'X\hat{\boldsymbol{\beta}}_B^{(1)} = X'X(\hat{\boldsymbol{\beta}}_B^{(0)} + \boldsymbol{\gamma}) \tag{4.93}$$

$$= X'X\hat{\boldsymbol{\beta}}_B^{(0)} + X'X(X'X + B'B)^{-1}B'\boldsymbol{\delta},$$

and using (4.88) and (4.92) in (4.93) we have

$$X'X\hat{\boldsymbol{\beta}}_B^{(1)} = X'\mathbf{y}. \tag{4.93a}$$

A word of warning: Constraints should be kept in mind quite separately from the question of estimability. If we are interested in $\mathbf{q}'\boldsymbol{\beta}$ and ask whether or not it is estimable, then whether or not we have constraints of the form $B\boldsymbol{\beta} = \boldsymbol{\delta}$, we must still look for a vector $\mathbf{t}$ which is such that $\mathbf{t}'X = \mathbf{q}'$, and so on. If we can find such a $\mathbf{t}$, and if $B\boldsymbol{\beta} = \boldsymbol{\delta}$, then $\mathbf{q}'\boldsymbol{\beta}$ is subject to $B\boldsymbol{\beta} = \boldsymbol{\delta}$, merely altering the form of $\mathbf{q}'\boldsymbol{\beta}$.

For example, if there is a constraint ($q = 1$) on the β's which says that (suppose $k = 3$) $\beta_1 + \beta_2 + \beta_3 = 0$, and if X is such that $3\beta_1 + 2\beta_2 + \beta_3$ is estimable, then in view of the constraint, the function that is estimable is $3\beta_1 + 2\beta_2 + (-\beta_1 - \beta_2)$, or $2\beta_1 + \beta_2$, and so on.

Note too that (Problem 4.17)

$$E(X\hat{\boldsymbol{\beta}}_B^{(1)}) = X\boldsymbol{\beta} = \boldsymbol{\eta} \tag{4.94}$$

and that the regression sum of squares may be produced using one of our previous rules. To see the latter, we first remark that, since (4.91) provides a solution to the normal equations we know that SS_r, the regression sum of squares, may be determined as

$$\mathrm{SS}_r = \hat{\boldsymbol{\beta}}_B^{(0)\prime}X'\mathbf{y} = \mathbf{y}'(XX_{14}^-)'\mathbf{y} = \mathbf{y}'XX_{14}^-\mathbf{y}. \tag{4.95}$$

Similarly, it is easy to see that

$$\hat{\boldsymbol{\beta}}_B^{(1)\prime}X'\mathbf{y} = \mathrm{SS}_r = \mathbf{y}'XX_{14}^-\mathbf{y} \tag{4.96}$$

and the proof is left to the reader.

Finally, we note that when given constraints

$$B\boldsymbol{\beta} = \boldsymbol{\delta}, \tag{4.97}$$

this is, of course, equivalent to

$$D\boldsymbol{\beta} = \boldsymbol{\xi} \tag{4.97a}$$

where $D = CB$, $\boldsymbol{\xi} = C\boldsymbol{\delta}$, and C is any full rank $q \times q$ matrix. Hence, we could augment X with D and change the definition of $\hat{\boldsymbol{\beta}}_B^{(0)}$ and γ accordingly. This elementary fact is useful when trying to select an augmenting matrix that simplifies arithmetic—see Section 5.2 for an example.

In proceeding with the analysis of the non-full-rank situation, the work in this chapter provides two alternative methods: (i) impose identifiability constraints and proceed as in this section; or (ii) compute a g_{14} inverse of X and proceed as in Section 4.5.

There is a third method, one which we will explore in the next chapter, termed the "reduce to full rank" procedure. After doing the problems of this chapter, the reader is invited to pursue this in Chapter 5.

PROBLEMS

4.1. (a) Prove Theorem 4.2.4, given Theorem 4.2.3.

(b) Verify that if X is of order $N \times k$, where $r(X) = k < N$, then $A^+ = (X'X)^{-1}X'$ satisfies the four conditions of (4.12).

(c) Show that under the conditions of Theorem 4.2.6, $\tilde{H} = A_1^- A$ is idempotent and of rank $r(A) = r$, and that $I_q - \tilde{H}$ is idempotent of rank $q - r$.

4.2. (a) Prove Corollary 4.2.8.1.

(b) Suppose $X = [X_1 \mid X_2]$ is of order $N \times k$ with $r(X) = p < k$ where X_1 is of order $N \times k_1$, X_2 is $N \times k_2$, $k_1 + k_2 = k$. Consider $X'X$, and let $V = X_2'[I - \mathcal{R}^{(\text{nf})}(X_1)]X_2$, where $\mathcal{R}^{(\text{nf})}(X_1) = X_1(X_1'X_1)^- X_1'$. Take $X'X$ in partitioned form, that is,

$$X'X = \left(\begin{array}{c|c} X_1'X_1 & X_1'X_2 \\ \hline X_2'X_1 & X_2'X_2 \end{array}\right).$$

Then show that the matrix W, given by

$$W = \left(\begin{array}{c|c} (X_1'X_1)^- + (X_1'X_1)^- X_1'X_2V^- X_2'X_1(X_1'X_1)^- & -(X_1'X_1)^- X_1'X_2V^- \\ \hline -V^- X_2'X_1(X_1'X_1)^- & V^- \end{array}\right)$$

is a g_1 inverse of $X'X$, so that $(X'X)W(X'X) = X'X$. [Note that

this result is the analogue of Problem 2.27(c) for the matrix $X'X$ as given above.]
Hints: If $G = (U'U)^-$ for some matrix U, then from Theorem 4.2.8 and Corollary 4.2.8.1, we have $U = UGU'U$ and $U' = U'UGU'$.

(c) Show too that, if in W we replace $(X_1'X_1)^-$ by $(X_1'X_1)_{12}^-$ and V^- by V_{12}^-, the resulting matrix is a g_{12} inverse of $X'X$, that is, $W(X'X)W = W$.

4.3. For the case X of order $N \times k$ with $r(X) = p < k$, verify (4.45).

4.4. Deduce that the regression sum of squares for the non-full-rank case can be expressed as in (4.46*h*).

4.5. Verify results (4.47*a*), (4.47*b*), and (4.47*c*), and (4.48*a*), (4.48*b*), and (4.48*c*).

4.6. (a) Starting from the relation $\mathbf{q}'\hat{\boldsymbol{\beta}}^0 = \mathbf{q}'GX'\mathbf{y}$, show that (4.58*b*) holds, where $\mathbf{q} = X'\mathbf{t}$ and $G = (X'X)_1^-$.

(b) Starting from the relation $V(\mathbf{q}'\hat{\boldsymbol{\beta}}^0) = \sigma^2\mathbf{t}'XX_{14}^-\mathbf{t}$, prove (4.58*b*) holds.

4.7. Deduce (4.59*c*).

4.8. If X is of order $N \times k$, $r(X) = p < k$, show that $\hat{\boldsymbol{\eta}}$ is orthogonal to $\mathbf{e}$.

4.9. Substantiate (i), (ii), and (iii) of (4.60*b*).

4.10. Verify the statement: "X_1 is orthogonal to $X_{2\cdot 1}$, defined in (4.62)."

4.11. Construct the relevant analysis of variance table when $E(\mathbf{y}) = X_1\boldsymbol{\beta}_1 + X_2\boldsymbol{\beta}_2$—for $\boldsymbol{\beta}_2$ of interest—$X = (X_1 \mid X_2)$ is not of full rank, and X_1 is a $N \times k_1$ matrix not of full rank, and so on. Supply an EMS column and derive its entries, under the usual assumptions for $V(\mathbf{y})$. If X is such that $\boldsymbol{\beta}_2$ is estimable, explain how you would test H_0: $\boldsymbol{\beta}_2 = \mathbf{0}$, if *AD*y holds.

4.12. Verify that the matrix specified in (4.75*b*) is a g_{14} inverse of the matrix specified in (4.75).

4.13. (a) By augmenting the matrix $X_{2\cdot 1}$ given (4.75) with $B = \frac{1}{2}\,(1, 1)$, find a g_1 inverse of $X_{2\cdot 1}'X_{2\cdot 1}$, and hence show that a g_{14} inverse of $X_{2\cdot 1}$ is as given in (4.75*b*).

(b) One could also find the Moore–Penrose inverse of $X_{2\cdot 1}$, and hence a g_{14} inverse, by using Theorem 4.2.4—do so.

4.14. Verify (4.76*a*). *Hint:* See (4.69*b*).

4.15. (a) Prove algebraically that SS_e of (4.77*a*) is as stated in (4.77*b*).
(b) If $\beta_1 - \beta_2 = 0$, deduce (4.79).

4.16. Using matrix notation, write SS_W of (4.77*b*) as a quadratic form in $\mathbf{y}$, and find the trace of the (idempotent) matrix involved.

4.17. Verify (4.94).

4.18. (a) Suppose A is of order $N \times k$ and rank p, $p < k < N$. Suppose we may write

$$A = \left[\begin{array}{c|c} A_{11} & A_{12} \\ \hline A_{21} & A_{22} \end{array}\right],$$

where A_{11} is of order $p \times p$ and of full rank, and

$$\left[\begin{array}{c} A_{12} \\ \hline A_{22} \end{array}\right] = \left[\begin{array}{c} A_{11} \\ \hline A_{21} \end{array}\right] L,$$

for some matrix L, of order $p \times (k-p)$, with A_{12}, A_{21}, and A_{22} of orders $p \times (k-p)$, $(N-p) \times p$, and $(N-p) \times (k-p)$, respectively; that is, the last $k-p$ columns of A are linearly dependent on the first (linearly independent) p columns of A.

Show that the $k \times N$ matrix D, where

$$D = \left[\begin{array}{c|c} A_{11}^{-1} & \bigcirc_{p \times (N-p)} \\ \hline \bigcirc_{(k-p) \times p} & \bigcirc_{(k-p) \times (N-p)} \end{array}\right],$$

is a g_1 inverse of A. The $\bigcirc$ matrices are matrices of zeros.

(b) If A is of order $N \times k$ and of rank $p < k < N$, and if A cannot be written as in part (a), then it is always possible to find P_1, a $N \times N$ permutation matrix (to be used to permute rows) and P_2, a $k \times k$ permutation matrix to be used to permute columns) so that

$$P_1 A P_2 = \left[\begin{array}{c|c} \tilde{A}_{11} & \tilde{A}_{12} \\ \hline \tilde{A}_{21} & \tilde{A}_{22} \end{array}\right],$$

where $\tilde{A}_{11}$ is of order $p \times p$ and of full rank, and

$$\left[\begin{array}{c} \tilde{A}_{12} \\ \hline \tilde{A}_{22} \end{array}\right] = \left[\begin{array}{c} \tilde{A}_{11} \\ \hline \tilde{A}_{21} \end{array}\right] \tilde{L}.$$

Hence, from part (a), the $k \times N$ matrix $\tilde{D}$, where

$$\tilde{D} = \left[\begin{array}{c|c} \tilde{A}_{11}^{-1} & \bigcirc_{p\times(n-p)} \\ \hline \bigcirc_{(k-p)\times p} & \bigcirc_{(k-p)\times(n-p)} \end{array}\right]$$

is a g_1 inverse of P_1AP_2, which implies that $P_2\tilde{D}P_1$ is a g_1 inverse of A. Show this.

4.19. (a) Suppose X is the 4×3 matrix given by

$$X = \begin{bmatrix} 1 & 1 & 0 \\ 1 & 1 & 0 \\ 1 & 0 & 1 \\ 1 & 0 & 1 \end{bmatrix}.$$

Find G—a g_1 inverse of $X'X$—by using Problem 4.18. Verify that $X'XGX'X = X'X$.

(b) Now find GX', and verify that XGX' is symmetric and that $XGX'X = X$, so that indeed $GX' = X_{14}^{-}$.

(c) Suppose X is as in part (a) and that we are given $\boldsymbol{\eta} = X\boldsymbol{\beta} = \mathbf{0}$. Construct a $\boldsymbol{\beta}$, where $\boldsymbol{\beta} \neq \mathbf{0}$, which is such that $X\boldsymbol{\beta} = \mathbf{0}$. Comment.

4.20. (a) Show that if $\theta_1 = \mathbf{q}_1'\boldsymbol{\beta}$ and $\theta_2 = \mathbf{q}_2'\boldsymbol{\beta}$ are estimable, then $c_1\theta_1 + c_2\theta_2$ is estimable [$\mathbf{q}_j'$ and $\mathbf{c} = (c_1, c_2)'$ are vectors of known constants].

(b) Show that, if $AE(\mathbf{y})$, $AV(\mathbf{y})$, and $AD(\mathbf{y})$ hold, and if $\mathbf{q}'\boldsymbol{\beta}$ is estimable, the interval

$$\left[\mathbf{q}'\hat{\boldsymbol{\beta}}^0 \pm t_{N-p;\,\alpha/2}\left(\frac{\text{SS}_e(\mathbf{q}'G\mathbf{q})}{N-p}\right)^{1/2}\right]$$

is a $100(1-\alpha)\%$ confidence interval for $\theta = \mathbf{q}'\boldsymbol{\beta}$. Give details.

4.21. (a) Let $H = GX'X$, where $G = (X'X)_1^-$. Prove that $Q\boldsymbol{\beta}$ is estimable if and only if $QH = Q$. ($Q\boldsymbol{\beta}$ is estimable if for all j, $\mathbf{q}_j'\boldsymbol{\beta}$ is estimable, where $\mathbf{q}_j'$ is the jth row of Q.)

(b) Use this to show that $\boldsymbol{\eta} = X\boldsymbol{\beta}$ is always estimable.

(c) Consider the situation implied by (4.70). Find, using Problem 4.18, $G = (X'X)^-$ and use it to test whether or not $\beta_1 - \beta_2$ is estimable by inquiring whether or not (4.58c) holds.

4.22. Suppose $E(\mathbf{y}) = X\boldsymbol{\beta}$, where X is $N \times k$ with $r(X) = p < k < N$, $V(\mathbf{y}) = \sigma^2 I_N$, with $AD(\mathbf{y})$ holding. Draw up the analysis of variance table with usual headings—Source, Degrees of Freedom, Sum of Squares, and Expected Mean Square—when $\hat{\boldsymbol{\beta}}^0 = X_{14}^{-}\mathbf{y}$ is used. Prove that your stated entries of the EMS column are correct. Consulting the EMS column, note that we are able to construct a test of $\boldsymbol{\eta} = \mathbf{0}$ even if $\boldsymbol{\beta}$ is not estimable. (Indeed $\boldsymbol{\eta} = \mathbf{0}$ does not imply that $\boldsymbol{\beta} = \mathbf{0}$ if X is not of full rank.) What is the test that is indicated by inspecting the EMS column? State the procedure of level α, giving justification for the distributional results used. Also, indicate why $X\boldsymbol{\beta}$ is always estimable so that $\boldsymbol{\eta} = \mathbf{0}$ is a testable hypothesis. (The reader should also consult Section 5.4 of Chapter 5.)

4.23. Suppose $E(y_{ij}) = \mu + \alpha_j$, $\Sigma_{j=1}^3 \alpha_j = 0$, and that $i = 1,\ldots,n_j$. Let $N = \Sigma n_j$ and suppose $V(\mathbf{y}) = \sigma^2 I_N$, with $\mathbf{y} = (y_{11}, y_{21}, y_{12}, y_{22}, y_{13}, y_{23})$. State X and the rank of X. Use the constraint to find an unbiased estimator $\hat{\boldsymbol{\beta}}$ of $\boldsymbol{\beta} = (\mu, \alpha_1, \alpha_2, \alpha_3)'$. Show that $\hat{\boldsymbol{\beta}}$ obeys the constraint.

APPENDIX A4.1. PROOF OF THEOREM 4.1.1

The proof given in this appendix is due to Searle (1971).

(i) *Sufficiency.* In this part we assume $AGA = A$; then $AGA\mathbf{v} = A\mathbf{v}$. Thus, when $A\mathbf{v} = \mathbf{u}$, the foregoing implies that $AG\mathbf{u} = \mathbf{u}$; that is, $A[G\mathbf{u}] = \mathbf{u}$ so that $\mathbf{v} = G\mathbf{u}$ is a solution of $A\mathbf{v} = \mathbf{u}$, proving the sufficiency part.

(ii) *Necessity.* Here we assume that the consistent equations $A\mathbf{v} = \mathbf{u}$ have a solution $\mathbf{v} = G\mathbf{u}$, G independent of $\mathbf{u}$. Consider the equations $A\mathbf{v} = \mathbf{a}_j$, where $\mathbf{a}_j$ is the jth column of A. These equations have a solution, namely a vector of zeros, except for the jth component which has the value 1. Thus the equations $A\mathbf{v} = \mathbf{a}_j$ are consistent (see the remark made before the statement of Theorem 4.1.1). But by the assumption for the necessity part, consistent equations $A\mathbf{v} = \mathbf{a}_j$ have a solution $\mathbf{v} = G\mathbf{a}_j$, which is to say $A(G\mathbf{a}_j) = \mathbf{a}_j$, for all j, so that $AGA = A$, proving the necessity part.

APPENDIX A4.2.

In the course of this chapter, we have repeatedly used the result that if $PX'X = QX'X$, then $PX' = QX'$. We now establish this as follows.

Lemma A4.2.1. *If $W'W = \bigcirc$, then $W = \bigcirc$.*

Proof. $W'W = \bigcirc$ means, in particular, that the sums of squares of components of each column are zero, so that the components themselves are zero, proving the lemma.

Lemma A4.2.2. *If $PX'X = QX'X$, then $PX' = QX'$.*

Proof. We have that

$$\bigcirc = PX'X - QX'X$$

so that

$$\begin{aligned}\bigcirc &= (PX'X - QX'X)(P - Q)' \\ &= (PX' - QX')(PX' - QX')'\end{aligned}$$

and we apply Lemma A4.2.1 and obtain

$$PX' - QX' = \bigcirc;$$

that is, if $PX'X = QX'X$, then $PX' = QX'$.

APPENDIX A4.3. PROOF OF THEOREM 4.2.9

Let $U = (X'X + B'B)^{-1}B'$. U is of order $k \times q$ and obviously of rank $r(B') = r(B) = q$. We will first show that

(A4.3.1) $$XU = \bigcirc_{N \times q} \quad \text{and} \quad BU = I_q,$$

where $\bigcirc_{N \times q}$ is a $N \times q$ matrix of zeros, and I_q is the identity of order q. Since $r(X) = k - q = p$, there are p linearly independent rows in X and $N - p$ rows that are linearly dependent on the p linearly independent rows. Suppose then that we write

(A4.3.2) $$PX = \begin{bmatrix} X_1 \\ \hline X_2 \end{bmatrix},$$

where the $p \times k$ matrix X_1 contains the p linearly independent rows of X, so that

(A4.3.2a) $$r(X_1) = p \quad \text{and} \quad X_2 = KX_1, \qquad \text{for some } K,$$

where K is a $(N-p)\times p$ matrix, X_2 is of order $(N-p)\times k$ and is such that its $N-p$ rows are linearly dependent on the rows of X_1, and where P is a $N\times N$ permutation matrix ($PP'=P'P=I$) used to permute rows of X so that indeed we may write PX as in (A4.3.2).

Now consider the $k\times k=(p+q)\times k$ matrix

$$\left[\begin{array}{c} X_1 \\ \hline B \end{array}\right]. \tag{A4.3.3}$$

Since $r(X_1)=p$ and $r(B)=q$, with the rows of B linearly independent of X_1, it is clear that the matrix in (A4.3.3) is of full rank k and hence possesses an inverse. Let

$$\left[\begin{array}{c} X_1 \\ \hline B \end{array}\right]^{-1}=\left[V_1 \mid V_2\right], \tag{A4.3.3a}$$

where V_1 is of order $k\times p$ and V_2 is of order $k\times q$, $p+q=k$, that is,

$$I=\left[\begin{array}{c} X_1 \\ \hline B \end{array}\right]\left[V_1 \mid V_2\right] \tag{A4.3.3b}$$

or

$$\left[\begin{array}{c|c} X_1V_1 & X_1V_2 \\ \hline BV_1 & BV_2 \end{array}\right]=\left[\begin{array}{c|c} I_p & \bigcirc_{p\times q} \\ \hline \bigcirc_{q\times p} & I_q \end{array}\right]. \tag{A4.3.3c}$$

Hence, in particular,

$$\text{(i)}\quad X_1V_2=\bigcirc_{p\times q} \qquad \text{and} \qquad \text{(ii)}\quad BV_2=I_q. \tag{A4.3.4}$$

But

$$PX=\left[\begin{array}{c} X_1 \\ \hline X_2 \end{array}\right]=\left[\begin{array}{c} X_1 \\ \hline KX_1 \end{array}\right], \tag{A4.3.5}$$

and we have

$$PXV_2=\left[\begin{array}{c} X_1V_2 \\ \hline KX_1V_2 \end{array}\right]=\left[\begin{array}{c} \bigcirc_{p\times q} \\ \hline K\bigcirc_{p\times q} \end{array}\right] \tag{A4.3.5a}$$

so that

$$XV_2 = \bigcirc_{N\times q}. \tag{A4.3.5b}$$

Thus, using (A4.3.5*b*),

$$(X'X + B'B)V_2 = B'BV_2 \tag{A4.3.6}$$

so that from (A4.3.4), part (ii), we have

$$(X'X + B'B)V_2 = B'. \tag{A4.3.6a}$$

Now it is easy to see that

$$r\left[\frac{X}{B}\right] = r(X'X + B'B) = k$$

so that we have

$$V_2 = (X'X + B'B)^{-1}B' = U. \tag{A4.3.6b}$$

Returning to (A4.3.5*b*) we have

$$XU = \bigcirc_{N\times q}, \tag{A4.3.7}$$

which establishes the result (4.44*a*) of Theorem 4.2.9. Also, from part (ii) of (A4.3.4), and (A.4.3.6*b*), we have

$$BU = I_q. \tag{A4.3.7a}$$

Transposing in (A4.3.7) and pre-multiplying by $-B'$, we have

$$-B'B(X'X + B'B)^{-1}X' = \bigcirc_{k\times N}. \tag{A4.3.8}$$

We now add to the left-hand side and right-hand side of (A4.3.8), the left- and right-hand sides, respectively, of the identity

$$(X'X + B'B)(X'X + B'B)^{-1}X' = X', \tag{A4.3.8a}$$

obtaining

$$[(X'X + B'B) - B'B](X'X + B'B)^{-1}X' = X' \tag{A4.3.9}$$

or

$$(X'X)(X'X + B'B)^{-1}X' = X', \tag{A4.3.9a}$$

and post-multiplying each side of (A4.3.9a) by X we have

$$X'X(X'X + B'B)^{-1}X'X = X'X, \tag{A4.3.10}$$

that is, $(X'X + B'B)^{-1}$ is a g_1 inverse of $X'X$, proving the first part of the result (4.44) of Theorem 4.2.9. Now from (A4.3.7a), we have upon pre- and post-multiplying both sides by B' and B, respectively, that

$$B'B(X'X + B'B)^{-1}B'B = B'B; \tag{A4.3.11}$$

that is, $(X'X + B'B)^{-1}$ is also a generalized inverse of $B'B$, and this proves the second part of (4.44) of Theorem 4.2.9.

We remark here that $(X'X + B'B)^{-1}X'$ and $(X'X + B'B)^{-1}B'$ are g_{14} inverses of X and B, respectively, as can be seen from the theorems in Section 4.2.

5 Some Non-Full-Rank Linear Models and the General Linear Hypothesis

Very often an experimenter is in a situation where appropriate use can be made of the model

$$E(y) = \sum_{j=1}^{k} \beta_j x_j, \tag{5.1}$$

where the independent variables x_j may take on only the values zero and one, corresponding to "absence" or "presence" of certain experimental factors. The literature terms such x_j dummy variables, and many authors term models of the type (5.1), where the x_j are dummy variables, *linear models*, as opposed to *regression models*, where, for the latter, the x_j may have values other than 0 or 1. In addition, linear models often lead to the general formulation $E(\mathbf{y}) = X\boldsymbol{\beta}$, where X is not of full rank.

Very important examples of the above are the so-called analysis of variance situations of the m-way classification models, and this chapter is devoted to a description of the analysis of two such models and some special tools that help with the analysis. The problems at the end of the chapter introduce, and ask for the analysis, of other analysis of variance models. The last section of this chapter, Section 5.4, has a discussion of the general linear hypothesis for the non-full-rank situation, justifying the implicit tests used in the discussion of the analysis of variance models and equipping us with the tools needed for deriving test procedures in any non-full-rank situation.

5.1. SOME USEFUL LEMMAS

We will be analyzing a linear model $E(\mathbf{y}) = X\boldsymbol{\beta}$, where the matrix X is $(N \times k)$ and of rank $r(X) = p < k$. This means that $X'X$ is not of full rank, with all that that implies, and to help us to arrive at regression sum of squares, residual sum of squares, and so on, we may use one of several methods. One method is implied by Theorem 4.2.9 (see below) and uses an augmented matrix $A = (X' \mid B')$. Another method is based on computing a g_{14} inverse of X and is also discussed in Chapter 4. In this chapter, we discuss yet another method, based on the following lemmas.

Lemma 5.1.1. *Consider the $p \times (p + q)$ matrix W, where*

$$W = [I_p \mid V], \tag{5.2}$$

with I_p the identity matrix of order p and V any matrix with p rows, and say, q columns, $q \geq 1$. Then the rank of W is p.

The proof is trivial and left to the reader.

Lemma 5.1.2. *Suppose X is of order $N \times k$, with rank $r(X) = p < k$, that is, X is not of full rank. Let $k - p = q$. Then, there exists a $k \times q$ matrix U, such that the rank of U is given by $r(U) = q < k$ and*

$$XU = \mathrm{O}, \tag{5.3}$$

that is, U is of full column rank and satisfies (5.3).

Proof. [We remark that Theorem 4.2.9 does show the existence of a matrix U satisfying (5.3)—see relation (4.44a) and Appendix A4.3. We give a less special proof here.] Suppose that we may write, after permutation of columns,

$$XP' = [X_1 \mid X_2], \tag{5.4}$$

where X_1 is $N \times (k - q) = N \times p$, X_1 is of full rank, and P' is a $k \times k$ permutation matrix that permutes columns. (See Section 14 of Appendix A2 of Chapter 2.) Then because $r(X) = p$, X_2 is made up of linear combinations of X_1, that is,

$$X_2 = X_1 K, \tag{5.5}$$

where K is of order $p \times q = (k - q) \times q$. Since P is invertible, we may now let a $k \times q$ matrix U be such that

$$PU = \begin{pmatrix} -K \\ \hdashline I_q \end{pmatrix} \tag{5.6}$$

(we recall that $p + q = k$). Then applying Lemma 5.1.1 to $(PU)'$, we have

$$r((PU)') = q \quad \text{or} \quad r(PU) = q \quad \text{or} \quad r(U) = q, \tag{5.6a}$$

and we have that

$$(XP')(PU) = XU = (X_1 \mid X_2)\begin{pmatrix} -K \\ \hdashline I_q \end{pmatrix} = -X_1K + X_2 \tag{5.7}$$

so that from (5.5),

$$XU = \bigcirc, \qquad U = P'\begin{pmatrix} -K \\ \hdashline I_q \end{pmatrix} \tag{5.7a}$$

which was to be proven.

A geometric interpretation of the result (5.3) of Lemma 5.1.2 is the following. Suppose we consider the *row space* of X, that is, the linear subspace generated by the span of the rows of X. In the notation first used in Chapter 3, this is the subspace $S(X')$, that is, the subspace generated by linear combinations of the columns of X', which of course are the rows of X. Since the rank of X is the same as the rank of X', $S(X')$ is of dimension p and is contained in R^k, and we will write $S_p(X') \subset R^k$.

Suppose now we inquire into the orthogonal complement of $S_p(X')$—say $S_p^{\perp}(X')$. Its dimension is $k - p = q$. Let $\mathbf{u}_j$ be q linearly independent $k \times 1$ vectors such that $\{\mathbf{u}_1, \ldots, \mathbf{u}_q\}$ forms a basis for this space, that is, $\mathbf{v} \in S_p^{\perp}(X')$ implies $\mathbf{v} = \sum_{i=1}^{q} \alpha_i \mathbf{u}_i$. Then, since each $\mathbf{u}_i$ is orthogonal to vectors in $S_p(X')$, we have

$$\mathbf{x}_r'\mathbf{u}_s = 0, \qquad \text{for all } r, s \tag{5.8}$$

where $\mathbf{x}_r'$ is the rth row of X. From the above we immediately have

$$XU = \bigcirc, \tag{5.8a}$$

where $U = (\mathbf{u}_1, \ldots, \mathbf{u}_q)$. Hence, Lemma 5.1.2 is merely recording the fact that vectors belonging to $S_p^{\perp}(X')$, the orthogonal complement of the row space of X, are indeed orthogonal to rows of X.

In general, we may make use of Lemma 5.1.2 as follows. We are dealing with the usual linear model, of non-full rank. Write

$$E(\mathbf{y}) = X\boldsymbol{\beta} = XHH^{-1}\boldsymbol{\beta} = XH\boldsymbol{\tau}, \qquad \boldsymbol{\tau} = H^{-1}\boldsymbol{\beta}, \tag{5.9}$$

where $r(X) = p < k$,

$$H = \left(V \mid U\right) \tag{5.9a}$$

with H a $k \times k$ nonsingular matrix, V a $k \times p$ matrix of full column rank, and U a $k \times q = k \times (k - p)$ matrix of full column rank satisfying (5.3).

Now if we rewrite (5.9) as

$$E(\mathbf{y}) = X\left(V \mid U\right)\begin{pmatrix} \boldsymbol{\tau}_1 \\ \hline \boldsymbol{\tau}_2 \end{pmatrix}, \tag{5.10}$$

where $\boldsymbol{\tau}_1$ is of order $p \times 1$ and $\boldsymbol{\tau}_2$ is of order $(k - p) \times 1$, then we have

$$E(\mathbf{y}) = XV\boldsymbol{\tau}_1 = W\boldsymbol{\tau}_1, \tag{5.11}$$

where W is of order $N \times p$ and such that

$$W = XV, \qquad r(W) = r(X) = p. \tag{5.12}$$

Hence, we have *reduced* (5.8), *a non-full-rank model to a full-rank model* (5.11) and (5.12), and as XH constitutes a nonsingular transformation of X, regressing $\mathbf{y}$ on W leads to the correct regression and residual sums of squares, and we may enter these in an appropriate analysis of variance table such as Table 5.1.1. We may do this because (5.11) is a "full-rank situation," and hopefully, we know how to deal with this.

Of course, we are still left with the problem, glossed over in the above, of finding V and U, that is, of constructing

$$H = \left(V \mid U\right), \qquad XU = O, \quad r(H) = k. \tag{5.13}$$

We wish to find $k \times 1$ vectors u_j, such that

$$X\mathbf{u}_j = \mathbf{0}, \qquad j = 1, \ldots, q \tag{5.14}$$

and p $k \times 1$ column vectors $\mathbf{v}_t$, such that

$$H = \left[\mathbf{v}_1 \mid \cdots \mid \mathbf{v}_p \mid \mathbf{u}_1 \mid \cdots \mid \mathbf{u}_q\right] \tag{5.15}$$

is of order $k \times k$ and of rank $r(H) = k$. There are, of course, an infinity of

TABLE 5.1.1. Analysis of Variance for (5.8)–(5.11)

Source	Sum of Squares	Degrees of Freedom	Expected Sum of Squares
Regression (due to X)	$\hat{\boldsymbol{\eta}}'\hat{\boldsymbol{\eta}} = \mathbf{y}'\mathcal{R}(W)\mathbf{y}$	p	$p\sigma^2 + \boldsymbol{\beta}'X'X\boldsymbol{\beta} = p\sigma^2 + \boldsymbol{\eta}'\boldsymbol{\eta}$
Residual	$\mathbf{y}'\mathbf{y} - \hat{\boldsymbol{\eta}}'\hat{\boldsymbol{\eta}} = \mathbf{y}'[I - \mathcal{R}(W)]\mathbf{y}$	$N - p$	$(N-p)\sigma^2$
Total	$\mathbf{y}'\mathbf{y}$	N	$N\sigma^2 + \boldsymbol{\beta}'X'X\boldsymbol{\beta}$

$\mathcal{R}(W) = W(W'W)^{-1}W'$

such matrices. It turns out, however, that often, in constructing V, an identity matrix for the upper left-hand corner is used to advantage. We illustrate with the following example.

EXAMPLE 5.1.1.

Suppose the 4×1 vector $\mathbf{y}$ of observations has

$$V(\mathbf{y}) = \sigma^2 I, \tag{5.16}$$

with

$$E(\mathbf{y}) = \beta_0 \mathbf{1}_4 + \beta_1 \mathbf{x}_1 + \beta_2 \mathbf{x}_2 = X\boldsymbol{\beta}, \tag{5.16a}$$

where the 4×3 matrix X is such that

$$X = (\mathbf{1}_4 \mid \mathbf{x}_1 \mid \mathbf{x}_2) = \begin{pmatrix} 1 & -1 & -1 \\ 1 & 1 & 1 \\ 1 & -1 & -1 \\ 1 & 1 & 1 \end{pmatrix}. \tag{5.16b}$$

Clearly, in this example, since $\mathbf{x}_1 = \mathbf{x}_2$, we cannot find an unbiased estimator of $(\beta_1, \beta_2)'$. On applying Lemma 5.1.2, we have

$$r(X) = 2 = p; \quad q = k - p = 3 - 2 = 1; \quad U = \mathbf{u}_1; \tag{5.17}$$

where $U = \mathbf{u}_1$ is a $k \times q = 3 \times 1$ vector such that

$$X\mathbf{u}_1 = \mathbf{0} \quad \text{or} \quad \begin{pmatrix} 1 & -1 & -1 \\ 1 & 1 & 1 \\ 1 & -1 & -1 \\ 1 & 1 & 1 \end{pmatrix} \begin{pmatrix} u_{11} \\ u_{21} \\ u_{31} \end{pmatrix} = \begin{pmatrix} 0 \\ 0 \\ 0 \\ 0 \end{pmatrix}. \tag{5.17a}$$

Since in each line of X the elements of the second and third columns are equal, we try as a solution

$$(5.18)\qquad \mathbf{u}_1 = \begin{pmatrix} 0 \\ -1 \\ 1 \end{pmatrix}$$

and, indeed, it is easy to verify that (5.18) satisfies equation (5.17a). We now try an "H matrix" with I_2 in its upper left-hand columns as follows.

$$(5.19)\qquad H = \left(\begin{array}{cc|c} 1 & 0 & 0 \\ 0 & 1 & -1 \\ \hline 0 & 0 & 1 \end{array}\right)$$

which means, as is easily verified,

$$(5.19a)\qquad H^{-1} = \left(\begin{array}{cc|c} 1 & 0 & 0 \\ 0 & 1 & 1 \\ \hline 0 & 0 & 1 \end{array}\right).$$

We may now rewrite (5.16a) as

$$(5.20)\qquad E(\mathbf{y}) = XHH^{-1}\boldsymbol{\beta}, \qquad H = [V \,|\, \mathbf{u}_1] = \left(\begin{array}{cc|c} 1 & 0 & 0 \\ 0 & 1 & -1 \\ 0 & 0 & 1 \end{array}\right)$$

and find

$$(5.21)\qquad E(\mathbf{y}) = W\boldsymbol{\tau}_1, \quad W = \begin{pmatrix} 1 & -1 \\ 1 & 1 \\ 1 & -1 \\ 1 & 1 \end{pmatrix}, \quad \boldsymbol{\tau}_1 = \begin{pmatrix} \tau_{11} \\ \tau_{21} \end{pmatrix} = \begin{pmatrix} \beta_0 \\ \beta_1 + \beta_2 \end{pmatrix}.$$

We note that, in this example, $\mathbf{w}_1$ is orthogonal to $\mathbf{w}_2$, so the analysis of variance table takes the form, given in Table 5.1.2, especially useful if only the statement that the effects "β_1 and β_2 are such that $\beta_1 + \beta_2$ is zero" is to be examined. Indeed it is easy to show that $\beta_1 + \beta_2$ is estimable, an exercise left to the reader (recall Definition 4.4.1 and Theorem 4.4.2). Of course, this is expected since, from (5.21) (or using the fact that $\mathbf{x}_1 = \mathbf{x}_2$), (5.16$a$) can be rewritten as

$$(5.21a)\qquad E(\mathbf{y}) = \beta_0\mathbf{1}_4 + (\beta_1 + \beta_2)\mathbf{x}_1,$$

where $W = (\mathbf{1}_4 \,|\, \mathbf{x}_1)$ is a 4×2 matrix of full rank 2, and so on.

TABLE 5.1.2. Analysis of Variance for Example 5.1.1

Source	Sum of Squares	Degrees of Freedom	Expected Sum of Squares
Due to mean ($\mathbf{w}_1 = \mathbf{1}_4$)	$(\mathbf{w}_1'\mathbf{y})^2/\mathbf{w}_1'\mathbf{w}_1 = 4\bar{y}^2$	1 } $p = 2$	$\sigma^2 + 4\tau_{11}^2 = \sigma^2 + 4\beta_0^2$
Extra, due to effect τ_{21} (i.e., due to β_1, β_2)	$(\mathbf{w}_2'\mathbf{y})^2/\mathbf{w}_2'\mathbf{w}_2$	1 }	$\sigma^2 + 4\tau_{21}^2 = \sigma^2 + 4(\beta_1 + \beta_2)^2$
Residual	$\mathbf{y}'[I - \mathfrak{R}(\mathbf{w}_1) - \mathfrak{R}(\mathbf{w}_2)]\mathbf{y}$	$N - p = 2$	$2\sigma^2$
Total	$\mathbf{y}'\mathbf{y}$	$N = 4$	$4\sigma^2 + 4[\beta_0^2 + (\beta_1 + \beta_2)^2]$

Of course, there will be examples when it is not easiest to have I in the upper left-hand corner of H. For example, suppose

$$X = \begin{bmatrix} 1 & 1 & 1 \\ 1 & 1 & -1 \\ 1 & 1 & 1 \\ 1 & 1 & -1 \end{bmatrix}.$$

Then, a natural choice for $U = \mathbf{u}_1$ is $(-1, 1, 0)'$, and a natural choice for H is

$$H = \begin{bmatrix} 0 & 0 & -1 \\ 1 & 0 & 1 \\ 0 & 1 & 0 \end{bmatrix}. \tag{5.21b}$$

5.2. ONE-WAY ANALYSIS OF VARIANCE

The one-way analysis of variance situation occurs frequently in experimentation. Indeed, we have encountered examples of the one-way solution previously in this text. (See, for example, Sections 3.4 and 4.5.) In this section, we sketch alternative formulations of the model and discuss the analysis using two methods, one of which brings in the question of constraints (sometimes called restraints). We suppose that n_j independent observations y_{ij}, $i = 1, \ldots, n_j$, $j = 1, \ldots, c$, are taken using factor level j (e.g., the factor could be drugs and level j could represent a treatment that uses drug j, there being c drugs available; or, as another example, the factor would be a single drug and level j would represent the amount of the single

TABLE 5.2.1.

Factor Levels	1	2	$\cdots$	j	$\cdots$	c
	y_{11}	y_{12}		y_{1j}		y_{1c}
	y_{21}	$\vdots$				$\vdots$
	$\vdots$	$y_{n_2 2}$		$\vdots$		$y_{n_c c}$
	$y_{n_1 1}$					
				$y_{n_j j}$		

drug in solution in units of percent). It is convenient to tabulate the y_{ij} so obtained as in Table 5.2.1. Assume that

$$\mathbf{y} = \left(y_{11},\ldots,y_{n_1 1}, y_{12},\ldots,y_{n_2 2},\ldots,y_{1c},\ldots,y_{n_c c}\right)', \tag{5.22}$$

where

$$E(y_{ij}) = \mu + \tau_j, \qquad i = 1,\ldots,n_j, \quad j = 1,\ldots,c \tag{5.23}$$

and

$$V(\mathbf{y}) = \sigma^2 I_N, \qquad N = \sum_{j=1}^{c} n_j. \tag{5.23a}$$

Assumption (5.23) is made because if the so-called *differential* effects τ_j due to levels j are zero, there is no reason to suppose $E(y_{ij}) = 0$. We now let

$$\boldsymbol{\beta} = (\mu, \tau_1,\ldots,\tau_c)', \tag{5.24}$$

and using the notation (5.22) and (5.24) and defining dummy variables x_{ij} by

$$x_{ij} = \begin{cases} 1, & \text{if } y_{ij} \text{ is the } i\text{th observation obtained using level } j \\ 0, & \text{otherwise} \end{cases} \tag{5.25}$$

we write the model in matrix form as

$$E(\mathbf{y}) = X\boldsymbol{\beta}, \qquad X = \left(\mathbf{1}_N \mid \mathbf{x}_1 \mid \mathbf{x}_2 \mid \cdots \mid \mathbf{x}_c\right). \tag{5.26}$$

Here X is of order $N \times (c+1)$ and

$$(5.27) \qquad X = \left[\begin{array}{c|ccccc} 1 & 1 & & & & \\ \vdots & \vdots & & & & \\ 1 & 1 & & & & \\ \hline 1 & & 1 & & & \\ \vdots & & \vdots & & & \\ 1 & & 1 & & & \\ \hline \vdots & & & & & \\ \hline 1 & & & & & 1 \\ \vdots & & & & & \vdots \\ 1 & & & & & 1 \end{array}\right]$$

(the blank spaces are to be read as zeros). That is, the $N \times 1$ vector $\mathbf{x}_j$ $(j = 1,\ldots,c)$ has ones for the n_j components numbered

$$(5.27a) \qquad \mathcal{N}_j + 1,\ \mathcal{N}_j + 2,\ldots,\mathcal{N}_j + n_j$$

and zeros for the values of the other components, where

$$(5.27b) \qquad \mathcal{N}_j = \sum_{t=0}^{j-1} n_t, \qquad j = 1, 2, \ldots, c.$$

(We define $n_0 = 0$.) Note that $\mathbf{x}_j$ is column $j+1$ of the X matrix, and note too that the first column of X, $\mathbf{1}_N$, is a $N \times 1$ vector of N ones.

The presence of a one in $\mathbf{x}_j$ indicates that the corresponding observation is present, that is, the corresponding observation was generated using factor level j.

We often wish to test the statement

$$(5.28) \qquad \tau_1 = \cdots = \tau_c$$

and so we now find the (extra) regression sum of squares due to $\mathbf{x}_1,\ldots,\mathbf{x}_c$ in the model, that is, due to the effects of the c levels of the factor. In the old notation,

$$(5.29) \qquad k = c + 1, \quad p = r(X) = c, \quad q = (c+1) - c = 1.$$

Pursuing the method of transformation on X (see Section 5.1), we wish to construct a matrix of order $(c + 1) \times (c + 1)$, say H, where, since $q = 1$,

$$H = \left(V \mid \mathbf{u}_1\right) \tag{5.30}$$

such that

$$X\mathbf{u}_1 = \mathbf{0}. \tag{5.30a}$$

We will try an H that has as its upper left-hand corner an identity matrix of order c, namely,

$$H = \left[V \mid \mathbf{u}_1\right] = \left[\begin{array}{cccc|c} 1 & & & & -1 \\ & 1 & & & 1 \\ & & \ddots & & \vdots \\ & & & 1 & 1 \\ \hline 0 & & \cdots & 0 & 1 \end{array}\right]. \tag{5.31}$$

Then

$$H^{-1} = \left[\begin{array}{cccc|c} 1 & & & & 1 \\ & 1 & & & -1 \\ & & \ddots & & \vdots \\ & & & 1 & -1 \\ \hline 0 & & \cdots & 0 & 1 \end{array}\right]. \tag{5.31a}$$

That is, we are choosing

$$V = \left[\begin{array}{ccc} 1 & & \\ & \ddots & \\ & & 1 \\ \hline 0 & \cdots & 0 \end{array}\right] \quad \text{and} \quad \mathbf{u}_1 = \begin{pmatrix} -1 \\ 1 \\ \vdots \\ 1 \end{pmatrix}. \tag{5.31b}$$

Now

$$XH = X\left[V \mid \mathbf{u}_1\right] = \left[\mathbf{1}_N \mid \mathbf{x}_1 \mid \mathbf{x}_2 \mid \cdots \mid \mathbf{x}_{c-1} \mid \mathbf{0}\right] \tag{5.32}$$

and

$$(5.32a) \qquad H^{-1}\boldsymbol{\beta} = H^{-1}\begin{pmatrix} \mu \\ \tau_1 \\ \vdots \\ \tau_c \end{pmatrix} = \begin{pmatrix} \mu + \tau_c \\ \tau_1 - \tau_c \\ \tau_2 - \tau_c \\ \vdots \\ \tau_{c-1} - \tau_c \\ \tau_c \end{pmatrix} = \begin{pmatrix} \mu + \tau_c \\ \gamma_1 \\ \vdots \\ \gamma_{c-1} \\ \tau_c \end{pmatrix},$$

where

$$(5.32b) \qquad \gamma_t = \tau_t - \tau_c, \qquad t = 1,\ldots,c-1.$$

Hence, using (5.32) and (5.32a), we may write (5.26) as

$$(5.33) \qquad E(\mathbf{y}) = XHH^{-1}\boldsymbol{\beta} = W\begin{pmatrix} \gamma_0 \\ \gamma_1 \\ \vdots \\ \gamma_{c-1} \end{pmatrix},$$

where the $N \times c$ matrix W is given by

$$(5.34) \quad W = XV = [\mathbf{1}_N \mid \mathbf{x}_1 \mid \mathbf{x}_2 \mid \cdots \mid \mathbf{x}_{c-1}], \qquad r(W) = c,$$

and

$$(5.35) \qquad \gamma_0 = \mu + \tau_c, \quad \gamma_t = \tau_t - \tau_c, \qquad t = 1,\ldots,c-1.$$

We note that (5.28) implies

$$(5.36) \qquad \gamma_t = 0, \qquad t = 1,\ldots,c-1$$

and vice versa. Since H is a nonsingular transformation of X, we know that the total regression sum of squares is the same whether we use (5.26) or (5.33), that the regression sum of squares due to the inclusion of $\mathbf{1}_N$ in the model (i.e., due to μ) is the same, and that the extra regression sum of squares due to the effects $\gamma_1,\ldots,\gamma_{c-1}$ (or $\tau_1,\ldots,\tau_{c-1}, \tau_c$) is the same. Because W *is of full rank*, we may proceed by eliminating the grand mean and regress residuals so obtained on $W_{2\cdot 1}$, where the latter is obtained by regressing $[\mathbf{x}_1 \cdots \mathbf{x}_{c-1}]$ on $\mathbf{1}_N$, and so on, and obtain the extra regression sum of squares due to $\gamma_1,\ldots,\gamma_{c-1}$. Since the statement (5.28) is equivalent to

TABLE 5.2.2. Analysis of Variance for the One-Way Classified Data of Table 5.2.1

Source	Sum of Squares	Degrees of Freedom		Expected Sum of Squares
Mean	$N\bar{y}^2$	1	$c = r(X) = r(W)$	
Extra due to $\gamma_1, \ldots, \gamma_{c-1}$ (or $\tau_1, \ldots, \tau_c$)	$\sum_{j=1}^{c} n_j(\bar{y}_{\cdot j} - \bar{y})^2$	$c - 1$		
Residual	$\sum_{j=1}^{c} \sum_{i=1}^{n_j} (y_{ij} - \bar{y}_{\cdot j})^2$	$N - p = N - c$		
Total	$\mathbf{y}'\mathbf{y}$	$N = \sum_{j=1}^{c} n_j$		

(5.36), we construct the usual analysis of variance Table 5.2.2. Problem 5.2 asks for all the details of the above [recall that $\bar{y} = N^{-1}\Sigma\Sigma\, y_{ij} = N^{-1}\Sigma\, n_j \bar{y}_{\cdot j}$ and $\bar{y}_{\cdot t} = (1/n_t)\Sigma_{i=1}^{n_t} y_{it}$]. Problem 5.2 also asks for the entries of the last column in Table 5.2.2.

It is interesting to note that it is very easy to prove the algebraic identity implied by the sum of squares (SS) column of Table 5.2.2. In fact, from the SS column we find that

$$(5.37)\quad \Sigma\Sigma\,(y_{ij} - \bar{y})^2 = \Sigma\Sigma y_{ij}^2 - N\bar{y}^2 = \Sigma\Sigma\,(y_{ij} - \bar{y}_{\cdot j})^2 + \Sigma\, n_j(\bar{y}_{\cdot j} - \bar{y})^2.$$

This is sometimes referred to as the *breakdown of the total sum of squares of deviations* of y_{ij}'s from the grand mean $\bar{y}$, namely, $\Sigma\Sigma\,(y_{ij} - \bar{y})^2$, into a part called the *within sum of squares* and a part called the *between sum of squares*.

If we assume that $AD(\mathbf{y})$ holds, (5.37) has two independent (Problem 5.3) parts, one that reflects information about σ^2, the error, (SS_W) and one that reflects information about the differences of the τ_j (SS_B), where

$$(5.37a)\qquad SS_W = \Sigma\Sigma\,(y_{ij} - \bar{y}_{\cdot j})^2, \qquad SS_B = \Sigma\, n_j(\bar{y}_{\cdot j} - \bar{y})^2.$$

Now it is interesting to analyze (5.26) using the method implied by Theorem 4.6.1. We have

$$(5.38)\qquad E(\mathbf{y}) = (\mathbf{1}_N \mid \mathbf{x}_1 \mid \cdots \mid \mathbf{x}_c)\begin{pmatrix} \mu \\ \tau_1 \\ \vdots \\ \tau_c \end{pmatrix},$$

and we are seeking the regression sum of squares due to $\mathbf{x}_1,\ldots,\mathbf{x}_c$. We can begin by eliminating the mean, which of course involves regressing $\mathbf{y}$ on $\mathbf{1}_N$ and taking residuals $\mathbf{e}_1$ and which is easily seen to be such that

$$\mathbf{e}_1 = \left(y_{11} - \bar{y},\ldots,y_{n_1 1} - \bar{y},\ldots,y_{1c} - \bar{y},\ldots,y_{n_c c} - \bar{y}\right)', \tag{5.39}$$

where

$$\bar{y} = N^{-1}\sum_{j=1}^{c}\sum_{i=1}^{n_j} y_{ij}. \tag{5.39a}$$

We then regress $\mathbf{e}_1$ on $X_{2\cdot 1} = Z$, where (see Problem 5.4)

$$Z = \begin{bmatrix} 1 - n_1/N & -n_2/N & \cdots & -n_c/N \\ \vdots & \vdots & & \vdots \\ 1 - n_1/N & -n_2/N & \cdots & -n_c/N \\ \hdashline -n_1/N & 1 - n_2/N & & -n_c/N \\ \vdots & \vdots & & \vdots \\ -n_1/N & 1 - n_2/N & \cdots & -n_c/N \\ \hdashline -n_1/N & -n_2/N & \cdots & -n_c/N \\ \vdots & \vdots & & \vdots \\ -n_1/N & -n_2/N & \cdots & -n_c/N \\ \hdashline \vdots & \vdots & & \vdots \\ \hdashline -n_1/N & -n_2/N & \cdots & 1 - n_c/N \\ \vdots & \vdots & & \vdots \\ -n_1/N & -n_2/N & \cdots & 1 - n_c/N \end{bmatrix}, \tag{5.40}$$

that is, the jth column of Z has components $-n_j/N$ everywhere except in places $\mathcal{N}_j + 1,\ldots,\mathcal{N}_j + n_j$, for which the components have the value $1 - n_j/N$, and where $\mathcal{N}_j$ is defined in (5.27b). Note that

$$Z'\mathbf{1}_N = \mathbf{0} \quad \text{or} \quad \mathbf{1}'_N Z = \mathbf{0}'. \tag{5.41}$$

For example,

$$\begin{aligned} \mathbf{1}'_N\mathbf{z}_1 &= n_1\left(1 - \frac{n_1}{N}\right) + (N - n_1)\left(-\frac{n_1}{N}\right) \\ &= \frac{n_1(N - n_1) - n_1(N - n_1)}{N} \\ &= 0. \end{aligned} \tag{5.42}$$

We also note that (Problem 5.4)

$$\sum_{j=1}^{c} \mathbf{z}_j = \mathbf{0} \tag{5.43}$$

so that the rank of Z is $c-1$, and we cannot routinely regress $\mathbf{e}_1$ on $X_{2\cdot 1} = Z$, since $Z'Z$ is not invertible. It is of interest to examine $Z'Z$ at this point, and it is easy to see that (Problem 5.4)

$$\mathbf{z}_j'\mathbf{z}_j = n_j\left(1 - \frac{n_j}{N}\right), \quad \mathbf{z}_i'\mathbf{z}_j = -\frac{n_i n_j}{N}, \qquad i \neq j. \tag{5.44}$$

Then

(5.45)

$$Z'Z = \begin{pmatrix} \mathbf{z}_1' \\ \hline \mathbf{z}_2' \\ \hline \vdots \\ \hline \mathbf{z}_c' \end{pmatrix} (\mathbf{z}_1 | \cdots | \mathbf{z}_c)$$

$$= \begin{pmatrix} n_1(1 - n_1/N) & -n_1 n_2/N & \cdots & -n_1 n_c/N \\ -n_2 n_1/N & n_2(1 - n_2/N) & \cdots & -n_2 n_c/N \\ \vdots & & \ddots & \vdots \\ -n_c n_1/N & \cdots & & n_c(1 - n_c/N) \end{pmatrix}$$

or

$$Z'Z = \begin{pmatrix} n_1 & & \\ & \ddots & \\ & & n_c \end{pmatrix} - \frac{1}{N}\left[\begin{pmatrix} n_1 \\ \vdots \\ n_c \end{pmatrix} (n_1, \ldots, n_c)\right]. \tag{5.45a}$$

Thus

$$\begin{pmatrix} n_1 & & \\ & \ddots & \\ & & n_c \end{pmatrix} = Z'Z + \frac{1}{N}\left[\begin{pmatrix} n_1 \\ \vdots \\ n_c \end{pmatrix} (n_1, \ldots, n_c)\right] \tag{5.45b}$$

$$= Z'Z + B'B,$$

where

$$B = \frac{1}{\sqrt{N}}(n_1,\ldots,n_c). \tag{5.45c}$$

At this stage we use the model

$$E(\mathbf{e}_1) = Z\boldsymbol{\tau} \tag{5.46}$$

where the $N \times c$ matrix Z is of rank $r(Z) = c - 1$. Now $c - (c - 1) = 1$, so that if we augment Z by a $1 \times c$ matrix to obtain A, where

$$A = \begin{pmatrix} Z \\ \hdashline B \end{pmatrix}, \qquad B = \frac{1}{\sqrt{N}}(n_1,\ldots,n_c), \tag{5.47}$$

then A is of full rank and [see (5.45b)]

$$(Z'Z + B'B)^{-1}Z' = \begin{pmatrix} 1/n_1 & & \\ & \ddots & \\ & & 1/n_c \end{pmatrix} Z' = Z_{14}^{-}, \tag{5.48}$$

that is, (5.48) is a g_{14} inverse of Z. Hence, the (extra) regression sum of squares due to $\tau_1,\ldots,\tau_c$ is

$$\mathbf{e}_1'ZZ_{14}^{-}\mathbf{e}_1 = \mathbf{e}_1'Z \begin{pmatrix} 1/n_1 & & \\ & \ddots & \\ & & 1/n_c \end{pmatrix} Z'\mathbf{e}_1, \tag{5.49}$$

and it is easily shown that (Problem 5.5)

$$Z'\mathbf{e}_1 = (n_1(\bar{y}_{\cdot 1} - \bar{y}),\ldots,n_c(\bar{y}_{\cdot c} - \bar{y}))' \tag{5.49a}$$

so that (5.49) has the value

$$(Z'\mathbf{e}_1)' \begin{pmatrix} 1/n_1 & & \\ & \ddots & \\ & & 1/n_c \end{pmatrix} (Z'\mathbf{e}_1) = \sum_{j=1}^{c} n_j(\bar{y}_{\cdot j} - \bar{y})^2. \tag{5.49b}$$

Furthermore, the residual sum of squares is

$$\begin{aligned}\mathbf{e}_1'\mathbf{e}_1 - \mathbf{e}_1'ZZ_{14}^{-}\mathbf{e}_1 &= \Sigma\Sigma\,(y_{ij} - \bar{y})^2 - \Sigma\, n_j(\bar{y}_{\cdot j} - \bar{y})^2 \\ &= \Sigma\Sigma\,(y_{ij} - \bar{y}_{\cdot j})^2 = \mathrm{SS}_W \end{aligned} \tag{5.50}$$

as before. The ingredients may be entered into an analysis of variance table.

We turn now to the question of constraints. To make the sample grand mean unbiased for μ (the theoretical grand mean), we often add the constraint

$$\sum_{j=1}^{c} n_j\tau_j = 0 \quad \text{or} \quad (n_1,\ldots,n_c)\boldsymbol{\tau} = 0 \tag{5.51}$$

for the model stated in (5.26), (5.27), and (5.23*a*). The addition of this assumption to our model has the advantage of uniquely defining μ and the τ_j's and is equivalent to

$$\frac{1}{\sqrt{N}}(n_1,\ldots,n_c)\boldsymbol{\tau} = \frac{1}{\sqrt{N}}\mathbf{n}'\boldsymbol{\tau} = B\boldsymbol{\tau} = 0. \tag{5.52}$$

[We choose B as $(1/\sqrt{N})\mathbf{n}'$ instead of $\mathbf{n}'$ to make the arithmetic simpler—see Section 4.6 and (5.45*a*)–(5.50).] A simple numerical example illustrates the uniqueness point. Suppose it is known that level averages are as in Table 5.2.3, and that $n_1 = 1$, $n_2 = 2$, and $n_3 = 3$. The ε_{ij}'s are assumed to be independent—mean 0 and variance σ^2. What we are saying is that

$$\begin{aligned} \mu + \tau_1 \qquad\qquad &= 71, \\ \mu \qquad + \tau_2 \qquad &= 73, \\ \mu \qquad\qquad + \tau_3 &= 78. \end{aligned} \tag{5.53}$$

If there are no assumptions about the τ_j's, then obviously the system (5.53) has an infinity of solutions, so that μ and the (differential) effects τ_j would not be uniquely defined. Now if (5.51) holds, that is, if

$$\tau_1 + 2\tau_2 + 3\tau_3 = 0, \tag{5.54}$$

TABLE 5.2.3.

$E(y_{i1}) = 71$	$E(y_{i2}) = 73$	$E(y_{i3}) = 78$
$y_{11} = \mu + \tau_1 + \varepsilon_{11}$	$y_{12} = \mu + \tau_2 + \varepsilon_{12}$	$y_{13} = \mu + \tau_3 + \varepsilon_{13}$
	$y_{22} = \mu + \tau_2 + \varepsilon_{22}$	$y_{23} = \mu + \tau_3 + \varepsilon_{23}$
		$y_{33} = \mu + \tau_3 + \varepsilon_{33}$

then (5.53) has the unique solution

$$(5.55) \qquad \mu = \frac{451}{6}, \quad \tau_1 = -\frac{25}{6}, \quad \tau_2 = -\frac{13}{6}, \quad \tau_3 = \frac{17}{6}.$$

Furthermore, a model which postulates that

$$(5.56) \qquad E(y_{ij}) = \mu + \tau_j, \qquad i = 1,\ldots,n_j, \quad j = 1,\ldots,c$$

can always be written in the forms (5.22) and (5.23) subject to the constraint (5.51), because if (5.51) does not hold, we can then write

$$(5.57) \qquad \begin{aligned} E(y_{ij}) &= (\mu + \bar{\tau}) + \tau_j - \bar{\tau}, \qquad \bar{\tau} = \frac{1}{N}\sum_{j=1}^{c} n_j\tau_j \\ &= \mu' + \tau_j'. \end{aligned}$$

Now $\Sigma n_j\tau_j' = \Sigma n_j(\tau_j - \bar{\tau}) = 0$. Note that $\tau_j' = \tau_j - \bar{\tau}$; this has motivated the use of the term *differential effects*, and the differential effects obey the constraint $\Sigma n_j\tau_j' = 0$. We note that the constraint $\Sigma n_j\tau_j' = 0$ is defined in terms of the experiment to be conducted. Indeed, if $n_1 = n_2 = \cdots = n_c = n$, so that $N = nc$, we have that the constraint then takes the form $\Sigma_{j=1}^{c}\tau_j' = 0$.

If (5.52) is part of the model, then (see Section 4.6)

$$(5.58) \qquad \hat{\boldsymbol{\tau}}_B^0 = (Z'Z + B'B)^{-1}Z'\mathbf{e}_1$$

is unbiased for $\boldsymbol{\tau}$, since $B\boldsymbol{\tau} = 0$. Using (5.45*b*) and (5.49*a*),

$$(5.59) \qquad \hat{\boldsymbol{\tau}}_B^0 = (\bar{y}_{\cdot 1} - \bar{y}, \ldots, \bar{y}_{\cdot c} - \bar{y})',$$

and we note that $\hat{\boldsymbol{\tau}}_B^0$ satisfies (5.52), that is,

$$(5.59a) \qquad B\hat{\boldsymbol{\tau}}_B^0 = 0,$$

since the left-hand side of (5.59a) is

$$B\hat{\boldsymbol{\tau}}_B^0 = \frac{1}{\sqrt{N}}(n_1,\ldots,n_c)(\bar{y}_{\cdot 1} - \bar{y},\ldots,\bar{y}_{\cdot c} - \bar{y})'$$

$$= \frac{1}{\sqrt{N}}(\Sigma n_j \bar{y}_{\cdot j} - \bar{y}\Sigma n_j) \tag{5.59b}$$

$$= \frac{1}{\sqrt{N}}(N\bar{y} - N\bar{y}) = 0.$$

5.3. THE TWO-WAY CLASSIFICATION

5.3a. No Interaction

The goal of many experiments is to determine the effect of more than one factor on a response variable y. For example, we may wish to determine the effect of varying the pressure and temperature on the production of a certain plastic. This may be accomplished by selecting $r = 2$ levels of the factor called pressure, say 14.8 and 15.0 lb/in.^{2}, and $c = 3$ cooking temperatures for the levels of the other factor—temperature—say 92, 94, and 96°C. In general, if budget and time allow, we may select r levels of the first factor and c levels of the second factor, and, using all rc combinations of the levels of both factors, make rc observations y_{ij}, $i = 1,\ldots,r$, $j = 1,\ldots,c$, where y_{ij} is the observation obtained when using the ith level of factor 1 and the jth level of factor 2. We may tabulate the observations in a two-way table because we need a double system to classify the observations here (unlike the one-way classification of the previous section) since we are varying two factors. Note the "balance" here. Any row in the resulting rectangular array has as many observations (c) as any other row, and any column has as many observations (r) as any other column. We say that the place of the entry y_{ij} is the (ij)th cell, and in this experiment there is one observation per cell. We sometimes refer to the *combination of the ith level of A and the jth level of B as a treatment*; again, in this experiment there is one observation per treatment.

TABLE 5.3.1. Observations Obtained by Varying Two Factors, *A* and *B* (*r* Levels of *A* and *c* Levels of *B*)

	B Levels							
A Levels	1	2	$\cdots$	j	$\cdots$	c	Row (*A*) Totals	Row Means
1	y_{11}	y_{12}	$\cdots$	y_{1j}	$\cdots$	y_{1c}	$T_{1\cdot}$	$\bar{y}_{1\cdot}$
2	y_{21}	y_{22}	$\cdots$	y_{2j}	$\cdots$	y_{2c}	$T_{2\cdot}$	$\bar{y}_{2\cdot}$
$\vdots$	$\vdots$	$\vdots$		$\vdots$		$\vdots$	$\vdots$	$\vdots$
i	y_{i1}	y_{i2}	$\cdots$	y_{ij}	$\cdots$	y_{ic}	$T_{i\cdot}$	$\bar{y}_{i\cdot}$
$\vdots$	$\vdots$	$\vdots$		$\vdots$		$\vdots$	$\vdots$	$\vdots$
r	y_{r1}	y_{r2}	$\cdots$	y_{rj}	$\cdots$	y_{rc}	$T_{r\cdot}$	$\bar{y}_{r\cdot}$
Column (B) Totals	$T_{\cdot 1}$	$T_{\cdot 2}$	$\cdots$	$T_{\cdot j}$	$\cdots$	$T_{\cdot c}$	$T_{\cdot\cdot}$	$\bar{y}_{\cdot\cdot}$
Column Means	$\bar{y}_{\cdot 1}$	$\bar{y}_{\cdot 2}$	$\cdots$	$\bar{y}_{\cdot j}$	$\cdots$	$\bar{y}_{\cdot c}$	$\bar{y}_{\cdot\cdot}$	

Often, the combined effect of factor A and factor B on the response variable y is the sum of the effects. Therefore, we often use the model

$$(5.60) \qquad E(y_{ij}) = \mu + \alpha_i + \gamma_j, \qquad i = 1,\ldots,r, \quad j = 1,\ldots,c$$

$$(5.60a) \qquad V(y_{ij}) = \sigma^2, \qquad \operatorname{cov}(y_{ij}, y_{i'j'}) = 0$$

$$(5.61) \qquad \sum_{i=1}^{r} \alpha_i = 0, \qquad \sum_{j=1}^{c} \gamma_j = 0,$$

where α_i is the effect of using level i of factor A, and γ_j is the effect of using level j of factor B. The constraints are imposed for uniqueness of μ, the α_i's, and the γ_j's. Suppose we had not imposed them; that is, suppose we assumed at the outset that

$$(5.62) \qquad E(y_{ij}) = \mu' + \alpha_i' + \gamma_j'.$$

Then we may rewrite (5.62) as

$$(5.62a) \qquad \begin{aligned} E(y_{ij}) &= \mu' + \bar{\alpha}' + \bar{\gamma}' + (\alpha_i' - \bar{\alpha}') + (\gamma_j' - \bar{\gamma}') \\ &= \mu + \alpha_i + \gamma_j \end{aligned}$$

where $\bar{\alpha}' = (1/r)\Sigma\alpha_i'$, $\bar{\gamma}' = (1/c)\Sigma\gamma_j'$, $\mu = \mu' + \bar{\alpha}' + \bar{\gamma}'$, $\alpha_i = \alpha_i' - \bar{\alpha}'$, and

$\gamma_j = \gamma'_j - \bar{\gamma}'$. Note that α_i and γ_j satisfy (5.61). The reader is invited to construct the obvious extension to the example mentioned in Section 5.2 at Table 5.2.3, and so on.

We again use the dot notation—which inserts into a subscript a dot if summation is over the corresponding subscript—so we have

$$\text{(5.63)} \qquad \begin{aligned} T_{i\cdot} &= \sum_{j=1}^{c} y_{ij}, \qquad i = 1,\ldots,r \\ T_{\cdot j} &= \sum_{i=1}^{r} y_{ij}, \qquad j = 1,\ldots,c. \end{aligned}$$

For purposes of general discussion, $T_{i\cdot}$ is called the ith row total, but in a specific experiment, it is referred to as the total of all the observations taken using the ith level of factor A. Also,

$$\text{(5.63}a\text{)} \qquad \bar{y}_{i\cdot} = \frac{T_{i\cdot}}{c}, \qquad \bar{y}_{\cdot j} = \frac{T_{\cdot j}}{r}$$

are row and column averages, respectively. We also define

$$\text{(5.63}b\text{)} \qquad T_{\cdot\cdot} = T = \sum_{i=1}^{r} \sum_{j=1}^{c} y_{ij} = \sum_{i=1}^{r} T_{i\cdot} = \sum_{j=1}^{c} T_{\cdot j} \qquad \text{(grand total)}$$

and

$$\text{(5.63}c\text{)} \qquad \begin{aligned} \bar{y}_{\cdot\cdot} = \bar{y} &= \frac{T}{rc} = \frac{1}{rc}\sum_{i} T_{i\cdot} = \frac{1}{rc}\sum_{j} T_{\cdot j} \\ &= \frac{1}{r}\sum_{i} \bar{y}_{i\cdot} = \frac{1}{c}\sum_{j} \bar{y}_{\cdot j} \qquad \text{(grand mean)}. \end{aligned}$$

Denoting

$$\text{(5.64)} \qquad \mathbf{y} = (y_{11},\ldots,y_{1c}, y_{21},\ldots,y_{2c},\ldots,y_{r1},\ldots,y_{rc})'$$

and

$$\text{(5.65)} \qquad \boldsymbol{\beta} = (\mu, \alpha_1,\ldots,\alpha_r, \gamma_1,\ldots,\gamma_c)'$$

(note that the total number of observations is $N = rc$ and $k = r + c + 1$), we may write (5.60) in matrix form as

$$E(\mathbf{y}) = X\boldsymbol{\beta}, \tag{5.66}$$

where X is of order $N \times (r+c+1)$, $N = rc$, and

$$X = \left[\begin{array}{c|cccc|cccc}
1 & 1 & & & & 1 & & & \\
1 & 1 & & & & & 1 & & \\
\vdots & \vdots & & & & & & \ddots & \\
1 & 1 & & & & & & & 1 \\
\hline
1 & & 1 & & & 1 & & & \\
 & & & & & & 1 & & \\
\vdots & & \vdots & & & & & \ddots & \\
1 & & 1 & & & & & & 1 \\
\hline
\vdots & & & \vdots & & & & \vdots & \\
\hline
1 & & & & 1 & 1 & & & \\
 & & & & & & 1 & & \\
\vdots & & & & \vdots & & & \ddots & \\
1 & & & & 1 & & & & 1
\end{array}\right] \tag{5.66a}$$

$$= \left[\mathbf{1}_N \,\vert\, \mathbf{x}_1 \,\vert\, \cdots \,\vert\, \mathbf{x}_r \,\vert\, \mathbf{x}_{r+1} \,\vert\, \cdots \,\vert\, \mathbf{x}_{r+c}\right].$$

Blank spaces are to be read as zeros in the above. What we are saying is that the $N \times 1$ vector $\mathbf{x}_j$, for $j = 1,\ldots,r$, has components zero except for the c places $(j-1)c + 1,\ldots, jc$, which are ones; and, similarly $\mathbf{x}_{r+t}$ is a $N \times 1$ vector of zeros, except for r ones located at components t, $c+t$, $2c + t,\ldots,(r-1)c + t$, $t = 1,\ldots,c$.

We note that the rank of X is

$$r(x) = p = r + c - 1 < k = r + c + 1, \tag{5.67}$$

since the sum of $\mathbf{x}_1,\ldots,\mathbf{x}_r$ is $\mathbf{1}_N$ and the sum of $\mathbf{x}_{r+1},\ldots,\mathbf{x}_{r+c}$ is also $\mathbf{1}_N$. We proceed as usual: The parameters of interest are the differential effects $(\alpha_1,\ldots,\alpha_r, \gamma_1,\ldots,\gamma_c)$, and so we eliminate the (grand) mean μ as a first step. [Note that if there is no effect on the response variable y when using A and B, that is, if changes in the levels of factor A and factor B do not affect the response variable so that $\alpha_i = \alpha_j$, $i \neq j = 1,\ldots,r$ and $\gamma_s = \gamma_t$, $s \neq t = 1,\ldots,c$, then the differential effects are all zero in view of the constraints (5.61), so that we have $E(y_{ij}) = \mu$. It is for this reason that μ is often referred to as the (unknown) grand mean.] So, regressing $\mathbf{y}$ on $\mathbf{1}_N$, we find on taking residuals that

$$\mathbf{e}_1 = \mathbf{y} - \bar{y}\mathbf{1}_N, \tag{5.68}$$

and we now deal with the model

$$(5.69) \qquad E(\mathbf{e}_1) = X_{2\cdot 1}\boldsymbol{\delta}, \qquad \boldsymbol{\delta} = (\boldsymbol{\alpha}', \boldsymbol{\gamma}')',$$

where the $N \times (r + c)$ matrix $X_{2\cdot 1}$ is found by regressing $\mathbf{x}_1, \ldots, \mathbf{x}_r, \mathbf{x}_{r+1}, \ldots, \mathbf{x}_{r+c}$ on $\mathbf{1}$ and taking residuals, and

$$(5.69a) \qquad \boldsymbol{\delta} = (\alpha_1, \ldots, \alpha_r, \gamma_1, \ldots, \gamma_c)'.$$

Denoting $X_{2\cdot 1}$ by Z, we have $E(\mathbf{e}) = Z\boldsymbol{\delta}$, where Z is as given in (5.70):

(5.70)

$$
E\left[\begin{array}{c}
e_{11}\\ \vdots \\ e_{1c}\\ \hline e_{21}\\ \vdots\\ e_{2c}\\ \hline e_{31}\\ \vdots\\ e_{3c}\\ \hline \vdots \\ \hline e_{r1}\\ \vdots\\ e_{rc}
\end{array}\right]
=
\left[\begin{array}{cccc|cccc}
1-1/r & -1/r & \cdots & -1/r & 1-1/c & -1/c & \cdots & -1/c\\
 & & & & -1/c & 1-1/c & & \\
\vdots & \vdots & & \vdots & & -1/c & \ddots & \vdots\\
 & & & & \vdots & \vdots & & -1/c\\
1-1/r & -1/r & \cdots & -1/r & -1/c & -1/c & \cdots & 1-1/c\\
\hline
-1/r & 1-1/r & \cdots & -1/r & 1-1/c & -1/c & \cdots & -1/c\\
 & & & & -1/c & 1-1/c & & \\
\vdots & \vdots & \vdots & \vdots & & -1/c & \ddots & \vdots\\
 & & & & \vdots & \vdots & & -1/c\\
-1/r & 1-1/r & \cdots & -1/r & -1/c & -1/c & \cdots & 1-1/c\\
\hline
-1/r & -1/r & \cdots & -1/r & 1-1/c & -1/c & \cdots & -1/c\\
 & & & & -1/c & 1-1/c & & \\
\vdots & \vdots & \vdots & \vdots & & -1/c & \ddots & \vdots\\
 & & & & \vdots & \vdots & & -1/c\\
-1/r & -1/r & \cdots & -1/r & -1/c & -1/c & \cdots & 1-1/c\\
\hline
\vdots & \vdots & & \vdots & \vdots & \vdots & & \vdots\\
\hline
-1/r & -1/r & \cdots & 1-1/r & 1-1/c & -1/c & \cdots & -1/c\\
 & & & & -1/c & 1-1/c & & \\
\vdots & \vdots & & \vdots & & -1/c & \ddots & \vdots\\
 & & & & \vdots & \vdots & & -1/c\\
-1/r & -1/r & \cdots & 1-1/r & -1/c & -1/c & \cdots & 1-1/c
\end{array}\right]
\times \left[\begin{array}{c}\boldsymbol{\alpha}\\ \hline \boldsymbol{\gamma}\end{array}\right].
$$

Problem 5.6 asks for verification of the form of Z. Let

$$(5.71) \qquad Z = \left(Z_1 \mid Z_2 \right) = \left(\mathbf{z}_1 \cdots \mathbf{z}_r \mid \mathbf{z}_{r+1} \cdots \mathbf{z}_{r+c} \right)$$

with Z_1 of order $N \times r$ and Z_2 of order $N \times c$. Then we find, in addition to $\mathbf{1}'\mathbf{z}_s = 0$, $s = 1,\ldots,r+c$, that

$$(5.71a) \qquad \mathbf{z}_i'\mathbf{z}_j = 0, \qquad i = 1,\ldots,r, \quad j = r+1,\ldots,r+c$$

and indeed $Z'Z$ is of order $(r+c) \times (r+c)$ and

(5.71b)

$$Z'Z = \left[\begin{array}{c|c} {}^{c}\!\begin{bmatrix} 1-1/r & -1/r & \cdots & -1/r \\ -1/r & 1-1/r & & \vdots \\ \vdots & \vdots & \ddots & -1/r \\ -1/r & -1/r & \cdots & 1-1/r \end{bmatrix} & \bigcirc \\ \hline \bigcirc' & {}^{r}\!\begin{bmatrix} 1-1/c & -1/c & \cdots & -1/c \\ -1/c & & & \vdots \\ \vdots & & \ddots & -1/c \\ -1/c & \cdots & & 1-1/c \end{bmatrix} \end{array} \right]$$

$$= \left[\begin{array}{c|c} Z_1'Z_1 & \bigcirc \\ \hline \bigcirc' & Z_2'Z_2 \end{array} \right].$$

Problem 5.6 asks for details.

Indeed, as can easily be verified (Problem 5.6),

$$(5.72) \qquad r(Z) = r(Z'Z) = r + c - 2.$$

Since $Z'Z$ is as in (5.71b), and since at this stage we are working with the model $E(\mathbf{e}) = Z\boldsymbol{\delta}$, the normal equations take the form

$$(5.73) \qquad \begin{bmatrix} Z_1'Z_1\hat{\boldsymbol{\alpha}}^{(0)} \\ Z_2'Z_2\hat{\boldsymbol{\gamma}}^{(0)} \end{bmatrix} = \begin{bmatrix} Z_1'\mathbf{e}_1 \\ Z_2'\mathbf{e}_1 \end{bmatrix},$$

and we may analyze

$$(5.73a) \qquad Z_1'Z_1\hat{\boldsymbol{\alpha}}^{(0)} = Z_1'\mathbf{e}_1 \quad \text{and} \quad Z_2'Z_2\hat{\boldsymbol{\gamma}}^{(0)} = Z_2'\mathbf{e}_1$$

separately. To find $\hat{\boldsymbol{\alpha}}^{(0)}$, we will use the method of Theorem 4.2.9. Note that [see (5.71)]

$$Z_1'Z_1 = c\left[I_r - \frac{1}{r}\mathbf{1}_r\mathbf{1}_r'\right] \tag{5.74}$$

or

$$\frac{1}{c}Z_1'Z_1 = I_r - \frac{1}{r}\mathbf{1}_r\mathbf{1}_r'. \tag{5.74a}$$

It is easy to check that the right-hand side of (5.74a) is idempotent of rank $r-1$, that is, not of full rank. We rearrange (5.74) to obtain

$$Z_1'Z_1 + \frac{c}{r}\mathbf{1}_r\mathbf{1}_r' = cI_r. \tag{5.75}$$

Now recalling the one-way example of the previous section, we augment Z_1 with B_1, where

$$B_1 = \left(\frac{c}{r}\right)^{1/2}\mathbf{1}_r'. \tag{5.75a}$$

Then, since one of the constraints is $\mathbf{1}_r'\boldsymbol{\alpha} = 0$ or equivalently $B_1\boldsymbol{\alpha} = 0$, we have (see Section 4.6, with $\boldsymbol{\delta} = \mathbf{0}$)

$$\begin{aligned}\hat{\boldsymbol{\alpha}}^{(0)} &= (Z_1'Z_1 + B_1'B_1)^{-1}Z_1'\mathbf{e}_1 \\ &= (cI_r)^{-1}Z_1'\mathbf{e}_1.\end{aligned} \tag{5.75b}$$

But (Problem 5.7)

$$Z_1'\mathbf{e}_1 = \left(c(\bar{y}_{1\cdot} - \bar{y}), \ldots, c(\bar{y}_{r\cdot} - \bar{y})\right)' \tag{5.75c}$$

so, from (5.75b),

$$\hat{\boldsymbol{\alpha}}^{(0)} = (\bar{y}_{1\cdot} - \bar{y}, \ldots, \bar{y}_{r\cdot} - \bar{y})'. \tag{5.75d}$$

Note that

$$B_1\hat{\boldsymbol{\alpha}}^{(0)} = 0. \tag{5.75e}$$

Clearly the (extra) regression sum of squares due to $\mathbf{x}_1, \ldots, \mathbf{x}_r$ is

(5.75f)

$$\mathbf{e}_1' Z_1 (Z_1' Z_1 + B_1' B_1)^{-1} Z_1' \mathbf{e}_1 = \frac{1}{c}\left(c(\bar{y}_{1\cdot} - \bar{y}), \ldots, c(\bar{y}_{r\cdot} - \bar{y})\right) \begin{bmatrix} c(\bar{y}_{1\cdot} - \bar{y}) \\ \vdots \\ c(\bar{y}_{r\cdot} - \bar{y}) \end{bmatrix}$$

$$= \frac{c^2}{c} \sum_i (\bar{y}_{i\cdot} - \bar{y})^2 = \sum_i c(\bar{y}_{i\cdot} - \bar{y})^2.$$

Similarly, $Z_2' Z_2$ is of rank $c - 1$, and the choice of

(5.76) $$B_2 = \left(\frac{r}{c}\right)^{1/2} \mathbf{1}_c'$$

leads to

(5.76a) $$\hat{\boldsymbol{\gamma}}^{(0)} = (\bar{y}_{\cdot 1} - \bar{y}, \ldots, \bar{y}_{\cdot c} - \bar{y})'$$

so that

(5.76b) $$B_2 \hat{\boldsymbol{\gamma}}^{(0)} = 0$$

and the (extra) regression sum of squares due to $\mathbf{x}_{r+1}, \ldots, \mathbf{x}_{r+c}$ is

(5.76c) $$\mathbf{e}_1' Z_2 (Z_2' Z_2 + B_2' B_2)^{-1} Z_2' \mathbf{e}_1 = \sum_j r(\bar{y}_{\cdot j} - \bar{y})^2.$$

We collect all the elements discussed above in an analysis of variance table: In matrix form we have Table 5.3.2, in simple algebraic form Table 5.3.3.

Under normality, it is straightforward to show that the various sums of squares listed in Table 5.3.2 are independent of the residual sum of squares [Problem 5.7(d)].

The algebraic form of Table 5.3.2 is presented in Table 5.3.3. Problem 5.7(e) asks for details. Note the analysis of variance breakdown implied by the sum of squares column of Table 5.3.3. We have

(5.77)

$$\Sigma\Sigma(y_{ij} - \bar{y})^2 = \mathbf{y}'\mathbf{y} - rc\bar{y}^2$$

$$= c\sum_1^r (\bar{y}_{i\cdot} - \bar{y})^2 + r\sum_1^c (\bar{y}_{\cdot j} - \bar{y})^2 + \Sigma\Sigma(y_{ij} - \bar{y}_{i\cdot} - \bar{y}_{\cdot j} + \bar{y})^2$$

$$= \quad \mathrm{SS}_A \quad + \quad \mathrm{SS}_B \quad + \quad \mathrm{SS}_e$$

TABLE 5.3.2. Analysis of Variance for the Data of Table 5.3.1 when (5.60), (5.60*a*), and (5.61) Hold

Source	Degrees of Freedom		Sum of Squares	Expected Sum of Squares
Mean	1	$p = r(X)$ $= r + c - 1$	$\mathbf{y}'\mathbf{1}_N(\mathbf{1}'_N\mathbf{1}_N)^{-1}\mathbf{1}'_N\mathbf{y} = N\bar{y}^2$	
Rows	$r - 1$		$\mathbf{e}'_1 Z_1(Z'_1Z_1 + B'_1B_1)^{-1}Z'_1\mathbf{e}_1$ $= \mathbf{e}'_1\mathcal{R}^{(\text{nf})}_{B_1}(Z_1)\mathbf{e}_1$ $= \mathbf{y}'\mathcal{R}^{(\text{nf})}_{B_1}(Z_1)\mathbf{y}$	
Columns	$c - 1$		$\mathbf{e}'_1 Z_2(Z'_2Z_2 + B'_2B_2)^{-1}Z'_2\mathbf{e}_1$ $= \mathbf{e}'_1\mathcal{R}^{(\text{nf})}_{B_2}(Z_2)\mathbf{e}'_1 = \mathbf{y}'\mathcal{R}^{(\text{nf})}_{B_2}(Z_2)\mathbf{y}$	
Residual	$rc - (r + c - 1)$ $= (r - 1)(c - 1)$		$\mathbf{e}'_1[I_N - \mathcal{R}^{(\text{nf})}_{B_1}(Z_1) - \mathcal{R}^{(\text{nf})}_{B_2}(Z_2)]\mathbf{e}_1$ $= (\mathbf{y} - \bar{y}\mathbf{1}_N)'(\mathbf{y} - \bar{y}\mathbf{1}_N)$ $-\mathbf{y}'[\mathcal{R}^{(\text{nf})}_{B_1}(Z_1) + \mathcal{R}^{(\text{nf})}_{B_2}(Z_2)]\mathbf{y}$	
Total	rc		$\mathbf{y}'\mathbf{y}$	

$$\mathcal{R}^{(\text{nf})}_{B_i}(Z_i) = Z_i(Z'_iZ_i + B'_iB_i)^{-1}Z'_i \qquad \mathbf{1}'_N Z_i = \mathbf{0}', \quad N = rc$$

Problem 5.7(e) asks for a direct proof. It is easy to see that SS_A, the between rows sum of squares, gives information about the differences of the α_i's, while SS_B, the between columns sum of squares, gives information about the differences in γ_j's, apart from σ^2, and indeed that SS_e, the residual sum of squares, reflects information about σ^2 only. Problem 5.7 also asks the reader to find the entries of the last column of Table 5.3.3 (or Table 5.3.2), which are expected values of SS_A, SS_B, and SS_e. Motivated by these latter results and assuming normality, we can construct an F-test for various standard hypotheses, and the reader is asked to do this also in Problem 5.7(f).

TABLE 5.3.3.

Source	Degrees of Freedom	Sum of Squares	Expected Sum of Squares
Mean	1	$rc\bar{y}^2$	
Rows	$r - 1$	$c\sum_1^r(\bar{y}_{i\cdot} - \bar{y})^2$	
Columns	$c - 1$	$r\sum_1^c(\bar{y}_{\cdot j} - \bar{y})^2$	
Residual	$(r - 1)(c - 1)$	$\Sigma\Sigma(y_{ij} - \bar{y}_{i\cdot} - \bar{y}_{\cdot j} + \bar{y})^2$	
Total	rc	$\mathbf{y}'\mathbf{y}$	

5.3b. Interaction

An experiment such as described and analyzed in Section 5.3a is often called a $(r \times c)$-*factorial* experiment—sometimes referred to as a *completely randomized block* experiment. The reader has doubtless noticed that the Section 5.3a was entitled "No Interaction." By this we mean that it is expected that an observation, say y_{ij}, will possibly be affected by the use of factor-A level i, as measured by α_i, and the use of factor-B level j, as measured by γ_j, and that these affect the response additively. That is,

$$E(y_{ij}) = \mu + \alpha_i + \gamma_j. \tag{5.78}$$

Using (5.78), we have

$$E(y_{ij} - y_{i'j}) = \alpha_i - \alpha_{i'}, \qquad j = 1,\ldots,c, \tag{5.79}$$

that is, the differences in response to different levels of A, using the same level j of B, is expected to be the same for all levels of B $(j = 1,\ldots,c)$.

A simple example introduced at the beginning of Section 5.3(a) illustrates the situation. Suppose the A factor is pressure, used at the two levels 14.8 and 15.0 lb/in.^{2}, while the B factor is temperature, used at the three levels 92, 94, and 96°C. Suppose we know that

$$\begin{aligned} \mu = 100, \qquad \alpha_1 &= -4, \quad \gamma_1 = -12, \\ \alpha_2 &= 4, \quad \gamma_2 = -4, \\ & \quad \gamma_3 = 16. \end{aligned} \tag{5.80}$$

From (5.78), we find $E(y_{ij})$, $i = 1, 2$, $j = 1, 2, 3$, and enter them in Table 5.3.4.

TABLE 5.3.4. Values of $E(y_{ij})$, $i = 1, 2, j = 1, 2, 3$

	Temperatures		
Pressures	1	2	3
1	84	92	112
2	92	100	120

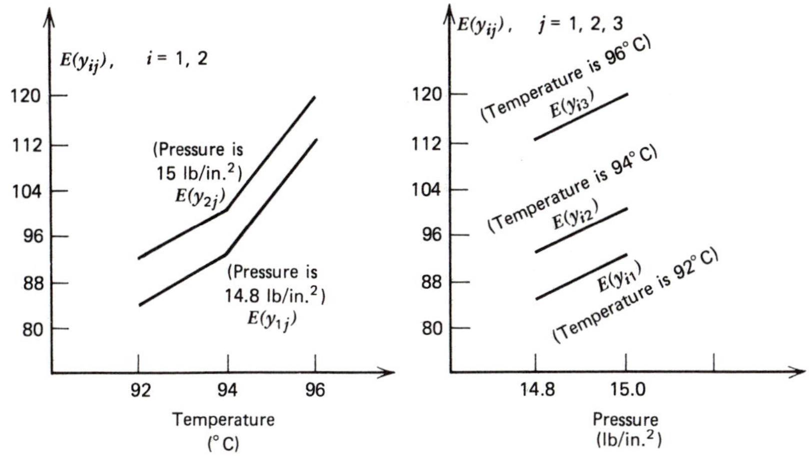

FIGURE 5.3.1.

We may plot the data in two ways, as shown in Figure 5.3.1. Note that the lines joining the plotted points are parallel. Indeed [see (5.79(*a*)],

$$E(y_{21}-y_{11}) = 92 - 84 = 8 = \alpha_2 - \alpha_1 = 4 - (-4) = 8,$$

(5.80*a*) $$E(y_{22}-y_{12}) = 100 - 92 = 8 = \alpha_2 - \alpha_1 = 4 - (-4) = 8,$$

$$E(y_{23}-y_{13}) = 120 - 112 = 8 = \alpha_2 - \alpha_1 = 4 - (-4) = 8.$$

Similarly,

$$E(y_{13}-y_{11}) = 112 - 84 = 28 = \gamma_3 - \gamma_1 = 16 - (-12) = 28,$$

(5.80*b*)

$$E(y_{23}-y_{21}) = 120 - 92 = 28 = \gamma_3 - \gamma_1 = 16 - (-12) = 28,$$

and so on.

Often, the model (5.78) is not appropriate, that is, the differences in response to different levels of A, using the same level of B, are *not* expected to be the same. We then say that the levels of A *interact* with the levels of B, and the appropriate model here is

(5.81) $$E(y_{ij}) = \mu + \alpha_i + \gamma_j + \lambda_{ij}.$$

TABLE 5.3.5. Values of $E(y_{ij})$

86	95	107
90	97	125

To gain insight into the effect of the so-called *interaction* effects, suppose that in our previous example (5.81) holds with

$$\lambda_{11} = +2, \quad \lambda_{12} = +3, \quad \lambda_{13} = -5,$$
$$\lambda_{21} = -2, \quad \lambda_{22} = -3, \quad \lambda_{23} = +5. \tag{5.82}$$

(Note that $\Sigma_{i=1}^{2}\lambda_{ij} = \Sigma_{j=1}^{3}\lambda_{ij} = 0$.) Then Table 5.3.4 is replaced by Table 5.3.5.

When we graph these (Figure 5.3.2), we find, of course, that the "parallelism" behavior is no longer present, because *A interacts* with *B* [see (5.82)]. The α_i and γ_j sometimes are called *main effects*, and to repeat, the λ_{ij} are called *interaction* or *first-order interaction effects*.

Usually, when an experimenter is unsure about the question of interaction of the factors in the experiment, a model of the type (5.81) is postulated and more than one observation per treatment taken. (A treatment is the combination of a level of factor *A* and a level of factor *B* used to generate

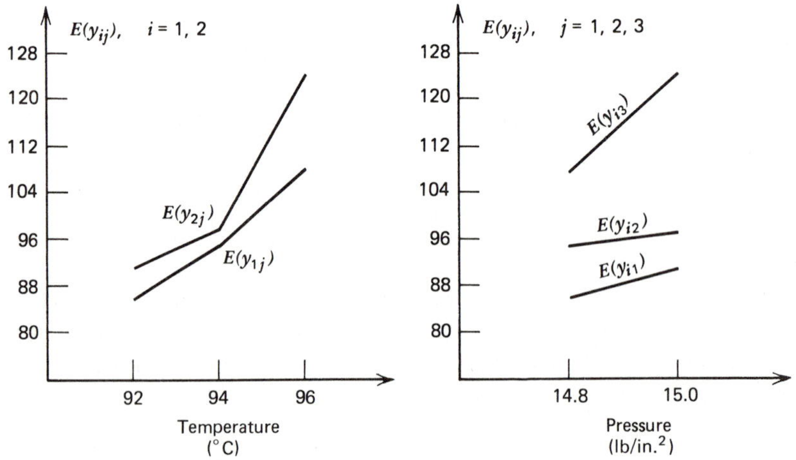

FIGURE 5.3.2.

an observation and denoted typically by A_iB_j.) We will see later that this is in accordance with the principle in Section 3.6, and another rationale for replication is given below (5.83). Hence, for each treatment we will generate l observations say y_{ijt}, $t = 1,\ldots,l$, when the ith level of A and the jth level of B are used. We may conveniently tabulate the observations in a two-way array such as Table 5.3.6.

Our model is

$$y_{ijt} = \mu + \alpha_i + \gamma_j + \lambda_{ij} + \varepsilon_{ijt}, \qquad i = 1,\ldots,r, \quad j = 1,\ldots,c,$$

(5.83)
$$\sum_{i=1}^{r} \alpha_i = 0, \quad \sum_{j=1}^{c} \gamma_j = 0, \quad \sum_i \lambda_{ij} = 0, \qquad j = 1,\ldots,c,$$

$$\sum_j \lambda_{ij} = 0, \qquad i = 1,\ldots,r,$$

where $t = 1,\ldots,l$, and

(5.83*a*)
$$E(\varepsilon_{ijt}) = 0, \quad V(\varepsilon_{ijt}) = \sigma^2, \quad \operatorname{cov}(\varepsilon_{ijt}, \varepsilon_{i'j't'}) = 0.$$

TABLE 5.3.6.

Factor A	Factor B: 1	2	$\cdots$	j	$\cdots$	c	$T_{i\cdot\cdot}$
1	y_{111} ⋮ y_{11l}	y_{121} ⋮ y_{12l}	$\cdots$	y_{1j1} ⋮ y_{1jl}	$\cdots$	y_{1c1} ⋮ y_{1cl}	$T_{1\cdot\cdot}$
2	y_{211} ⋮ y_{21l}	y_{221} ⋮ y_{22l}	$\cdots$	y_{2j1} ⋮ y_{2jl}	$\cdots$	y_{2c1} ⋮ y_{2cl}	$T_{2\cdot\cdot}$
⋮	⋮	⋮	$\cdots$	⋮	$\cdots$	⋮	⋮
i	y_{i11} ⋮ y_{i1l}	y_{i21} ⋮ y_{i2l}	$\cdots$	y_{ij1} ⋮ y_{ijl}	$\cdots$	y_{ic1} ⋮ y_{icl}	$T_{i\cdot\cdot}$
⋮	⋮	⋮	$\cdots$	⋮	$\cdots$	⋮	⋮
r	y_{r11} ⋮ y_{r1l}	y_{r21} ⋮ y_{r2l}	$\cdots$	y_{rj1} ⋮ y_{rjl}	$\cdots$	y_{rc1} ⋮ y_{rcl}	$T_{r\cdot\cdot}$
$T_{\cdot j\cdot}$	$T_{\cdot 1\cdot}$	$T_{\cdot 2\cdot}$	$\cdots$	$T_{\cdot j\cdot}$	$\cdots$	$T_{\cdot c\cdot}$	$T_{\cdot\cdot\cdot} = T$

Examining the constraints used in the model (5.83) shows that the number of linearly independent parameters used in describing the expectations $E(y_{ijt})$ is

$$(r-1)(c-1)+(r-1)+(c-1)+1=rc,$$

where $(r-1)(c-1)$ is the number of independent λ_{ij}'s, $r-1$ is the number of independent α's, $c-1$ is the number of independent γ's, and the addition of 1 is necessary for the parameter μ. Hence, if we were to use an experimental plan of the type implied by Table 5.3.1, which calls for only rc observations generated by taking one observation for each treatment A_iB_j, we would need the rc degrees of freedom to estimate the rc linearly independent parameters, say μ, $\alpha_1,\ldots,\alpha_{r-1}$, $\gamma_1,\ldots,\gamma_{c-1}$, $\lambda_{11},\ldots,\lambda_{1,c-1}$, $\lambda_{21},\ldots,\lambda_{2,c-1},\ldots,\lambda_{r-1,1},\ldots,\lambda_{r-1,c-1}$, and would have no degrees of freedom left over for error. Thus, as mentioned before, some replication is called for, and as model (5.83) suggests, we are discussing the case of balanced replication, that is, *for each treatment* A_iB_j, we generate l observations y_{ijt}, $t=1,\ldots,l$.

We note also that the constraints of (5.83) are imposed, just as in previous work, to ensure uniqueness of definition for the parameters μ, α_i, γ_j, and λ_{ij}. We also note here that a model which implies the additivity of the "main effects" of A and the "main effects" of B along with the interaction effect of A and B can always be written as in (5.83). That is, if we have

$$E(y_{ijt})=\mu'+\alpha_i'+\gamma_j'+\lambda_{ij}', \tag{5.83b}$$

then we may write

(5.83c)

$$E(y_{ijt})=\left[\mu'+\bar{\alpha}'+\bar{\gamma}'+\bar{\lambda}'_{..}\right]+\left[(\alpha_i'-\bar{\alpha}')+(\bar{\lambda}'_{i.}-\bar{\lambda}'_{..})\right]$$
$$+\left[(\gamma_j'-\bar{\gamma}')+(\bar{\lambda}'_{.j}-\bar{\lambda}'_{..})\right]+\left[\lambda_{ij}'-\bar{\lambda}'_{i.}-\bar{\lambda}'_{.j}+\bar{\lambda}'_{..}\right],$$

where

(5.83d)

$$\bar{\alpha}'=r^{-1}\sum_{i=1}^{r}\alpha_i',\quad \bar{\lambda}'_{i.}=c^{-1}\sum_{j=1}^{c}\lambda_{ij}',\quad \lambda'_{.j}=r^{-1}\sum_{i=1}^{r}\lambda_{ij}',$$
$$\bar{\lambda}'_{..}=(rc)^{-1}\sum_{i=1}^{r}\sum_{j=1}^{c}\lambda_{ij},\quad \text{and so on.}$$

We now define

$$(5.83e)\quad \begin{aligned} &\mu = \mu' + \bar{\alpha}' + \bar{\gamma}' + \bar{\lambda}'_{..}, \quad \alpha_i = \left[(\alpha'_i - \bar{\alpha}') + (\bar{\lambda}'_{i.} - \bar{\lambda}'_{..})\right] \\ &\gamma_j = \left[(\gamma'_j - \bar{\gamma}') + (\bar{\lambda}'_{.j} - \bar{\lambda}'_{..})\right], \quad \lambda_{ij} = \lambda'_{ij} - \bar{\lambda}'_{i.} - \bar{\lambda}'_{.j} + \bar{\lambda}'_{..} \end{aligned}$$

and as the reader may easily verify,

$$(5.84)\qquad \sum_i \alpha_i = 0, \quad \sum_j \gamma_j = 0, \quad \sum_i \lambda_{ij} = \sum_j \lambda_{ij} = 0.$$

One important point emerges in the above exercise—an exercise needed, of course, for uniqueness. For example, we note that the effect α_i is composed of main effects α'_i, through $(\alpha'_i - \bar{\alpha}')$, and interaction effects of the levels of B with A_i, through $\bar{\lambda}'_{i.} - \bar{\lambda}'_{..}$. Hence, in practice, after constructing the analysis of variance which we discuss below (see Table 5.3.7), interaction effects are always examined first. If the data support the statement that interaction effects are zero, we can then examine statements about the α_i and γ_j, because when interaction effects are zero, $\alpha_i = \alpha'_i - \bar{\alpha}'$ and $\gamma_j = \gamma'_j - \bar{\gamma}'$, that is, the α_i are differential main effects due only to row effects, and so on.

To see all this more precisely, we need the following notation: Let

$$T_{ij.} = \sum_{t=1}^{l} y_{ijt}, \qquad i = 1,\dots,r, \quad j = 1,\dots,c \text{ (cell or treatment totals)},$$

$$T_{i..} = \sum_{j=1}^{c} T_{ij.} = \sum_{j=1}^{c}\sum_{t=1}^{l} y_{ijt}, \qquad i = 1,\dots,r \text{ (row totals)},$$

$$T_{.j.} = \sum_{i=1}^{r} T_{ij.} = \sum_{i=1}^{r}\sum_{t=1}^{l} y_{ijt}, \qquad j = 1,\dots,c \text{ (column totals)},$$

$$T_{...} = T = \sum_{i=1}^{r} T_{i..} = \sum_{j=1}^{c} T_{.j.} = \sum_i\sum_j\sum_t y_{ijt} \qquad \text{(grand total)},$$

and, correspondingly,

$$\bar{y}_{ij.} = \frac{T_{ij.}}{l} \qquad \text{(cell means)},$$

$$(5.84a)\quad \bar{y}_{i..} = \frac{T_{i..}}{cl} \qquad \text{(row means)}, \quad \bar{y}_{.j.} = T_{.j.}/rl \qquad \text{(column means)},$$

$$\bar{y}_{...} = \bar{y} = \frac{T}{rcl} = \frac{T}{N}, \qquad N = rcl.$$

Denoting

$$\mathbf{y} = (y_{111}, \ldots, y_{11l}, y_{121}, \ldots, y_{12l}, \ldots, y_{1c1}, \ldots, y_{1cl}, \tag{5.85}$$
$$y_{211}, \ldots, y_{21l}, \ldots, y_{rc1}, \ldots, y_{rcl})'$$

and

(5.85a)

$$\boldsymbol{\beta} = (\mu, \alpha_1, \ldots, \alpha_r, \gamma_1, \ldots, \gamma_c, \lambda_{11}, \lambda_{12}, \ldots, \lambda_{1c}, \lambda_{21}, \ldots, \lambda_{2c}, \ldots, \lambda_{r1}, \ldots, \lambda_{rc})',$$

we may write the model in familiar matrix notation form. This is asked for in Problem 5.8. But, with the results of Section 5.3(a) in our mind, we proceed by first breaking up the total sum of squares of deviations into a between rows sum of squares (SS_A), a between columns sum of squares (SS_B), and at that point examine what is left over. Using natural notation,

$$\mathbf{e}_1'\mathbf{e}_1 = \Sigma\Sigma\Sigma(y_{ijt} - \bar{y})^2 \tag{5.86}$$
$$= \Sigma\Sigma\Sigma\left[(y_{ijt} - \bar{y}_{i\cdot\cdot} - \bar{y}_{\cdot j\cdot} + \bar{y}) + (\bar{y}_{i\cdot\cdot} - \bar{y}) + (\bar{y}_{\cdot j\cdot} - \bar{y})\right]^2.$$

We have inserted $(\bar{y}_{i\cdot\cdot} - \bar{y})$ and $(\bar{y}_{\cdot j\cdot} - \bar{y})$ in order to develop terms which, when squared and summed, would give terms analogous to SS_A and SS_B of Section 5.3a. Indeed, using the relations (5.84a), it may be shown that (Problem 5.9)

(5.86a)

$$\mathbf{e}_1'\mathbf{e}_1 = \Sigma\Sigma\Sigma(y_{ijt} - \bar{y}_{i\cdot\cdot} - \bar{y}_{\cdot j\cdot} + \bar{y})^2 + lc\sum_i(\bar{y}_{i\cdot\cdot} - \bar{y})^2 + lr\sum_j(\bar{y}_{\cdot j\cdot} - \bar{y})^2$$
$$= \Sigma\Sigma\Sigma(y_{ijt} - \bar{y}_{i\cdot\cdot} - \bar{y}_{\cdot j\cdot} + \bar{y})^2 + SS_A + SS_B.$$

We now wish to break down the first term on the right-hand side of (5.86a). We are in an analogous position to that described in Section 3.6: We have rc groups, each with l observations, so that

$$\frac{1}{l-1}\sum_{t=1}^{l}(y_{ijt} - \bar{y}_{ij\cdot})^2 \tag{5.87}$$

is unbiased for σ^2 for each (i, j). Hence,

$$\text{(5.87}a\text{)} \qquad \frac{1}{rc(l-1)} \sum_i \sum_j \sum_t (y_{ijt} - \bar{y}_{ij\cdot})^2$$

is unbiased for σ^2, and using this as a clue we rewrite the first term on the right-hand side of (5.86a) as

$$\text{(5.88)} \qquad \begin{aligned} &\Sigma\Sigma\Sigma(y_{ijt} - \bar{y}_{i\cdot\cdot} - \bar{y}_{\cdot j\cdot} + \bar{y})^2 \\ &\qquad = \Sigma\Sigma\Sigma\left[(y_{ijt} - \bar{y}_{ij\cdot}) + (\bar{y}_{ij\cdot} - \bar{y}_{i\cdot\cdot} - \bar{y}_{\cdot j\cdot} + \bar{y})\right]^2. \end{aligned}$$

[Incidentally, if the model were $E(y_{ijt}) = \mu + \alpha_i + \gamma_j$, then it is easy to see that the residual sum of squares that would be obtained is the left-hand side of (5.88).] Now using (5.84a), we find

$$\text{(5.89)} \qquad \begin{aligned} &\Sigma\Sigma\Sigma(y_{ijt} - \bar{y}_{i\cdot\cdot} - \bar{y}_{\cdot j\cdot} + \bar{y})^2 \\ &\qquad = \Sigma\Sigma\Sigma(y_{ijt} - \bar{y}_{ij\cdot})^2 + l\Sigma\Sigma(\bar{y}_{ij\cdot} - \bar{y}_{i\cdot\cdot} - \bar{y}_{\cdot j\cdot} + \bar{y})^2 \\ &\qquad = \text{SS}_e + \text{SS}_{I_{AB}}. \end{aligned}$$

(I_{AB} stands for interaction of the A and B factors.) We know from (5.87a) that SS_e gives information only about error, and we note that the general term of $\text{SS}_{I_{AB}}$ is the square of

$$\text{(5.90)} \qquad \bar{y}_{ij\cdot} - \bar{y}_{i\cdot\cdot} - \bar{y}_{\cdot j\cdot} + \bar{y}.$$

Using (5.83),

(5.90a)

$$\begin{aligned} \bar{y}_{ij\cdot} - \bar{y}_{i\cdot\cdot} - \bar{y}_{\cdot j\cdot} + \bar{y} &= \left(\lambda_{ij} - \bar{\lambda}_{i\cdot} - \bar{\lambda}_{\cdot j} + \bar{\lambda}\right) + \left(\bar{\varepsilon}_{ij\cdot} - \bar{\varepsilon}_{i\cdot\cdot} - \bar{\varepsilon}_{\cdot j\cdot} + \bar{\varepsilon}\right) \\ &= \lambda_{ij} + \left(\bar{\varepsilon}_{ij\cdot} - \bar{\varepsilon}_{i\cdot\cdot} - \bar{\varepsilon}_{\cdot j\cdot} + \bar{\varepsilon}\right), \end{aligned}$$

in view of the constraints given in (5.83), so that indeed $\text{SS}_{I_{AB}}$ will give information about the interactions, apart from error.

Note the form of $\text{SS}_{I_{AB}}$ and compare it with the error term in Section 5.3a. In Section 5.3a, we assumed no interaction, that is, that the λ_{ij} are zero, and in fact, if $\lambda_{ij} \neq 0$, then the error term for Section 5.3a will reflect

TABLE 5.3.7. Analysis of Variance for the Data of Table 5.3.6, when the Model (5.83) Applies ($n = rcl$)

Source	Degrees of Freedom	Sum of Squares	Expected Sum of Squares
Mean	1	$N\bar{y}^2$	
Rows	$r-1$	SS_A	
Columns	$c-1$	SS_B	
Interaction	$(r-1)(c-1)$	$SS_{I_{AB}}$	
Error	$rc(l-1)$	SS_e	
Total	rcl	$\mathbf{y}'\mathbf{y}$	

information about σ^2 and the λ_{ij}, and hence could not be used to compare SS_A, SS_B, and so on. See Problem 5.10 for further points and details.

Back to the experiment of this section, we have

$$\mathbf{e}_1'\mathbf{e}_1 = SS_A + SS_B + SS_{I_{AB}} + SS_e. \tag{5.91}$$

Recall that $\mathbf{e}_1$ is obtained by elimination of the mean, and hence the total degrees of freedom associated with (5.91) is $N - 1 = rcl - 1$. The sum of squares SS_A and SS_B have $r - 1$ and $c - 1$ degrees of freedom, respectively.

Furthermore, it is clear from the definition of SS_e that the degrees of freedom associated with it are $rc \times (l - 1)$. Hence, we have that the degrees of freedom associated with $SS_{I_{AB}}$ are

$$(N-1) - (r-1) - (c-1) - rc(l-1) = rc - r - c + 1 \tag{5.91a}$$

or

$$(r-1)(c-1). \tag{5.91b}$$

If $l = 1$, then $\bar{y}_{ij\cdot} = y_{ij}$ and $SS_e = 0$, and we find ourselves in the same situation as Section 5.3a, and we would construct an analysis of variance table that resembles Table 5.3.3, with SS_e of that table equal to $SS_{I_{AB}}$ with $l = 1$. This would be inadequate if the model (5.83) should be used with $l = 1$, since then the residual line of Table 5.3.3 would not be a valid error term (Problem 5.11). Because we wish to use a model that entertains the possibility of interaction, we proceed as in Section 3.6: We take more than one observation per treatment (here we do this for all treatments and take equal numbers per treatment for balance). Doing this yields the analysis sketched above, and the analysis of variance table is as in Table 5.3.7. (Problem 5.12 asks for the relevant tests under normality, proofs of independence of the quadratic forms, and the expected sum of squares column.)

Using the results (Problem 5.12) of the expected sum of squares column as a guide, we can construct F-statistics to use as test statistics of various hypotheses, such as H_0: $\boldsymbol{\alpha} = \mathbf{0}$ or H_0': $\boldsymbol{\gamma} = \mathbf{0}$ or H_0'': $\boldsymbol{\lambda} = \mathbf{0}$ [see (5.83)]. These F-tests turn out to be the tests derived from the likelihood ratio criterion for the non-full-rank case—see Section 5.4 (and Problem 5.12). We would of course test H_0'' first, and only if this test is accepted, would we consider testing H_0 and/or H_0'.

It is interesting to note that we may indeed treat the situation of this subsection as a "lack-of-fit" problem, as alluded to previously [see after (5.86a)]. Suppose indeed that data are taken under the assumptions leading to Table 5.3.6, and that the *model fitted* is [see (5.60), (5.60a), and (5.61)]

$$(5.92) \qquad E(y_{ijt}) = \mu + \alpha_i + \gamma_j, \qquad t = 1, \ldots, l, \text{ and so on.}$$

It is easy to see from Section 5.3a that the residual sum of squares for this situation is given by the first term on the right-hand side of (5.86a). Furthermore, we are in the position where we suspect lack of fit and, indeed, that the model (see (5.83).

$$(5.92a) \quad E(y_{ijt}) = \mu + \alpha_i + \gamma_j + \lambda_{ij}, \qquad t = 1, \ldots, l, \text{ and so on,}$$

is the correct model. The design used calls for replication, with l observations ($l > 1$) for each of the rc treatments (called groups in Section 3.6). Using the principles found when discussing lack of fit in Section 3.6, we know that a basis for estimating error is

$$(5.92b) \qquad \mathrm{SS}_W = \sum_{i=1}^{r} \sum_{j=1}^{c} \sum_{t=1}^{l} (y_{ijt} - \bar{y}_{ij\cdot})^2$$

and furthermore, that the lack of fit term is given by the difference of the residual sum of squares and SS_W. That is,

$$(5.92c) \quad \mathrm{SS}_L = \Sigma\Sigma\Sigma(y_{ijt} - \bar{y}_{i\cdot\cdot} - \bar{y}_{\cdot j\cdot} + \bar{y}_{\cdot\cdot\cdot})^2 - \Sigma\Sigma\Sigma(y_{ijt} - \bar{y}_{ij\cdot})^2,$$

and it is easy to see that we have

$$(5.92d) \qquad \mathrm{SS}_L = \Sigma\Sigma\Sigma(\bar{y}_{ij\cdot} - \bar{y}_{i\cdot\cdot} - \bar{y}_{\cdot j\cdot} + \bar{y}_{\cdot\cdot\cdot})^2 = \mathrm{SS}_I.$$

Hence, the interaction sum of squares is the lack of fit term for this problem.

5.4. GENERAL LINEAR HYPOTHESIS: NON-FULL-RANK CASE

Often we wish to examine the statement (compare Section 3.7)

$$H_0: B\boldsymbol{\beta} = \mathbf{b}, \tag{5.93}$$

where $\boldsymbol{\beta}$ is of order $k \times 1$ and B is a $s \times k$ matrix of rank s, and our model is

$$\mathbf{y} = X\boldsymbol{\beta} + \boldsymbol{\varepsilon}, \tag{5.94}$$

with the rank of the $N \times k$ matrix X given by $r(X) = p$, where we assume

$$s \leq p < k. \tag{5.94a}$$

We also assume that H_0 is a *testable* hypothesis (see Section 4.4 of Chapter 4), which is to say that the s functions $B\boldsymbol{\beta}$ are *estimable*. Since this is so, we know that there exists a matrix T, which is of order $s \times N$ and such that $T\mathbf{y}$ is an unbiased estimator of $B\boldsymbol{\beta}$ (for all possible values of $\boldsymbol{\beta}$) so that

$$B = TX. \tag{5.95}$$

Now recall that we may reduce a non-full-rank model to a full-rank model by finding a $k \times q$ matrix U $(q = k - p)$ of rank q such that

$$XU = \bigcirc_{N\times q}, \tag{5.96}$$

and constructing a $k \times k$ nonsingular matrix H, where

$$H = \left[V \;\vdots\; U\right] \tag{5.96a}$$

with V a $k \times p$ matrix of rank $r(V) = p$. Then

$$\begin{aligned} E(\mathbf{y}) &= XHH^{-1}\boldsymbol{\beta} = \left[XV \;\vdots\; \bigcirc_{N\times q}\right](\boldsymbol{\phi}_1' \;\vdots\; \boldsymbol{\phi}_2')' \\ &= XV\boldsymbol{\phi}_1, \end{aligned} \tag{5.97}$$

where

$$\boldsymbol{\phi} = (\boldsymbol{\phi}_1' \;\vdots\; \boldsymbol{\phi}_2')' = H^{-1}\boldsymbol{\beta} \tag{5.97a}$$

and $\boldsymbol{\phi}_1$ is $p \times 1$, and so on.

We also note that

(5.98) $$B\boldsymbol{\beta} = BHH^{-1}\boldsymbol{\beta} = \left[\,BV \;\vdots\; BU\,\right]\begin{bmatrix}\boldsymbol{\phi}_1 \\ \hdashline \boldsymbol{\phi}_2\end{bmatrix},$$

that is, H_0 given in (5.93) is now of the form

(5.98a) $$B\boldsymbol{\beta} = BV\boldsymbol{\phi}_1 + BU\boldsymbol{\phi}_2 = \mathbf{b}.$$

But $B = TX$ and $XU = \bigcirc$, so that (5.98a) is of the form

(5.98b) $$BV\boldsymbol{\phi}_1 = \mathbf{b}.$$

We note too that since

(5.99)

$$BH = \left[\,BV \;\vdots\; BU\,\right] = \left[\,TXV \;\vdots\; TXU\,\right] = \left[\,TXV \;\vdots\; \bigcirc\,\right] = \left[\,BV \;\vdots\; \bigcirc\,\right]$$

and H is nonsingular, $r(BH) = r(B) = s$, so that from (5.99) we have

(5.99a) $$r(BV) = s.$$

That is,

(5.99b) $$BV\boldsymbol{\phi}_1 = \mathbf{b}$$

represents s linearly independent constraints on $\boldsymbol{\phi}_1$. In summary then, (5.93), (5.94), and (5.94a) are equivalent to (5.97)–(5.98b); that is, we have reduced the non-full-rank problem (5.93) and (5.94) to the full-rank problem

(5.100) $$E(\mathbf{y}) = W\boldsymbol{\phi}_1, \qquad H_0\colon BV\boldsymbol{\phi}_1 = \mathbf{b},$$

where $W = XV$ is of order $N \times p$ and of full rank p [see (5.99)], and $BV\boldsymbol{\phi}_1 = \mathbf{b}$ puts s linearly independent constraints on the $s \times 1$ vector $\boldsymbol{\phi}_1$. Assuming that

(5.101) $$\boldsymbol{\varepsilon} = N(\mathbf{0}, \sigma^2 I)$$

we may apply the full-rank method of Section 3.7 and find the likelihood ratio tests, and so on. We illustrate with a simple example, at this point.

EXAMPLE 5.4.1.

Suppose we have a one-way analysis of variance with two groups and two observations per group, so that we may write (see Section 5.2)

$$\mathbf{y} = X\boldsymbol{\beta} + \boldsymbol{\varepsilon}, \tag{5.102}$$

where $\mathbf{y}$ is of order 4×1, $\boldsymbol{\beta}$ is of order 3×1, $\boldsymbol{\varepsilon} = N(\mathbf{0}, \sigma^2 I_4)$, and

$$X = \begin{bmatrix} 1 & 1 & 0 \\ 1 & 1 & 0 \\ 1 & 0 & 1 \\ 1 & 0 & 1 \end{bmatrix}, \quad \boldsymbol{\beta} = (\mu, \beta_1, \beta_2)', \quad \beta_1 + \beta_2 = 0. \tag{5.102a}$$

We wish to test

$$H_0: \beta_1 = \beta_2 \tag{5.102b}$$

or, equivalently,

$$H_0: (0, 1, -1)\boldsymbol{\beta} = 0. \tag{5.102c}$$

Here $N = 4$, $k = 3$, $p = 2$, and $s = q = 1$, with $B = (0, 1, -1)$ and $\mathbf{b} = 0$, a scalar. It is also clear that $B\boldsymbol{\beta}$ is estimable–in fact, if we choose T as

$$\mathbf{t}' = \tfrac{1}{2}(1, 1, -1, -1), \tag{5.103}$$

we find that

$$TX = \mathbf{t}'X = (0, 1, -1) = B, \tag{5.103a}$$

so that

$$E(\mathbf{t}'\mathbf{y}) = B\boldsymbol{\beta} = \beta_1 - \beta_2. \tag{5.103b}$$

Therefore, H_0 of (5.102b) is testable. Proceeding as discussed earlier in this section, we may find H, a 3×3 matrix, such that

$$H = \left[V \mid U\right] = \left[V \mid \mathbf{u}\right], \tag{5.104}$$

where V is of order 3×2 and $U = \mathbf{u}$ is of order 3×1, and such that

$$X\mathbf{u} = \mathbf{0}_4. \tag{5.104a}$$

For example, we may choose

$$\mathbf{u} = (-1, 1, 1)' \tag{5.104b}$$

and

$$V = \begin{bmatrix} 1 & 0 \\ 0 & 1 \\ 0 & 0 \end{bmatrix}. \tag{5.104c}$$

Then

$$H = \begin{bmatrix} 1 & 0 & -1 \\ 0 & 1 & 1 \\ 0 & 0 & 1 \end{bmatrix}, \qquad H^{-1} = \begin{bmatrix} 1 & 0 & 1 \\ 0 & 1 & -1 \\ 0 & 0 & 1 \end{bmatrix}, \tag{5.104d}$$

and

$$XH = [\, XV \mid \mathbf{0}_4 \,]. \tag{5.104e}$$

We rewrite (5.102) and (5.102a) as

$$E(\mathbf{y}) = XHH^{-1}\boldsymbol{\beta}, \tag{5.105}$$

which is to say

$$E(\mathbf{y}) = W\boldsymbol{\phi}_1, \quad W = XV = \begin{bmatrix} 1 & 1 \\ 1 & 1 \\ 1 & 0 \\ 1 & 0 \end{bmatrix}, \quad \boldsymbol{\phi}_1 = \begin{bmatrix} \mu + \beta_2 \\ \beta_1 - \beta_2 \end{bmatrix} \tag{5.105a}$$

or

$$E(\mathbf{y}) = \mathbf{1}_4\gamma_0 + \mathbf{w}_1\gamma_1 \tag{5.105b}$$

with

$$\gamma_0 = \mu + \beta_2, \quad \gamma_1 = \beta_1 - \beta_2, \quad \mathbf{w}_1 = (1, 1, 0, 0)'. \tag{5.105c}$$

The "model" (5.105b) is of full rank, with W being a 4×2 matrix, of rank 2. Furthermore, note that H_0 is equivalent to

$$H_0\colon \gamma_1 = 0, \tag{5.106}$$

and applying the method of Section 3.7, we find an F-statistic for this problem. [See, for example, (3.222) and (3.227).] We have that the numerator of the F-statistic (3.222) for (3.227) is the regression sum of squares due to γ_1 divided by the degrees of freedom which is 1. Thus, the numerator is

$$\hat{\gamma}_1 \mathbf{w}'_{1\cdot 0} \mathbf{w}_{1\cdot 0} \hat{\gamma}_1 = SS_e(H_0) - SS_e, \tag{5.107}$$

where $SS_e = SS_e(\gamma_0, \gamma_1)$ denotes the residual sum of squares when fitting (5.105b), while $SS_e(H_0) = SS_e(\gamma_0)$ denotes the residual sum of squares that would result when fitting (5.105b), given that H_0 is true—that is, when fitting

$$E(\mathbf{y}) = \mathbf{1}_4 \gamma_0. \tag{5.107a}$$

It may be shown (Problem 5.15) that (5.107) becomes

$$(\bar{y}_{\cdot 1} - \bar{y}_{\cdot 2})^2 = 2 \sum_{j=1}^{2} (\bar{y}_{\cdot j} - \bar{y})^2 = SS_B, \tag{5.108}$$

where

$$\bar{y}_{\cdot j} = \frac{\sum_{i=1}^{2} y_{ij}}{2}, \quad \mathbf{y} = (y_1, y_2, y_3, y_4)'$$
$$= (y_{11}, y_{21}, y_{12}, y_{22})', \; \bar{y} = \frac{\sum_{i=1}^{4} \sum_{j=1}^{4} y_{ij}}{4}$$

(SS_B stands for the between sum of squares) and the denominator is

$$MS_e = \frac{SS_W}{2} = \frac{1}{2} \sum_{j=1}^{2} \sum_{i=1}^{2} (y_{ij} - \bar{y}_{\cdot j})^2. \tag{5.109}$$

As stated, Problem 5.15 asks for the remaining details. We note too that if the original hypothesis was H_0': $\beta_1 = \beta_2 + c$, c a specified constant, then by transforming $\mathbf{y}$ to $\mathbf{t}$, where

$$\mathbf{t} = \mathbf{y} - \mathbf{w}_1 c, \tag{5.110}$$

we would have that (5.105b) becomes

$$E(\mathbf{t}) = \mathbf{1}_4 \gamma_0 + \mathbf{w}_1 \tau, \qquad \tau = \beta_1 - \beta_2 - c = \gamma_1 - c \tag{5.110a}$$

so that H_0' would be equivalent to the statement $\tau = 0$, and we would proceed using $\mathbf{t}$ instead of $\mathbf{y}$ in the discussion that led to (5.108) and (5.109).

At this point we may well ask what would occur if the functions $B\boldsymbol{\beta}$ were not estimable, so that H_0: $B\boldsymbol{\beta} = \mathbf{b}$ is *not* testable. *We are assuming that B is of order $s \times k$ and of rank $r(B) = s$.* We first look at this question when $s = q = k - p$, where $r(X) = p$ and $q = k - p$.

Now we can always rewrite the non-full-rank model

$$E(\mathbf{y}) = X\boldsymbol{\beta} \tag{5.111}$$

in a manner completely analogous to the method used at the beginning of Section 3.7, so we assume we are dealing with (5.111) and that we wish to test

$$H_0': B\boldsymbol{\beta} = \mathbf{0}. \tag{5.112}$$

Now if $B\boldsymbol{\beta}$ are not estimable, then $B \neq TX$ for any T and since we are assuming that the rank of the $q \times k$ matrix B is q, where $q = k - p$ and $r(X) = p$, then to find $\mathrm{SS}_e(H_0')$ we may (see Corollary 4.6.1.1) augment the matrix X with the matrix B, so that

$$G = (X'X + B'B)^{-1} \tag{5.113}$$

exists and is a g_1 inverse of $X'X$, leading to

$$\mathrm{SS}_e(H_0') = \mathbf{y}'(I - XGX')\mathbf{y}, \qquad G = (X'X + B'B)^{-1}. \tag{5.114}$$

We may do the above since, under H_0', we are dealing with a model that is subject to the constraints (5.112). To find SS_e, we augment the non-full-rank matrix X by any $q \times k$ matrix such that the rank of the augmented matrix is k; B satisfies the conditions so we use it, leading to

$$\mathrm{SS}_e = \mathbf{y}'(I - XGX')\mathbf{y}, \qquad G = (X'X + B'B)^{-1}. \tag{5.115}$$

Thus, the numerator of the F-statistic would be

$$\mathrm{SS}_e(H_0') - \mathrm{SS}_e = 0, \tag{5.116}$$

that is, there is no test for H_0'. Now if B is of order $s \times k$ and $r(B) = s$, $s < k$, then we augment with B and a full-row-rank matrix B^*, which is of order $(q-s) \times k$, so that

$$A = \begin{bmatrix} X \\ B \\ B^* \end{bmatrix} \tag{5.117}$$

is of full rank k. We can do this as long as $B \neq TX$, which we know is the case by the assumption that $B\boldsymbol{\beta}$ are not estimable, and use a B^* that has rows independent of the rows of X and B and is of full row rank. We would then be led by a similar argument to (5.116) and we then have that there is no test for $B\boldsymbol{\beta} = \mathbf{b}$ when $B\boldsymbol{\beta}$ are not estimable.

Finally, we conclude this chapter and this book with the remark that the methods of Chapters 3, 4, and 5 enable us to deal with any linear model of the form

$$E(\mathbf{y}) = X\boldsymbol{\beta} + \boldsymbol{\varepsilon}, \qquad \boldsymbol{\varepsilon} = D\big(\mathbf{0}, \sigma^2 I_N\big), \tag{5.118}$$

no matter what the form of X may be. In particular, X may arise from a balanced design, such as the two-way analysis of variance design with replication, or X may arise from use of partially balanced designs, such as incomplete block designs not discussed in this book [for further reading, the reader is referred to the book by Box, Hunter, and Hunter (1978), and the references therein]. Also, X may arise due to the use of unbalanced designs, which may be of full or non-full rank. The point is that, depending on the experimenter's interest, we may proceed using the methods of Chapter 3 if X is of full rank, or using the methods of Chapters 4 and 5 if X is of non-full rank.

Problems

5.1. Verify (5.32) and (5.32a).

5.2. For the experimental setup of Section 5.2, we rewrote the model as in (5.33), namely, $E(\mathbf{y}) = W(\gamma_0, \gamma_1, \ldots, \gamma_{c-1})'$, with W given by (5.34), and so on. Detail the orthogonalization procedure necessary to find the extra regression sum of squares due to $\gamma_1, \ldots, \gamma_{c-1}$, given γ_0 in the model, and show that the analysis of variance table is equivalent to Table 5.2.2. Supply the entries for the expected sum of squares column, with derivations, for Table 5.2.2.

5.3. Referring to Problem 5.2, show that, if $AD(\mathbf{y})$ holds, the (extra) regression sum of squares due to $\gamma_1, \ldots, \gamma_{c-1}$ is independent of the residual sum of squares. [These were denoted by SS_B and SS_e respectively in (5.37a) and are the entries given in Table 5.2.2.]

5.4. (a) Show that $X_{2\cdot 1} = Z$ is as stated in (5.40).

(b) Verify (5.43) and (5.44).

(c) Using the model (5.46), show that $\tau_1 - \tau_c, \ldots, \tau_{c-1} - \tau_c$ are estimable [see Problem 4.21(a) of Chapter 4].

5.5. Verify that (5.48) leads to the correct (extra) regression sum of squares SS_B and the correct residual sum of squares SS_e mentioned in Problem 5.3.

5.6. (a) Verify that $X_{2\cdot 1} = Z$ of Section 5.3a, defined by the process of eliminating the means in (5.66) and (5.66a), is as stated in (5.70).

(b) Deduce (5.71b) and verify (5.72).

(c) Using (5.70), show that $\alpha_i - \alpha_r$, $i = 1, \ldots, r-1$, are estimable, and similarly, that $\gamma_j - \gamma_c$, $j = 1, \ldots, c-1$, are estimable. [Use the definition of estimability or consult Problem 4.21(a) of Chapter 4. It may also be that Problem 4.18(b) proves helpful to the reader, or Theorem 4.2.9.]

5.7. (a) Verify (5.75c) and (5.76a).

(b) Substantiate that the entries of Table 5.3.2 are correct and that the entries of the sum of squares column of Table 5.3.3 follow from the corresponding Table 5.3.2 entries.

(c) Give the entries for the expected sum of squares column of Table 5.3.3 with derivation.

(d) Show that the sum of squares due to the mean, rows, and columns are independent of the residual sum of squares under normality.

(e) Prove (5.77) directly.

(f) Using the results of (c), and assuming normality, construct a test utilizing F-statistics for (i) H_0: $\boldsymbol{\alpha} = \mathbf{0}$ and (ii) H_0': $\boldsymbol{\gamma} = \mathbf{0}$. You must prove that certain sums of squares are independent of each other, and so on.

5.8. Referring to (5.83) with $i = 1, 2$, $j = 1, 2$, and $t = 1, 2$, give the form of X. Determine the rank of X, giving the details justifying your answer.

5.9. (a) Verify (5.86a).

(b) If $E(y_{ijt}) = \mu + \alpha_i + \gamma_j$, where $\Sigma_i \alpha_i = \Sigma_j \gamma_j = 0$, show that $\Sigma\Sigma\Sigma(y_{ijt} - \bar{y}_{i\cdot\cdot} - \bar{y}_{\cdot j\cdot} + \bar{y})^2$ gives information about σ^2 only.

(c) If $E(y_{ijt}) = \mu + \alpha_i + \gamma_j + \lambda_{ij}$, where $\Sigma_i \alpha_i = 0$, $\Sigma_j \gamma_j = 0$, $\Sigma_i \lambda_{ij} = 0$, $j = 1,\ldots,c$, and $\Sigma_j \lambda_{ij} = 0$, $i = 1,\ldots,r$, then show that $\Sigma\Sigma\Sigma(y_{ijt} - \bar{y}_{ij\cdot})^2$ gives information about σ^2 only.

5.10. Treat the model of Section 5.3b as that of a "lack of fit" situation of Section 3.6. That is, fit a model with no interaction but run the experiment so that a lack of fit component may be extracted. What specifically is the lack of fit component sum of squares? What is its expected value? Give details.

5.11. If the model of Section 5.3a is not appropriate, but should take into account interaction, that is, $E(y_{ij}) = \mu + \alpha_i + \gamma_j + \lambda_{ij}$, with the usual constraints $\Sigma\alpha_i = \Sigma\gamma_j = \Sigma_i\lambda_{ij} = \Sigma_j\lambda_{ij} = 0$ holding, show that the residual sum of squares of Table 5.3.3 has an expectation which depends on the λ_{ij}'s, so that the residual sum of squares cannot be used as the basis for estimating σ^2.

5.12. (a) Verify that the entries for the sum of squares column of Table 5.3.7 are independent, under the assumption of normality of the y's, and so on, and derive the entries for the expected sum of squares column of Table 5.3.7. Use the results as a guide to construct a test of the hypotheses (i) $\boldsymbol{\lambda} = \mathbf{0}$; (ii) $\boldsymbol{\alpha} = \mathbf{0}$; (iii) $\boldsymbol{\gamma} = \mathbf{0}$.

(b) By using the method of Section 5.4, verify that the tests found in (a) are those given by the likelihood criterion and the corresponding F-test statistics.

5.13. Suppose $E(\mathbf{y}) = \Sigma_{j=1}^{5} \mathbf{x}_j \beta_j$, where $X = [\mathbf{x}_1 \mid \cdots \mid \mathbf{x}_5]$ is

$$X = \begin{bmatrix} 1 & -1 & -1 & 0 & -1 \\ 1 & 1 & 1 & 2 & 3 \\ 1 & -1 & 1 & 0 & 1 \\ 1 & 1 & 1 & 2 & 3 \\ 1 & -2 & 1 & -1 & 0 \\ 1 & 2 & 1 & 3 & 4 \end{bmatrix}.$$

(a) Determine the rank of X, say r.

(b) Find a matrix U of order $5 \times q$, such that $XU = \bigcirc$, where $q = k - r = 5 - r$.

(c) Find a matrix V of order $5 \times r$, such that $U'V = \bigcirc$.

(d) On the basis of (a)–(c), form a matrix, H, which will reduce the problem to one of full rank. Draw up the analysis of variance table.

5.14. Suppose $E(y) = \Sigma_{i=1}^{3} \beta_i x_i$, and that six experiments are run so that

$$X = \begin{bmatrix} 1 & -2 & 0 \\ 1 & -3 & -1 \\ 1 & 1 & 3 \\ 1 & 3 & 5 \\ 1 & -1 & 1 \\ 1 & 2 & 4 \end{bmatrix}.$$

(a) Show whether or not one can obtain linear unbiased estimators for β_1, β_2, and β_3.

(b) If interest is in the contribution due to x_2 and x_3, show how to determine the regression sum of squares due to x_2 and x_3.

(c) Use Problem 4.18(a) to find a g_1 inverse of $X'X$, say G, and construct the matrix $H = GX'X$. With this H, use Theorem 4.4.2 or Problem 4.21(a) to show that $\beta_2 + \beta_3$ and $\beta_1 + 2\beta_3$ are estimable.

5.15. (a) Verify that (5.107) may be expressed as in (5.108). Also, determine that MS_e is as given in (5.109) and construct the relevant analysis of variance table, stating the form for the test alluded to in (5.106)–(5.109).

(b) Follow the instructions of Problem 5.14(c) to show that $\beta_1 - \beta_2$ is estimable for the model (5.102) and (5.102a).

(c) Show also that $(\hat{\beta}_1, \hat{\beta}_2) = [(\bar{y}_{\cdot 1} - \bar{y}), (\bar{y}_{\cdot 2} - \bar{y})]$ is the form of the unbiased estimator of (β_1, β_2), given the model (5.102) and (5.102a).

5.16. Consider the situation diagrammed by Table 5.3.6 and summarized by (5.83) and (5.83a) where $r = 3$, $c = 4$, and $l = 2$. Find $\hat{\boldsymbol{\beta}}$ by using Theorem 4.6.1. If possible, find the regression sum of squares and the residual sum of squares and their expectations.

5.17. Suppose we were given $E(y_{ij})$ as in Table 5.3.5, where $E(y_{ij}) = \mu + \alpha_i + \gamma_j + \lambda_{ij}$. Show that this information would only lead to non-unique μ, (α_1, α_2), $(\gamma_1, \gamma_2, \gamma_3)$, and $(\lambda_{11}, \ldots, \lambda_{23})$, but that if, in addition, we have $\alpha_1 + \alpha_2 = 0$, $\gamma_1 + \gamma_2 + \gamma_3 = 0$, $\lambda_{11} + \lambda_{12} + \lambda_{13} = 0$, $\lambda_{21} + \lambda_{22} + \lambda_{23} = 0$, and $\lambda_{11} + \lambda_{21} = 0$, $\lambda_{12} + \lambda_{22} = 0$, then we would obtain unique values for these parameters satisfying $E(y_{ij}) = \mu + \alpha_i + \gamma_j + \lambda_{ij} = a_{ij}$, where the a_{ij} are as in Table 5.3.5. [*Note*: The last constraint (missing in the above), $\lambda_{13} + \lambda_{23} = 0$, is redundant since $0 = (1)(\lambda_{11} + \lambda_{12} + \lambda_{13}) + (1)(\lambda_{21} + \lambda_{22} + \lambda_{23}) + (-1)(\lambda_{11} + \lambda_{21}) + (-1)(\lambda_{12} + \lambda_{22}) = \lambda_{13} + \lambda_{23}$.

References

Abramowitz, M., and I. A. Stegun, Eds. (1965), *Handbook of Mathematical Functions*, National Bureau of Standards, Applied Mathematics Series, Vol. 55, 4th printing, Washington, D.C.

Box, G. E. P., and I. Guttman (1966), Some aspects of randomization, *J. R. Stat. Soc.* **28**, 543–558.

Box, G. E. P., W. G. Hunter, and J. S. Hunter (1978), *Statistics for Experimenters*, Wiley, New York.

Box, G. E. P., and G. C. Tiao (1973), *Bayesian Inference in Statistical Analysis*, Addison-Wesley, Boston.

Cochran, W. G. (1934), The distribution of quadratic forms in a normal system, with applications to the analysis of covariance, *Proc. Cambridge Philos. Soc.* **30**, 178–191.

Craig, A. T. (1943), Note on the independence of certain quadratic forms, *Ann. Math. Stat.* **14**, 195–197.

DeLury, D. B. (1950), *Values and Integrals of the Orthogonal Polynomials up to $n = 26$*, University of Toronto Press, Toronto, Ontario.

Draper, N. R., and H. Smith (1981), *Applied Regression Analysis*, 2nd ed., Wiley, New York.

Fisher, Sir R. A., and F. Yates (1974), *Statistical Tables for Biological, Agricultural, and Medical Research*, 6th ed., Longman, England.

Golub, G. (1968), Least squares, singular values and matrix approximations, *App. Math.* **13**, 44–51.

Golub, G. (1969), *Matrix Decompositions and Statistical Calculations*, Statistical Computation, Academic Press, New York, pp. 365–385.

Golub, G., and C. Reinsch (1970), Singular value decomposition and least-squares solutions, *Numer. Math.* **14**, 403–420.

Graybill, F. A. (1969), *Introduction to Matrices with Applications in Statistics*, Wadsworth, Belmont, CA.

Graybill, F. A., and G. Marsaglia (1957), Idempotent matrices and quadratic forms in the general linear hypothesis, *Ann. Math. Stat.* **28**, 678–686.

Kendall, M. A., and Alan Stuart (1966), *The Advanced Theory of Statistics*, Vol. 3, Hafner, New York.

Kleinbaum, D. G., and L. L. Kupper (1978), *Applied Regression Analysis and Other Multivariable Methods*, Duxbury, North Scituate, MA.

Lancaster, H. D. (1954), Traces and cumulants of quadratic forms in normal variables, *J. R. Stat. Soc. Ser. B* **16**, 247–254.

Lapczak, L. (1978), A Program to Calculate the Moore–Penrose Inverse based on the McGill University DSVD Program, available from the Department of Statistics, University of Toronto.

Loynes, R. M. (1966), On idempotent matrices, *Ann. Math. Stat.* **37**, 295–296.

Rao, C. R., and S. K. Mitra (1971), *Generalized Inverse of Matrices and its Applications*, Wiley, New York.

Scheffé, H. (1959), *The Analysis of Variance*, Wiley, New York.

Searle, S. R. (1966), *Matrix Algebra for the Biological Sciences*, Wiley, New York.

Searle, S. R. (1971), *Linear Models*, Wiley, New York.

Seber, G. A. F. (1977), *Linear Regression Analysis*, Wiley, New York.

Srivastava, M. S., and C. G. Khatri (1979), *An Introduction to Multivariate Statistics*, Elsevier North Holland, New York.

Tocher, K. D. (1952), The design and analysis of block experiments, *J. R. Stat. Soc. Ser. B* **14**, 45–91.

Appendix of Commonly Used Tables

TABLE I. Percentage Points of the χ^2_m Distribution*†

m \ α	.995	.990	.975	.950	.050	.025	.010	.005
1	392704×10^{-10}	157088×10^{-9}	982069×10^{-9}	393214×10^{-8}	3.84146	5.02389	6.63490	7.87944
2	.0100251	.0201007	.0506356	.102587	5.99147	7.37776	9.21034	10.5966
3	.0717212	.114832	.215795	.351846	7.81473	9.34840	11.3449	12.8381
4	.206990	.297110	.484419	.710721	9.48773	11.1433	13.2767	14.8602
5	.411740	.554300	.831211	1.145476	11.0705	12.8325	15.0863	16.7496
6	.675727	.872085	1.237347	1.63539	12.5916	14.4494	16.8119	18.5476
7	.989265	1.239043	1.68987	2.16735	14.0671	16.0128	18.4753	20.2777
8	1.344419	1.646482	2.17973	2.73264	15.5073	17.5346	20.0902	21.9550
9	1.734926	2.087912	2.70039	3.32511	16.9190	19.0228	21.6660	23.5893
10	2.15585	2.55821	3.24697	3.94030	18.3070	20.4831	23.2093	25.1882
11	2.60321	3.05347	3.81575	4.57481	19.6751	21.9200	24.7250	26.7569
12	3.07382	3.57056	4.40379	5.22603	21.0261	23.3367	26.2170	28.2995
13	3.56503	4.10691	5.00874	5.89186	22.3621	24.7356	27.6883	29.8194
14	4.07468	4.66043	5.62872	6.57063	23.6848	26.1190	29.1413	31.3193
15	4.60094	5.22935	6.26214	7.26094	24.9958	27.4884	30.5779	32.8013
16	5.14224	5.81221	6.90766	7.96164	26.2962	28.8454	31.9999	34.2672
17	5.69724	6.40776	7.56418	8.67176	27.5871	30.1910	33.4087	35.7185
18	6.26481	7.01491	8.23075	9.39046	28.8693	31.5264	34.8053	37.1564
19	6.84398	7.63273	8.90655	10.1170	30.1435	32.8523	36.1908	38.5822

20	7.43386	8.26040	9.59083	10.8508	31.4104	34.1696	37.5662	39.9968
21	8.03366	8.89720	10.28293	11.5913	32.6705	35.4789	38.9321	41.4010
22	8.64272	9.54249	10.9823	12.3380	33.9244	36.7807	40.2894	42.7956
23	9.26042	10.19567	11.6885	13.0905	35.1725	38.0757	41.6384	44.1813
24	9.88623	10.8564	12.4011	13.8484	36.4151	39.3641	42.9798	45.5585
25	10.5197	11.5240	13.1197	14.6114	37.6525	40.6465	44.3141	46.9278
26	11.1603	12.1981	13.8439	15.3791	38.8852	41.9232	45.6417	48.2899
27	11.8076	12.8786	14.5733	16.1513	40.1133	43.1944	46.9630	49.6449
28	12.4613	13.5648	15.3079	16.9279	41.3372	44.4607	48.2782	50.9933
29	13.1211	14.2565	16.0471	17.7083	42.5569	45.7222	49.5879	52.3356
30	13.7867	14.9535	16.7908	18.4926	43.7729	46.9792	50.8922	53.6720
40	20.7065	22.1643	24.4331	26.5093	55.7585	59.3417	63.6907	66.7659
50	27.9907	29.7067	32.3574	34.7642	67.5048	71.4202	76.1539	79.4900
60	35.5346	37.4848	40.4817	43.1879	79.0819	83.2976	88.3794	91.9517
70	43.2752	45.4418	48.7576	51.7393	90.5312	95.0231	100.425	104.215
80	51.1720	53.5400	57.1532	60.3915	101.879	106.629	112.329	116.321
90	59.1963	61.7541	65.6466	69.1260	113.145	118.136	124.116	128.299
100	67.3276	70.0648	74.2219	77.9295	124.342	129.561	135.807	140.169

* That is, values of $\chi^2_{m;\alpha}$, where m represents degrees of freedom and

$$\int_{\chi^2_{m;\alpha}}^{\infty} \frac{1}{2\Gamma(m/2)} \left(\frac{\chi^2}{2}\right)^{(m/2)-1} e^{-\chi^2/2}\, d\chi^2 = \alpha.$$

For $m < 100$, linear interpolation is adequate. For $m > 100$, $\sqrt{2\chi^2_m}$ is approximately normally distributed with mean $\sqrt{2m-1}$ and unit variance, so that percentage points may be obtained from Table II.

† From *Biometrika Tables for Statisticians*, Vol. 1 (2nd edition), Cambridge University Press (1958); edited by E. S. Pearson and H. O. Hartley; reproduced by permission of the publishers.

TABLE II. **Percentage Points of the t_m Distribution***†‡

m \ α	.25	.1	.05	.025	.01	.005
1	1.000	3.078	6.314	12.706	31.821	63.657
2	.816	1.886	2.920	4.303	6.965	9.925
3	.765	1.638	2.353	3.182	4.541	5.841
4	.741	1.533	2.132	2.776	3.747	4.604
5	.727	1.476	2.015	2.571	3.365	4.032
6	.718	1.440	1.943	2.447	3.143	3.707
7	.711	1.415	1.895	2.365	2.998	3.499
8	.706	1.397	1.860	2.306	2.896	3.355
9	.703	1.383	1.833	2.262	2.821	3.250
10	.700	1.372	1.812	2.228	2.764	3.169
11	.697	1.363	1.796	2.201	2.718	3.106
12	.695	1.356	1.782	2.179	2.681	3.055
13	.694	1.350	1.771	2.160	2.650	3.012
14	.692	1.345	1.761	2.145	2.624	2.977
15	.691	1.341	1.753	2.131	2.602	2.947
16	.690	1.337	1.746	2.120	2.583	2.921
17	.689	1.333	1.740	2.110	2.567	2.898
18	.688	1.330	1.734	2.101	2.552	2.878
19	.688	1.328	1.729	2.093	2.539	2.861
20	.687	1.325	1.725	2.086	2.528	2.845
21	.686	1.323	1.721	2.080	2.518	2.831
22	.686	1.321	1.717	2.074	2.508	2.819
23	.685	1.319	1.714	2.069	2.500	2.807
24	.685	1.318	1.711	2.064	2.492	2.797
25	.684	1.316	1.708	2.060	2.485	2.787
26	.684	1.315	1.706	2.056	2.479	2.779
27	.684	1.314	1.703	2.052	2.473	2.771
28	.683	1.313	1.701	2.048	2.467	2.763
29	.683	1.311	1.699	2.045	2.462	2.756
30	.683	1.310	1.697	2.042	2.457	2.750
40	.681	1.303	1.684	2.021	2.423	2.704
60	.679	1.296	1.671	2.000	2.390	2.660
120	.677	1.289	1.658	1.980	2.358	2.617
∞	.674	1.282	1.645	1.960	2.326	2.576

* That is, values of $t_{m;\alpha}$, where m equals degrees of freedom and

$$\int_{t_{m;\alpha}}^{\infty} \frac{\Gamma[(m+1)/2]}{\sqrt{\pi m}\,\Gamma(m/2)}\left(1+\frac{t^2}{m}\right)^{-(m+1)/2} dt = \alpha.$$

† From *Biometrika Tables for Statisticians*, Vol. 1 (2nd edition) Cambridge University Press (1958); edited by E. S. Pearson and H. O. Hartley; reproduced by permission of the publishers.

‡ Where necessary, interpolation should be carried out using the reciprocals of the degrees of freedom, and for this the function $120/m$ is convenient.

TABLE III. Percentage Points of the F_{m_1, m_2} Distribution*†

$\alpha = .10$

m_2 \ m_1	1	2	3	4	5	6	7	8	9
1	39.864	49.500	53.593	55.833	57.241	58.204	58.906	59.439	59.858
2	8.5263	9.0000	9.1618	9.2434	9.2926	9.3255	9.3491	9.3668	9.3805
3	5.5383	5.4624	5.3908	5.3427	5.3092	5.2847	5.2662	5.2517	5.2400
4	4.5448	4.3246	4.1908	4.1073	4.0506	4.0098	3.9790	3.9549	3.9357
5	4.0604	3.7797	3.6195	3.5202	3.4530	3.4045	3.3679	3.3393	3.3163
6	3.7760	3.4633	3.2888	3.1808	3.1075	3.0546	3.0145	2.9830	2.9577
7	3.5894	3.2574	3.0741	2.9605	2.8833	2.8274	2.7849	2.7516	2.7247
8	3.4579	3.1131	2.9238	2.8064	2.7265	2.6683	2.6241	2.5893	2.5612
9	3.3603	3.0065	2.8129	2.6927	2.6106	2.5509	2.5053	2.4694	2.4403
10	3.2850	2.9245	2.7277	2.6053	2.5216	2.4606	2.4140	2.3772	2.3473
11	3.2252	2.8595	2.6602	2.5362	2.4512	2.3891	2.3416	2.3040	2.2735
12	3.1765	2.8068	2.6055	2.4801	2.3940	2.3310	2.2828	2.2446	2.2135
13	3.1362	2.7632	2.5603	2.4337	2.3467	2.2830	2.2341	2.1953	2.1638
14	3.1022	2.7265	2.5222	2.3947	2.3069	2.2426	2.1931	2.1539	2.1220
15	3.0732	2.6952	2.4898	2.3614	2.2730	2.2081	2.1582	2.1185	2.0862
16	3.0481	2.6682	2.4618	2.3327	2.2438	2.1783	2.1280	2.0880	2.0553
17	3.0262	2.6446	2.4374	2.3077	2.2183	2.1524	2.1017	2.0613	2.0284
18	3.0070	2.6239	2.4160	2.2858	2.1958	2.1296	2.0785	2.0379	2.0047
19	2.9899	2.6056	2.3970	2.2663	2.1760	2.1094	2.0580	2.0171	1.9836
20	2.9747	2.5893	2.3801	2.2489	2.1582	2.0913	2.0397	1.9985	1.9649
21	2.9609	2.5746	2.3649	2.2333	2.1423	2.0751	2.0232	1.9819	1.9480
22	2.9486	2.5613	2.3512	2.2193	2.1279	2.0605	2.0084	1.9668	1.9327
23	2.9374	2.5493	2.3387	2.2065	2.1149	2.0472	1.9949	1.9531	1.9189
24	2.9271	2.5383	2.3274	2.1949	2.1030	2.0351	1.9826	1.9407	1.9063
25	2.9177	2.5283	2.3170	2.1843	2.0922	2.0241	1.9714	1.9292	1.8947
26	2.9091	2.5191	2.3075	2.1745	2.0822	2.0139	1.9610	1.9188	1.8841
27	2.9012	2.5106	2.2987	2.1655	2.0730	2.0045	1.9515	1.9091	1.8743
28	2.8939	2.5028	2.2906	2.1571	2.0645	1.9959	1.9427	1.9001	1.8652
29	2.8871	2.4955	2.2831	2.1494	2.0566	1.9878	1.9345	1.8918	1.8560
30	2.8807	2.4887	2.2761	2.1422	2.0492	1.9803	1.9269	1.8841	1.8498
40	2.8354	2.4404	2.2261	2.0909	1.9968	1.9269	1.8725	1.8289	1.7929
60	2.7914	2.3932	2.1774	2.0410	1.9457	1.8747	1.8194	1.7748	1.7380
120	2.7478	2.3473	2.1300	1.9923	1.8959	1.8238	1.7675	1.7220	1.6843
∞	2.7055	2.3026	2.0838	1.9449	1.8473	1.7741	1.7167	1.6702	1.6315

TABLE III (*Continued*)

$\alpha = .10$									
10	12	15	20	24	30	40	60	120	∞
60.195	60.705	61.220	61.740	62.002	62.265	62.529	62.794	63.061	63.328
9.3916	9.4081	9.4247	9.4413	9.4496	9.4579	9.4663	9.4746	9.4829	9.4913
5.2304	5.2156	5.2003	5.1845	5.1764	5.1681	5.1597	5.1512	5.1425	5.1337
3.9199	3.8955	3.8689	3.8443	3.8310	3.8174	3.8036	3.7896	3.7753	3.7607
3.2974	3.2682	3.2380	3.2067	3.1905	3.1741	3.1573	3.1402	3.1228	3.1050
2.9369	2.9047	2.8712	2.8363	2.8183	2.8000	2.7812	2.7620	2.7423	2.7222
2.7025	2.6681	2.6322	2.5947	2.5753	2.5555	2.5351	2.5142	2.4928	2.4708
2.5380	2.5020	2.4642	2.4246	2.4041	2.3830	2.3614	2.3391	2.3162	2.2926
2.4163	2.3789	2.3396	2.2983	2.2768	2.2547	2.2320	2.2085	2.1843	2.1592
2.3226	2.2841	2.2435	2.2007	2.1784	2.1554	2.1317	2.1072	2.0818	2.0554
2.2482	2.2087	2.1671	2.1230	2.1000	2.0762	2.0516	2.0261	1.9997	1.9721
2.1878	2.1474	2.1049	2.0597	2.0360	2.0115	1.9861	1.9597	1.9323	1.9036
2.1376	2.0966	2.0532	2.0070	1.9827	1.9576	1.9315	1.9043	1.8759	1.8462
2.0954	2.0537	2.0095	1.9625	1.9377	1.9119	1.8852	1.8572	1.8280	1.7973
2.0593	2.0171	1.9722	1.9243	1.8990	1.8728	1.8454	1.8168	1.7867	1.7551
2.0281	1.9854	1.9399	1.8913	1.8656	1.8388	1.8108	1.7816	1.7507	1.7182
2.0009	1.9577	1.9117	1.8624	1.8362	1.8090	1.7805	1.7506	1.7191	1.6856
1.9770	1.9333	1.8868	1.8368	1.8103	1.7827	1.7537	1.7232	1.6910	1.6567
1.9557	1.9117	1.8647	1.8142	1.7873	1.7592	1.7298	1.6988	1.6659	1.6308
1.9367	1.8924	1.8449	1.7938	1.7667	1.7382	1.7083	1.6768	1.6433	1.6074
1.9197	1.8750	1.8272	1.7756	1.7481	1.7193	1.6890	1.6569	1.6228	1.5862
1.9043	1.8593	1.8111	1.7590	1.7312	1.7021	1.6714	1.6389	1.6042	1.5668
1.8903	1.8450	1.7964	1.7439	1.7159	1.6864	1.6554	1.6224	1.5871	1.5490
1.8775	1.8319	1.7831	1.7302	1.7019	1.6721	1.6407	1.6073	1.5715	1.5327
1.8658	1.8200	1.7708	1.7175	1.6890	1.6589	1.6272	1.5934	1.5570	1.5176
1.8550	1.8090	1.7596	1.7059	1.6771	1.6468	1.6147	1.5805	1.5437	1.5036
1.8451	1.7989	1.7492	1.6951	1.6662	1.6356	1.6032	1.5686	1.5313	1.4906
1.8359	1.7895	1.7395	1.6852	1.6560	1.6252	1.5925	1.5575	1.5198	1.4784
1.8274	1.7808	1.7306	1.6759	1.6465	1.6155	1.5825	1.5472	1.5090	1.4670
1.8195	1.7727	1.7223	1.6673	1.6377	1.6065	1.5732	1.5376	1.4989	1.4564
1.7627	1.7146	1.6624	1.6052	1.5741	1.5411	1.5056	1.4672	1.4248	1.3769
1.7070	1.6574	1.6034	1.5435	1.5107	1.4755	1.4373	1.3952	1.3476	1.2915
1.6524	1.6012	1.5450	1.4821	1.4472	1.4094	1.3676	1.3203	1.2646	1.1926
1.5987	1.5458	1.4871	1.4206	1.3832	1.3419	1.2951	1.2400	1.1686	1.0000

TABLE III (*Continued*)

$\alpha = .05$

m_2 \ m_1	1	2	3	4	5	6	7	8	9
1	161.45	199.50	215.71	224.58	230.16	233.99	236.77	238.88	240.54
2	18.513	19.000	19.164	19.247	19.296	19.330	19.353	19.371	19.385
3	10.128	9.5521	9.2766	9.1172	9.0135	8.9406	8.8868	8.8452	8.8123
4	7.7086	6.9443	6.5914	6.3883	6.2560	6.1631	6.0942	6.0410	5.9988
5	6.6079	5.7861	5.4095	5.1922	5.0503	4.9503	4.8759	4.8183	4.7725
6	5.9874	5.1433	4.7571	4.5337	4.3874	4.2839	4.2066	4.1468	4.0990
7	5.5914	4.7374	4.3468	4.1203	3.9715	3.8660	3.7870	3.7257	3.6767
8	5.3177	4.4590	4.0662	3.8378	3.6875	3.5806	3.5005	3.4381	3.3881
9	5.1174	4.2565	3.8626	3.6331	3.4817	3.3738	3.2927	3.2296	3.1789
10	4.9646	4.1028	3.7083	3.4780	3.3258	3.2172	3.1355	3.0717	3.0204
11	4.8443	3.9823	3.5874	3.3567	3.2039	3.0946	3.0123	2.9480	2.8962
12	4.7472	3.8853	3.4903	3.2592	3.1059	2.9961	2.9134	2.8486	2.7964
13	4.6672	3.8056	3.4105	3.1791	3.0254	2.9153	2.8321	2.7669	2.7144
14	4.6001	3.7389	3.3439	3.1122	2.9582	2.8477	2.7642	2.6987	2.6458
15	4.5431	3.6823	3.2874	3.0556	2.9013	2.7905	2.7066	2.6408	2.5876
16	4.4940	3.6337	3.2389	3.0069	2.8524	2.7413	2.6572	2.5911	2.5377
17	4.4513	3.5915	3.1968	2.9647	2.8100	2.6987	2.6143	2.5480	2.4943
18	4.4139	3.5546	3.1599	2.9277	2.7729	2.6613	2.5767	2.5102	2.4563
19	4.3808	3.5219	3.1274	2.8951	2.7401	2.6283	2.5435	2.4768	2.4227
20	4.3513	3.4928	3.0984	2.8661	2.7109	2.5990	2.5140	2.4471	2.3928
21	4.3248	3.4668	3.0725	2.8401	2.6848	2.5727	2.4876	2.4205	2.3661
22	4.3009	3.4434	3.0491	2.8167	2.6613	2.5491	2.4638	2.3965	2.3419
23	4.2793	3.4221	3.0280	2.7955	2.6400	2.5277	2.4422	2.3748	2.3201
24	4.2597	3.4028	3.0088	2.7763	2.6207	2.5082	2.4226	2.3551	2.3002
25	4.2417	3.3852	2.9912	2.7587	2.6030	2.4904	2.4047	2.3371	2.2821
26	4.2252	3.3690	2.9751	2.7426	2.5868	2.4741	2.3883	2.3205	2.2655
27	4.2100	3.3541	2.9604	2.7278	2.5719	2.4591	2.3732	2.3053	2.2501
28	4.1960	3.3404	2.9467	2.7141	2.5581	2.4453	2.3593	2.2913	2.2360
29	4.1830	3.3277	2.9340	2.7014	2.5454	2.4324	2.3463	2.2782	2.2229
30	4.1709	3.3158	2.9223	2.6896	2.5336	2.4205	2.3343	2.2662	2.2107
40	4.0848	3.2317	2.8387	2.6060	2.4495	2.3359	2.2490	2.1802	2.1240
60	4.0012	3.1504	2.7581	2.5252	2.3683	2.2540	2.1665	2.0970	2.0401
120	3.9201	3.0718	2.6802	2.4472	2.2900	2.1750	2.0867	2.0164	1.9588
∞	3.8415	2.9957	2.6049	2.3719	2.2141	2.0986	2.0096	1.9384	1.8799

TABLE III (*Continued*)

α = .05									
10	12	15	20	24	30	40	60	120	∞
241.88	243.91	245.95	248.01	249.05	250.09	251.14	252.20	253.25	254.32
19.396	19.413	19.429	19.446	19.454	19.462	19.471	19.479	19.487	19.496
8.7855	8.7446	8.7029	8.6602	8.6385	8.6166	8.5944	8.5720	8.5494	8.5265
5.9644	5.9117	5.8578	5.8025	5.7744	5.7459	5.7170	5.6878	5.6581	5.6281
4.7351	4.6777	4.6188	4.5581	4.5272	4.4957	4.4638	4.4314	4.3984	4.3650
4.0600	3.9999	3.9381	3.8742	3.8415	3.8082	3.7743	3.7398	3.7047	3.6688
3.6365	3.5747	3.5108	3.4445	3.4105	3.3758	3.3404	3.3043	3.2674	3.2298
3.3472	3.2840	3.2184	3.1503	3.1152	3.0794	3.0428	3.0053	2.9669	2.9276
3.1373	3.0729	3.0061	2.9365	2.9005	2.8637	2.8259	2.7872	2.7475	2.7067
2.9782	2.9130	2.8450	2.7740	2.7372	2.6996	2.6609	2.6211	2.5801	2.5379
2.8536	2.7876	2.7186	2.6464	2.6090	2.5705	2.5309	2.4901	2.4480	2.4045
2.7534	2.6866	2.6169	2.5436	2.5055	2.4663	2.4259	2.3842	2.3410	2.2962
2.6710	2.6037	2.5331	2.4589	2.4202	2.3803	2.3392	2.2966	2.2524	2.2064
2.6021	2.5342	2.4630	2.3879	2.3487	2.3082	2.2664	2.2230	2.1778	2.1307
2.5437	2.4753	2.4035	2.3275	2.2878	2.2468	2.2043	2.1601	2.1141	2.0658
2.4935	2.4247	2.3522	2.2756	2.2354	2.1938	2.1507	2.1058	2.0589	2.0096
2.4499	2.3807	2.3077	2.2304	2.1898	2.1477	2.1040	2.0584	2.0107	1.9604
2.4117	2.3421	2.2686	2.1906	2.1497	2.1071	2.0629	2.0166	1.9681	1.9168
2.3779	2.3080	2.2341	2.1555	2.1141	2.0712	2.0264	1.9796	1.9302	1.8780
2.3479	2.2776	2.2033	2.1242	2.0825	2.0391	1.9938	1.9464	1.8963	1.8432
2.3210	2.2504	2.1757	2.0960	2.0540	2.0102	1.9645	1.9165	1.8657	1.8117
2.2967	2.2258	2.1508	2.0707	2.0283	1.9842	1.9380	1.8895	1.8380	1.7831
2.2747	2.2036	2.1282	2.0476	2.0050	1.9605	1.9139	1.8649	1.8128	1.7570
2.2547	2.1834	2.1077	2.0267	1.9838	1.9390	1.8920	1.8424	1.7897	1.7331
2.2365	2.1649	2.0889	2.0075	1.9643	1.9192	1.8718	1.8217	1.7684	1.7110
2.2197	2.1479	2.0716	1.9898	1.9464	1.9010	1.8533	1.8027	1.7488	1.6906
2.2043	2.1323	2.0558	1.9736	1.9299	1.8842	1.8361	1.7851	1.7307	1.6717
2.1900	2.1179	2.0411	1.9586	1.9147	1.8687	1.8203	1.7689	1.7138	1.6541
2.1768	2.1045	2.0275	1.9446	1.9005	1.8543	1.8055	1.7537	1.6981	1.6377
2.1646	2.0921	2.0148	1.9317	1.8874	1.8409	1.7918	1.7396	1.6835	1.6223
2.0772	2.0035	1.9245	1.8389	1.7929	1.7444	1.6928	1.6373	1.5766	1.5089
1.9926	1.9174	1.8364	1.7480	1.7001	1.6491	1.5943	1.5343	1.4673	1.3893
1.9105	1.8337	1.7505	1.6587	1.6084	1.5543	1.4952	1.4290	1.3519	1.2539
1.8307	1.7522	1.6664	1.5705	1.5173	1.4591	1.3940	1.3180	1.2214	1.0000

TABLE III (*Continued*)

$\alpha = .025$

m_2 \ m_1	1	2	3	4	5	6	7	8	9
1	647.79	799.50	864.16	899.58	921.85	937.11	948.22	956.66	963.28
2	38.506	39.000	39.165	39.248	39.298	39.331	39.355	39.373	39.387
3	17.443	16.044	15.439	15.101	14.885	14.735	14.624	14.540	14.473
4	12.218	10.649	9.9792	9.6045	9.3645	9.1973	9.0741	8.9796	8.9047
5	10.007	8.4336	7.7636	7.3879	7.1464	6.9777	6.8531	6.7572	6.6810
6	8.8131	7.2598	6.5988	6.2272	5.9876	5.8197	5.6955	5.5996	5.5234
7	8.0727	6.5415	5.8898	5.5226	5.2852	5.1186	4.9949	4.8994	4.8232
8	7.5709	6.0595	5.4160	5.0526	4.8173	4.6517	4.5286	4.4332	4.3572
9	7.2093	5.7147	5.0781	4.7181	4.4844	4.3197	4.1971	4.1020	4.0260
10	6.9367	5.4564	4.8256	4.4683	4.2361	4.0721	3.9498	3.8549	3.7790
11	6.7241	5.2559	4.6300	4.2751	4.0440	3.8807	3.7586	3.6638	3.5879
12	6.5538	5.0959	4.4742	4.1212	3.8911	3.7283	3.6065	3.5118	3.4358
13	6.4143	4.9653	4.3472	3.9959	3.7667	3.6043	3.4827	3.3880	3.3120
14	6.2979	4.8567	4.2417	3.8919	3.6634	3.5014	3.3799	3.2853	3.2093
15	6.1995	4.7650	4.1528	3.8043	3.5764	3.4147	3.2934	3.1987	3.1227
16	6.1151	4.6867	4.0768	3.7294	3.5021	3.3406	3.2194	3.1248	3.0488
17	6.0420	4.6189	4.0112	3.6648	3.4379	3.2767	3.1556	3.0610	2.9849
18	5.9781	4.5597	3.9539	3.6083	3.3820	3.2209	3.0999	3.0053	2.9291
19	5.9216	4.5075	3.9034	3.5587	3.3327	3.1718	3.0509	2.9563	2.8800
20	5.8715	4.4613	3.8587	3.5147	3.2891	3.1283	3.0074	2.9128	2.8365
21	5.8266	4.4199	3.8188	3.4754	3.2501	3.0895	2.9686	2.8740	2.7977
22	5.7863	4.3828	3.7829	3.4401	3.2151	3.0546	2.9338	2.8392	2.7628
23	5.7498	4.3492	3.7505	3.4083	3.1835	3.0232	2.9024	2.8077	2.7313
24	5.7167	4.3187	3.7211	3.3794	3.1548	2.9946	2.8738	2.7791	2.7027
25	5.6864	4.2909	3.6943	3.3530	3.1287	2.9685	2.8478	2.7531	2.6766
26	5.6586	4.2655	3.6697	3.3289	3.1048	2.9447	2.8240	2.7293	2.6528
27	5.6331	4.2421	3.6472	3.3067	3.0828	2.9228	2.8021	2.7074	2.6309
28	5.6096	4.2205	3.6264	3.2863	3.0625	2.9027	2.7820	2.6872	2.6106
29	5.5878	4.2006	3.6072	3.2674	3.0438	2.8840	2.7633	2.6686	2.5919
30	5.5675	4.1821	3.5894	3.2499	3.0265	2.8667	2.7460	2.6513	2.5746
40	5.4239	4.0510	3.4633	3.1261	2.9037	2.7444	2.6238	2.5289	2.4519
60	5.2857	3.9253	3.3425	3.0077	2.7863	2.6274	2.5068	2.4117	2.3344
120	5.1524	3.8046	3.2270	2.8943	2.6740	2.5154	2.3948	2.2994	2.2217
∞	5.0239	3.6889	3.1161	2.7858	2.5665	2.4082	2.2875	2.1918	2.1136

TABLE III (*Continued*)

$\alpha = .025$									
10	12	15	20	24	30	40	60	120	∞
968.63	976.71	984.87	993.10	997.25	1001.4	1005.6	1009.8	1014.0	1018.3
39.398	39.415	39.431	39.448	39.456	39.465	39.473	39.481	39.490	39.498
14.419	14.337	14.253	14.167	14.124	14.081	14.037	13.992	13.947	13.902
8.8439	8.7512	8.6565	8.5599	8.5109	8.4613	8.4111	8.3604	8.3092	8.2573
6.6192	6.5246	6.4277	6.3285	6.2780	6.2269	6.1751	6.1225	6.0693	6.0153
5.4613	5.3662	5.2687	5.1684	5.1172	5.0652	5.0125	5.9589	4.9045	4.8491
4.7611	4.6658	4.5678	4.4667	4.4150	4.3624	4.3089	4.2544	4.1989	4.1423
4.2951	4.1997	4.1012	3.9995	3.9472	3.8940	3.8398	3.7844	3.7279	3.6702
3.9639	3.8682	3.7694	3.6669	3.6142	3.5604	3.5055	3.4493	3.3918	3.3329
3.7168	3.6209	3.5217	3.4186	3.3654	3.3110	3.2554	3.1984	3.1399	3.0798
3.5257	3.4296	3.3299	3.2261	3.1725	3.1176	3.0613	3.0035	2.9441	2.8828
3.3736	3.2773	3.1772	3.0728	3.0187	2.9633	2.9063	2.8478	2.7874	2.7249
3.2497	3.1532	3.0527	2.9477	2.8932	2.8373	2.7797	2.7204	2.6590	2.5955
3.1469	3.0501	2.9493	2.8437	2.7888	2.7324	2.6742	2.6142	2.5519	2.4872
3.0602	2.9633	2.8621	2.7559	2.7006	2.6437	2.5850	2.5242	2.4611	2.3953
2.9862	2.8890	2.7875	2.6808	2.6252	2.5678	2.5085	2.4471	2.3831	2.3163
2.9222	2.8249	2.7230	2.6158	2.5598	2.5021	2.4422	2.3801	2.3153	2.2474
2.8664	2.7689	2.6667	2.5590	2.5027	2.4445	2.3842	2.3214	2.2558	2.1869
2.8173	2.7196	2.6171	2.5089	2.4523	2.3937	2.3329	2.2695	2.2032	2.1333
2.7737	2.6758	2.5731	2.4645	2.4076	2.3486	2.2873	2.2234	2.1562	2.0853
2.7348	2.6368	2.5338	2.4247	2.3675	2.3082	2.2465	2.1819	2.1141	2.0422
2.6998	2.6017	2.4984	2.3890	2.3315	2.2718	2.2097	2.1446	2.0760	2.0032
2.6682	2.5699	2.4665	2.3567	2.2989	2.2389	2.1763	2.1107	2.0415	1.9677
2.6396	2.5412	2.4374	2.3273	2.2693	2.2090	2.1460	2.0799	2.0099	1.9353
2.6135	2.5149	2.4110	2.3005	2.2422	2.1816	2.1183	2.0517	1.9811	1.9055
2.5895	2.4909	2.3867	2.2759	2.2174	2.1565	2.0928	2.0257	1.9545	1.8781
2.5676	2.4688	2.3644	2.2533	2.1946	2.1334	2.0693	2.0018	1.9299	1.8527
2.5473	2.4484	2.3438	2.2324	2.1735	2.1121	2.0477	1.9796	1.9072	1.8291
2.5286	2.4295	2.3248	2.2131	2.1540	2.0923	2.0276	1.9591	1.8861	1.8072
2.5112	2.4120	2.3072	2.1952	2.1359	2.0739	2.0089	1.9400	1.8664	1.7867
2.3882	2.2882	2.1819	2.0677	2.0069	1.9429	1.8752	1.8028	1.7242	1.6371
2.2702	2.1692	2.0613	1.9445	1.8817	1.8152	1.7440	1.6668	1.5810	1.4822
2.1570	2.0548	1.9450	1.8249	1.7597	1.6899	1.6141	1.5299	1.4327	1.3104
2.0483	1.9447	1.8326	1.7085	1.6402	1.5660	1.4835	1.3883	1.2684	1.0000

TABLE III (*Continued*)

$\alpha = .01$

m_2 \ m_1	1	2	3	4	5	6	7	8	9
1	4052.2	4999.5	5403.3	5624.6	5763.7	5859.0	5928.3	5981.6	6022.5
2	98.503	99.000	99.166	99.249	99.299	99.332	99.356	99.374	99.388
3	34.116	30.817	29.457	28.710	28.237	27.911	27.672	27.489	27.345
4	21.198	18.000	16.694	15.977	15.522	15.207	14.976	14.799	14.659
5	16.258	13.274	12.060	11.392	10.967	10.672	10.456	10.289	10.158
6	13.745	10.925	9.7795	9.1483	8.7459	8.4661	8.2600	8.1016	7.9761
7	12.246	9.5466	8.4513	7.8467	7.4604	7.1914	6.9928	6.8401	6.7188
8	11.259	8.6491	7.5910	7.0060	6.6318	6.3707	6.1776	6.0289	5.9106
9	10.561	8.0215	6.9919	6.4221	6.0569	5.8018	5.6129	5.4671	5.3511
10	10.044	7.5594	6.5523	5.9943	5.6363	5.3858	5.2001	5.0567	4.9424
11	9.6460	7.2057	6.2167	5.6683	5.3160	5.0692	4.8861	4.7445	4.6315
12	9.3302	6.9266	5.9526	5.4119	5.0643	4.8206	4.6395	4.4994	4.3875
13	9.0738	6.7010	5.7394	5.2053	4.8616	4.6204	4.4410	4.3021	4.1911
14	8.8616	6.5149	5.5639	5.0354	4.6950	4.4558	4.2779	4.1399	4.0297
15	8.6831	6.3589	5.4170	4.8932	4.5556	4.3183	4.1415	4.0045	3.8948
16	8.5310	6.2262	5.2922	4.7726	4.4374	4.2016	4.0259	3.8896	3.7804
17	8.3997	6.1121	5.1850	4.6690	4.3359	4.1015	3.9267	3.7910	3.6822
18	8.2854	6.0129	5.0919	4.5790	4.2479	4.0146	3.8406	3.7054	3.5971
19	8.1850	5.9259	5.0103	4.5003	4.1708	3.9386	3.7653	3.6305	3.5225
20	8.0960	5.8489	4.9382	4.4307	4.1027	3.8714	3.6987	3.5644	3.4567
21	8.0166	5.7804	4.8740	4.3688	4.0421	3.8117	3.6396	3.5056	3.3981
22	7.9454	5.7190	4.8166	4.3134	3.9880	3.7583	3.5867	3.4530	3.3458
23	7.8811	5.6637	4.7649	4.2635	3.9392	3.7102	3.5390	3.4057	3.2986
24	7.8229	5.6136	4.7181	4.2184	3.8951	3.6667	3.4959	3.3629	3.2560
25	7.7698	5.5680	4.6755	4.1774	3.8550	3.6272	3.4568	3.3239	3.2172
26	7.7213	5.5263	4.6366	4.1400	3.8183	3.5911	3.4210	3.2884	3.1818
27	7.6767	5.4881	4.6009	4.1056	3.7848	3.5580	3.3882	3.2558	3.1494
28	7.6356	5.4529	4.5681	4.0740	3.7539	3.5276	3.3581	3.2259	3.1195
29	7.5976	5.4205	4.5378	4.0449	3.7254	3.4995	3.3302	3.1982	3.0920
30	7.5625	5.3904	4.5097	4.0179	3.6990	3.4735	3.3045	3.1726	3.0665
40	7.3141	5.1785	4.3126	3.8283	3.5138	3.2910	3.1238	2.9930	2.8876
60	7.0771	4.9774	4.1259	3.6491	3.3389	3.1187	2.9530	2.8233	2.7185
120	6.8510	4.7865	3.9493	3.4796	3.1735	2.9559	2.7918	2.6629	2.5586
∞	6.6349	4.6052	3.7816	3.3192	3.0173	2.8020	2.6393	2.5113	2.4073

TABLE III (*Continued*)

$\alpha = .01$

10	12	15	20	24	30	40	60	120	∞
6055.8	6106.3	6157.3	6208.7	6234.6	6260.7	6286.8	6313.0	6339.4	6366.0
99.399	99.416	99.432	99.449	99.458	99.466	99.474	99.483	99.491	99.501
27.229	27.052	26.872	26.690	26.598	26.505	26.411	26.316	26.221	26.125
14.546	14.374	14.198	14.020	13.929	13.838	13.745	13.652	13.558	13.463
10.051	9.8883	9.7222	9.5527	9.4665	9.3793	9.2912	9.2020	9.1118	9.0204
7.8741	7.7183	7.5590	7.3958	7.3127	7.2285	7.1432	7.0568	6.9690	6.8801
6.6201	6.4691	6.3143	6.1554	6.0743	5.9921	5.9084	5.8236	5.7372	5.6495
5.8143	5.6668	5.5151	5.3591	5.2793	5.1981	5.1156	5.0316	4.9460	4.8588
5.2565	5.1114	4.9621	4.8080	4.7290	4.6486	4.5667	4.4831	4.3978	4.3105
4.8492	4.7059	4.5582	4.4054	4.3269	4.2469	4.1653	4.0819	3.9965	3.9090
4.5393	4.3974	4.2509	4.0990	4.0209	3.9411	3.8596	3.7761	3.6904	3.6025
4.2961	4.1553	4.0096	3.8584	3.7805	3.7008	3.6192	3.5355	3.4494	3.3608
4.1003	3.9603	3.8154	3.6646	3.5868	3.5070	3.4253	3.3413	3.2548	3.1654
3.9394	3.8001	3.6557	3.5052	3.4274	3.3476	3.2656	3.1813	3.0942	3.0040
3.8049	3.6662	3.5222	3.3719	3.2940	3.2141	3.1319	3.0471	2.9595	2.8684
3.6909	3.5527	3.4089	3.2588	3.1808	3.1007	3.0182	2.9330	2.8447	2.7528
3.5931	3.4552	3.3117	3.1615	3.0835	3.0032	2.9205	2.8348	2.7459	2.6530
3.5082	3.3706	3.2273	3.0771	2.9990	2.9185	2.8354	2.7493	2.6597	2.5660
3.4338	3.2965	3.1533	3.0031	2.9249	2.8442	2.7608	2.6742	2.5839	2.4893
3.3682	3.2311	3.0880	2.9377	2.8594	2.7785	2.6947	2.6077	2.5168	2.4212
3.3098	3.1729	3.0299	2 8796	2.8011	2.7200	2.6359	2.5484	2.4568	2.3603
3.2576	3.1209	2.9780	2.8274	2.7488	2.6675	2.5831	2.4951	2.4029	2.3055
3.2106	3.0740	2.9311	2.7805	2.7017	2.6202	2.5355	2.4471	2.3542	2.2559
3.1681	3.0316	2.8887	2.7380	2.6591	2.5773	2.4923	2.4035	2.3099	2.2107
3.1294	2.9931	2.8502	2.6993	2.6203	2.5383	2.4530	2.3637	2.2695	2.1694
3.0941	2.9579	2.8150	2.6640	2.5848	2.5026	2.4170	2.3273	2.2325	2.1315
3.0618	2.9256	2.7827	2.6316	2.5522	2.4699	2.3840	2.2938	2.1984	2.0965
3.0320	2.8959	2.7530	2.6017	2.5223	2.4397	2.3535	2.2629	2.1670	2.0642
3.0045	2.8685	2.7256	2.5742	2.4946	2.4118	2.3253	2.2344	2.1378	2.0342
2.9791	2.8431	2.7002	2.5487	2.4689	2.3860	2.2992	2.2079	2.1107	2.0062
2.8005	2.6648	2.5216	2.3689	2.2880	2.2034	2.1142	2.0194	1.9172	1.8047
2.6318	2.4961	2.3523	2.1978	2.1154	2.0285	1.9360	1.8363	1.7263	1.6006
2.4721	2.3363	2.1915	2.0346	1.9500	1.8600	1.7628	1.6557	1.5330	1.3805
2.3209	2.1848	2.0385	1.8783	1.7908	1.6964	1.5923	1.4730	1.3246	1.0000

TABLE III (*Continued*)

$\alpha = .005$

m_2 \ m_1	1	2	3	4	5	6	7	8	9
1	16211	20000	21615	22500	23056	23437	23715	23925	24091
2	198.50	199.00	199.17	199.25	199.30	199.33	199.36	199.37	199.39
3	55.552	49.799	47.467	46.195	45.392	44.838	44.434	44.126	43.882
4	31.333	26.284	24.259	23.155	22.456	21.975	21.622	21.352	21.139
5	22.785	18.314	16.530	15.556	14.940	14.513	14.200	13.961	13.772
6	18.635	14.544	12.917	12.028	11.464	11.073	10.786	10.566	10.391
7	16.236	12.404	10.882	10.050	9.5221	9.1554	8.8854	8.6781	8.5138
8	14.688	11.042	9.5965	8.8051	8.3018	7.9520	7.6942	7.4960	7.3386
9	13.614	10.107	8.7171	7.9559	7.4711	7.1338	6.8849	6.6933	6.5411
10	12.826	9.4270	8.0807	7.3428	6.8723	6.5446	6.3025	6.1159	5.9676
11	12.226	8.9122	7.6004	6.8809	6.4217	6.1015	5.8648	5.6821	5.5368
12	11.754	8.5096	7.2258	6.5211	6.0711	5.7570	5.5245	5.3451	5.2021
13	11.374	8.1865	6.9257	6.2335	5.7910	5.4819	5.2529	5.0761	4.9351
14	11.060	7.9217	6.6803	5.9984	5.5623	5.2574	5.0313	4.8566	4.7173
15	10.798	7.7008	6.4760	5.8029	5.3721	5.0708	4.8473	4.6743	4.5364
16	10.575	7.5138	6.3034	5.6378	5.2117	4.9134	4.6920	4.5207	4.3838
17	10.384	7.3536	6.1556	5.4967	5.0746	4.7789	4.5594	4.3893	4.2535
18	10.218	7.2148	6.0277	5.3746	4.9560	4.6627	4.4448	4.2759	4.1410
19	10.073	7.0935	5.9161	5.2681	4.8526	4.5614	4.3448	4.1770	4.0428
20	9.9439	6.9865	5.8177	5.1743	4.7616	4.4721	4.2569	4.0900	3.9564
21	9.8295	6.8914	5.7304	5.0911	4.6808	4.3931	4.1789	4.0128	3.8799
22	9.7271	6.8064	5.6524	5.0168	4.6088	4.3225	4.1094	3.9440	3.8116
23	9.6348	6.7300	5.5823	4.9500	4.5441	4.2591	4.0469	3.8822	3.7502
24	9.5513	6.6610	5.5190	4.8898	4.4857	4.2019	3.9905	3.8264	3.6949
25	9.4753	6.5982	5.4615	4.8351	4.4327	4.1500	3.9394	3.7758	3.6447
26	9.4059	6.5409	5.4091	4.7852	4.3844	4.1027	3.8928	3.7297	3.5989
27	9.3423	6.4885	5.3611	4.7396	4.3402	4.0594	3.8501	3.6875	3.5571
28	9.2838	6.4403	5.3170	4.6977	4.2996	4.0197	3.8110	3.6487	3.5186
29	9.2297	6.3958	5.2764	4.6591	4.2622	3.9830	3.7749	3.6130	3.4832
30	9.1797	6.3547	5.2388	4.6233	4.2276	3.9492	3.7416	3.5801	3.4505
40	8.8278	6.0664	4.9759	4.3738	3.9860	3.7129	3.5088	3.3498	3.2220
60	8.4946	5.7950	4.7290	4.1399	3.7600	3.4918	3.2911	3.1344	3.0083
120	8.1790	5.5393	4.4973	3.9207	3.5482	3.2849	3.0874	2.9330	2.8083
∞	7.8794	5.2983	4.2794	3.7151	3.3499	3.0913	2.8968	2.7444	2.6210

TABLE III (*Continued*)

$\alpha = .005$

10	12	15	20	24	30	40	60	120	∞
24224	24426	24630	24836	24940	25044	25148	25253	25359	25465
199.40	199.42	199.43	199.45	199.46	199.47	199.47	199.48	199.49	199.51
43.686	43.387	43.085	42.778	42.622	42.466	42.308	42.149	41.989	41.829
20.967	20.705	20.438	20.167	20.030	19.892	19.752	19.611	19.468	19.325
13.618	13.384	13.146	12.903	12.780	12.656	12.530	12.402	12.274	12.144
10.250	10.034	9.8140	9.5888	9.4741	9.3583	9.2408	9.1219	9.0015	8.8793
8.3803	8.1764	7.9678	7.7540	7.6450	7.5345	7.4225	7.3088	7.1933	7.0760
7.2107	7.0149	6.8143	6.6082	6.5029	6.3961	6.2875	6.1772	6.0649	5.9505
6.4171	6.2274	6.0325	5.8318	5.7292	5.6248	5.5186	5.4104	5.3001	5.1875
5.8467	5.6613	5.4707	5.2740	5.1732	5.0705	4.9659	4.8592	4.7501	4.6385
5.4182	5.2363	5.0489	4.8552	4.7557	4.6543	4.5508	4.4450	4.3367	4.2256
5.0855	4.9063	4.7214	4.5299	4.4315	4.3309	4.2282	4.1229	4.0149	3.9039
4.8199	4.6429	4.4600	4.2703	4.1726	4.0727	3.9704	3.8655	3.7577	3.6465
4.6034	4.4281	4.2468	4.0585	3.9614	3.8619	3.7600	3.6553	3.5473	3.4359
4.4236	4.2498	4.0698	3.8826	3.7859	3.6867	3.5850	3.4803	3.3722	3.2602
4.2719	4.0994	3.9205	3.7342	3.6378	3.5388	3.4372	3.3324	3.2240	3.1115
4.1423	3.9709	3.7929	3.6073	3.5112	3.4124	3.3107	3.2058	3.0971	2.9839
4.0305	3.8599	3.6827	3.4977	3.4017	3.3030	3.2014	3.0962	2.9871	2.8732
3.9329	3.7631	3.5866	3.4020	3.3062	3.2075	3.1058	3.0004	2.8908	2.7762
3.8470	3.6779	3.5020	3.3178	3.2220	3.1234	3.0215	2.9159	2.8058	2.6904
3.7709	3.6024	3.4270	3.2431	3.1474	3.0488	2.9467	2.8408	2.7302	2.6140
3.7030	3.5350	3.3600	3.1764	3.0807	2.9821	2.8799	2.7736	2.6625	2.5455
3.6420	3.4745	3.2999	3.1165	3.0208	2.9221	2.8198	2.7132	2.6016	2.4837
3.5870	3.4199	3.2456	3.0624	2.9667	2.8679	2.7654	2.6585	2.5463	2.4276
3.5370	3.3704	3.1963	3.0133	2.9176	2.8187	2.7160	2.6088	2.4960	2.3765
3.4916	3.3252	3.1515	2.9685	2.8728	2.7738	2.6709	2.5633	2.4501	2.3297
3.4499	3.2839	3.1104	2.9275	2.8318	2.7327	2.6296	2.5217	2.4078	2.2867
3.4117	3.2460	3.0727	2.8899	2.7941	2.6949	2.5916	2.4834	2.3689	2.2469
3.3765	3.2111	3.0379	2.8551	2.7594	2.6601	2.5565	2.4479	2.3330	2.2102
3.3440	3.1787	3.0057	2.8230	2.7272	2.6278	2.5241	2.4151	2.2997	2.1760
3.1167	2.9531	2.7811	2.5984	2.5020	2.4015	2.2958	2.1838	2.0635	1.9318
2.9042	2.7419	2.5705	2.3872	2.2898	2.1874	2.0789	1.9622	1.8341	1.6885
2.7052	2.5439	2.3727	2.1881	2.0890	1.9839	1.8709	1.7469	1.6055	1.4311
2.5188	2.3583	2.1868	1.9998	1.8983	1.7891	1.6691	1.5325	1.3637	1.0000

* That is, values of $F_{m_1,m_2;\,\alpha}$, where (m_1, m_2) is the pair of degrees of freedom in F_{m_1,m_2} and

$$\frac{\Gamma((m_1+m_2)/2)}{\Gamma(m_1/2)\Gamma(m_2/2)}\left(\frac{m_1}{m_2}\right)^{m_1/2}\int_{F_{m_1,m_2;\alpha}}^{\infty} F^{(m_1/2)-1}\left(1+\frac{m_1}{m_2}F\right)^{-(m_1+m_2)/2} dF = \alpha.$$

† From "Tables of percentage points of the Inverted Beta (F) Distribution," *Biometrika*, Vol. 33 (1943), pp. 73–88, by Maxine Merrington and Catherine M. Thompson; reproduced by permission of E. S. Pearson. If necessary, interpolation should be carried out using the reciprocals of the degrees of freedom.

TABLE IV. Values of the Orthogonal Polynomials* $w_j(x) = w_j^{(1)}(x)/\lambda_j$

	$N = 3$		$N = 4$			$N = 5$				$N = 6$				
	$w_1^{(1)}$	$w_2^{(1)}$	$w_1^{(1)}$	$w_2^{(1)}$	$w_3^{(1)}$	$w_1^{(1)}$	$w_2^{(1)}$	$w_3^{(1)}$	$w_4^{(1)}$	$w_1^{(1)}$	$w_2^{(1)}$	$w_3^{(1)}$	$w_4^{(1)}$	$w_5^{(1)}$
	−1	+1	−3	+1	−1	−2	+2	−1	+1	−5	+5	−5	+1	−1
	0	−2	−1	−1	+3	−1	−1	+2	−4	−3	−1	+7	−3	+5
	+1	+1	+1	−1	−3	0	−2	0	+6	−1	−4	+4	+2	−10
			+3	+1	+1	+1	−1	−2	−4	+1	−4	−4	+2	+10
						+2	+2	+1	+1	+3	−1	−7	−3	−5
										+5	+5	+5	+1	+1
$(\mathbf{w}_j^{(1)}(x))'(\mathbf{w}_j^{(1)}(x))$	2	6	20	4	20	10	14	10	70	70	84	180	28	252
λ_j	1	3	2	1	$\frac{10}{3}$	1	1	$\frac{5}{6}$	$\frac{35}{12}$	2	$\frac{3}{2}$	$\frac{5}{3}$	$\frac{7}{12}$	$\frac{21}{10}$

	$N = 7$					$N = 8$					$N = 9$†				
	$w_1^{(1)}$	$w_2^{(1)}$	$w_3^{(1)}$	$w_4^{(1)}$	$w_5^{(1)}$	$w_1^{(1)}$	$w_2^{(1)}$	$w_3^{(1)}$	$w_4^{(1)}$	$w_5^{(1)}$	$w_1^{(1)}$	$w_2^{(1)}$	$w_3^{(1)}$	$w_4^{(1)}$	$w_5^{(1)}$
	−3	+5	−1	+3	−1	−7	+7	−7	+7	−7					
	−2	0	+1	−7	+4	−5	+1	+5	−13	+23					
	−1	−3	+1	+1	−5	−3	−3	+7	−3	−17	0	−20	0	+18	0
	0	−4	0	+6	0	−1	−5	+3	+9	−15	+1	−17	−9	+9	+9
	+1	−3	−1	+1	+5	+1	−5	−3	+9	+15	+2	−8	−13	−11	+4
	+2	0	−1	−7	−4	+3	−3	−7	−3	+17	+3	+7	−7	−21	−11
	+3	+5	+1	+3	+1	+5	+1	−5	−13	−23	+4	+28	+14	+14	+4
						+7	+7	+7	+7	+7					
$(\mathbf{w}_j^{(1)}(x))'(\mathbf{w}_j^{(1)}(x))$	28	84	6	154	84	168	168	264	616	2184	60	2772	990	2002	468
λ_j	1	1	$\frac{1}{6}$	$\frac{7}{12}$	$\frac{7}{20}$	2	1	$\frac{2}{3}$	$\frac{7}{12}$	$\frac{7}{10}$	1	3	$\frac{5}{6}$	$\frac{7}{12}$	$\frac{3}{20}$

TABLE IV (*Continued*)

	$N = 10$					$N = 11$					$N = 12$				
	$w_1^{(1)}$	$w_2^{(1)}$	$w_3^{(1)}$	$w_4^{(1)}$	$w_5^{(1)}$	$w_1^{(1)}$	$w_2^{(1)}$	$w_3^{(1)}$	$w_4^{(1)}$	$w_5^{(1)}$	$w_1^{(1)}$	$w_2^{(1)}$	$w_3^{(1)}$	$w_4^{(1)}$	$w_5^{(1)}$
	+1	−4	−12	+18	+6	0	−10	0	+6	0	+1	−35	−7	+28	+20
	+3	−3	−31	+3	+11	+1	−9	−14	+4	+4	+3	−29	−19	+12	+44
	+5	−1	−35	−17	+1	+2	−6	−23	−1	+4	+5	−17	−25	−13	+29
	+7	+2	−14	−22	−14	+3	−1	−22	−6	−1	+7	+1	−21	−33	−21
	+9	+6	+42	+18	+6	+4	+6	−6	−6	−6	+9	+25	−3	−27	−57
						+5	+15	+30	+3	+3	+11	+55	+33	+33	+33
$(\mathbf{w}_j^{(1)}(x))'(\mathbf{w}_j^{(1)}(x))$	330	132	8580	2860	780	110	858	4290	286	156	572	12,012	5148	8008	15,912
λ_j	2	$\frac{1}{2}$	$\frac{5}{3}$	$\frac{5}{12}$	$\frac{1}{10}$	1	1	$\frac{5}{6}$	$\frac{1}{12}$	$\frac{1}{40}$	2	3	$\frac{2}{3}$	$\frac{7}{24}$	$\frac{3}{20}$

	$N = 13$					$N = 14$					$N = 15$				
	$w_1^{(1)}$	$w_2^{(1)}$	$w_3^{(1)}$	$w_4^{(1)}$	$w_5^{(1)}$	$w_1^{(1)}$	$w_2^{(1)}$	$w_3^{(1)}$	$w_4^{(1)}$	$w_5^{(1)}$	$w_1^{(1)}$	$w_2^{(1)}$	$w_3^{(1)}$	$w_4^{(1)}$	$w_5^{(1)}$
	0	−14	0	+84	0	+1	−8	−24	+108	+60	0	−56	0	+756	0
	+1	−13	−4	+64	+20	+3	−7	−67	+63	+145	+1	−53	−27	+621	+675
	+2	−10	−7	+11	+26	+5	−5	−95	−13	+139	+2	−44	−49	+251	+1000
	+3	−5	−8	−54	+11	+7	−2	−98	−92	+28	+3	−29	−61	−249	+751
	+4	+2	−6	−96	−18	+9	+2	−66	−132	−132	+4	−8	−58	−704	−44
	+5	+11	0	−66	−33	+11	+7	+11	−77	−187	+5	+19	−35	−869	−979
	+6	+22	+11	+99	+22	+13	+13	+143	+143	+143	+6	+52	+13	−429	−1144
											+7	+91	+91	+1001	+1001
$(\mathbf{w}_j^{(1)}(x))'(\mathbf{w}_j^{(1)}(x))$	182	2002	572	68,068	6188	910	728	97,240	136,136	235,144	280	37,128	39,780	6,466,460	10,581,480
λ_j	1	1	$\frac{1}{6}$	$\frac{7}{12}$	$\frac{7}{120}$	2	$\frac{1}{2}$	$\frac{5}{3}$	$\frac{7}{12}$	$\frac{7}{30}$	1	3	$\frac{5}{6}$	$\frac{35}{12}$	$\frac{21}{20}$

	$N = 16$					$N = 17$				
	$w_1^{(1)}$	$w_2^{(1)}$	$w_3^{(1)}$	$w_4^{(1)}$	$w_5^{(1)}$	$w_1^{(1)}$	$w_2^{(1)}$	$w_3^{(1)}$	$w_4^{(1)}$	$w_5^{(1)}$
	+1	−21	−63	+189	+45	0	−24	0	+36	0
	+3	−19	−179	+129	+115	+1	−23	−7	+31	+55
	+5	−15	−265	+23	+131	+2	−20	−13	+17	+88
	+7	−9	−301	−101	+77	+3	−15	−17	−3	+83
	+9	−1	−267	−201	−33	+4	−8	−18	−24	+36
	+11	+9	−143	−221	−143	+5	+1	−15	−39	−39
	+13	+21	+91	−91	−143	+6	+12	−7	−39	−104
	+15	+35	+455	+273	+143	+7	+25	+7	−13	−91
						+8	+40	+28	+52	+104
$(\mathbf{w}_j^{(1)}(x))'(\mathbf{w}_j^{(1)}(x))$	1360	5712	1,007,760	470,288	201,552	408	7752	3876	16,796	100,776
λ_j	2	1	$\frac{10}{3}$	$\frac{7}{12}$	$\frac{1}{10}$	1	1	$\frac{1}{6}$	$\frac{1}{12}$	$\frac{1}{20}$

	$N = 18$					$N = 19$				
	$w_1^{(1)}$	$w_2^{(1)}$	$w_3^{(1)}$	$w_4^{(1)}$	$w_5^{(1)}$	$w_1^{(1)}$	$w_2^{(1)}$	$w_3^{(1)}$	$w_4^{(1)}$	$w_5^{(1)}$
	+1	−40	−8	+44	+220	0	−30	0	+396	0
	+3	−37	−23	+33	+583	+1	−29	−44	+352	+44
	+5	−31	−35	+13	+733	+2	−26	−83	+227	+74
	+7	−22	−42	−12	+588	+3	−21	−112	+42	+79
	+9	−10	−42	−36	+156	+4	−14	−126	−168	+54
	+11	+5	−33	−51	−429	+5	−5	−120	−354	+3
	+13	+23	−13	−47	−871	+6	+6	−89	−453	−58
	+15	+44	+20	−12	−676	+7	+19	−28	−388	−98
	+17	+68	+68	+68	+884	+8	+34	+68	−68	−68
						+9	+51	+204	+612	+109
$(\mathbf{w}_j^{(1)}(x))'(\mathbf{w}_j^{(1)}(x))$	1938	23,256	23,256	28,424	6,953,544	570	13,566	213,180	2,288,132	89,148
λ_j	2	$\frac{3}{2}$	$\frac{1}{3}$	$\frac{1}{12}$	$\frac{3}{10}$	1	1	$\frac{5}{6}$	$\frac{7}{12}$	$\frac{1}{40}$

TABLE IV (*Continued*)

	$N = 20$					$N = 21$				
	$w_1^{(1)}$	$w_2^{(1)}$	$w_3^{(1)}$	$w_4^{(1)}$	$w_5^{(1)}$	$w_1^{(1)}$	$w_2^{(1)}$	$w_3^{(1)}$	$w_4^{(1)}$	$w_5^{(1)}$
	+1	−33	−99	+1188	+396	0	−110	0	+594	0
	+3	−31	−287	+948	+1076	+1	−107	−54	+540	+1404
	+5	−27	−445	+503	+1441	+2	−98	−103	+385	+2444
	+7	−21	−553	−77	+1351	+3	−83	−142	+150	+2819
	+9	−13	−591	−687	+771	+4	−62	−166	−130	+2354
	+11	−3	−539	−1187	−187	+5	−35	−170	−406	+1063
	+13	+9	−377	−1402	−1222	+6	−2	−149	−615	−788
	+15	+23	−85	−1122	−1802	+7	+37	−98	−680	−2618
	+17	+39	+357	−102	−1122	+8	+82	−12	−510	−3468
	+19	+57	+969	+1938	+1938	+9	+133	+114	0	−1938
						+10	+190	+285	+969	+3876
$\left(\mathbf{w}_j^{(1)}(x)\right)'\left(\mathbf{w}_j^{(1)}(x)\right)$	2660	17,556	4,903,140	22,881,320	31,201,800	770	201,894	432,630	5,720,330	121,687,020
λ_j	2	1	$\frac{10}{3}$	$\frac{35}{24}$	$\frac{7}{20}$	1	3	$\frac{5}{6}$	$\frac{7}{12}$	$\frac{21}{40}$

	$N = 22$					$N = 23$				
	$w_1^{(1)}$	$w_2^{(1)}$	$w_3^{(1)}$	$w_4^{(1)}$	$w_5^{(1)}$	$w_1^{(1)}$	$w_2^{(1)}$	$w_3^{(1)}$	$w_4^{(1)}$	$w_5^{(1)}$
	+1	−20	−12	+702	+390	0	−44	0	+858	0
	+3	−19	−35	+585	+1079	+1	−43	−13	+793	+65
	+5	−17	−55	+365	+1509	+2	−40	−25	+605	+116
	+7	−14	−70	+70	+1554	+3	−35	−35	+315	+141
	+9	−10	−78	−258	+1158	+4	−28	−42	−42	+132
	+11	−5	−77	−563	+363	+5	−19	−45	−417	+87
	+13	+1	−65	−775	−663	+6	−8	−43	−747	+12
	+15	+8	−40	−810	−1598	+7	+5	−35	−955	−77
	+17	+16	0	−570	−1938	+8	+20	−20	−950	−152
	+19	+25	+57	+57	−969	+9	+37	+3	−627	−171
	+21	+35	+133	+1197	+2261	+10	+56	+35	+133	−76
						+11	+77	+77	+1463	+209
$\left(\mathbf{w}_j^{(1)}(x)\right)'\left(\mathbf{w}_j^{(1)}(x)\right)$	3542	7084	96,140	8,748,740	40,562,340	1012	35,420	32,890	13,123,110	340,860
λ_j	2	$\frac{1}{2}$	$\frac{1}{3}$	$\frac{7}{12}$	$\frac{7}{30}$	1	1	$\frac{1}{6}$	$\frac{7}{12}$	$\frac{1}{60}$

TABLE IV ***(Continued)***

	N = 24					N = 25				
	$w_1^{(1)}$	$w_2^{(1)}$	$w_3^{(1)}$	$w_4^{(1)}$	$w_5^{(1)}$	$w_1^{(1)}$	$w_2^{(1)}$	$w_3^{(1)}$	$w_4^{(1)}$	$w_5^{(1)}$
	+1	−143	−143	+143	+715	0	−52	0	+858	0
	+3	−137	−419	+123	+2005	+1	−51	−77	+803	+275
	+5	−125	−665	+85	+2893	+2	−48	−149	+643	+500
	+7	−107	−861	+33	+3171	+3	−43	−211	+393	+631
	+9	−83	−987	−27	+2721	+4	−36	−258	+78	+636
	+11	−53	−1023	−87	+1551	+5	−27	−285	−267	+501
	+13	−17	−949	−137	−169	+6	−16	−287	−597	−236
	+15	+25	−745	−165	−2071	+7	−3	−259	−857	−119
	+17	+73	−391	−157	−3553	+8	+12	−196	−982	−488
	+19	+127	+133	−97	−3743	+9	+29	−93	−897	−753
	+21	+187	+847	+33	−1463	+10	+48	+55	−517	−748
	+23	−253	+1771	+253	+4807	+11	+69	+253	+253	−253
						+12	+92	+506	+1518	+1012
$(\mathbf{w}_j^{(1)}(x))'(\mathbf{w}_j^{(1)}(x))$	4600	394,680	17,760,600	394,680	177,928,920	1300	53,820	1,480,050	14,307,150	7,803,900
λ_j	2	3	$\frac{10}{3}$	$\frac{1}{12}$	$\frac{3}{10}$	1	1	$\frac{5}{6}$	$\frac{5}{12}$	$\frac{1}{20}$

*From Table XXIII of Fisher and Yates: *Statistical Tables for Biological, Agricultural and Medical Research*, 6th ed., 1974, published by Longman, England, by the kind permission of the publisher.

†Only the last $(N+1)/2$ components are tabled for $N \geq 9$. Note that the missing first components are mirror images if j is odd, and pure images if j is even. See, for example, the entries for $w_j^{(1)}$ for $N = 7$ and 8, and so on.

Index

Probability and Mathematical Statistics (Continued)

DAVID • Order Statistics, *Second Edition*
DAVISON • Multidimensional Scaling
DEMING • Sample Design in Business Research
DODGE and ROMIG • Sampling Inspection Tables, *Second Edition*
DOWDY and WEARDEN • Statistics for Research
DRAPER and SMITH • Applied Regression Analysis, *Second Edition*
DUNN • Basic Statistics: A Primer for the Biomedical Sciences, *Second Edition*
DUNN and CLARK • Applied Statistics: Analysis of Variance and Regression
ELANDT-JOHNSON • Probability Models and Statistical Methods in Genetics
ELANDT-JOHNSON and JOHNSON • Survival Models and Data Analysis
FLEISS • Statistical Methods for Rates and Proportions, *Second Edition*
FRANKEN, KÖNIG, ARNDT, and SCHMIDT • Queues and Point Processes
GALAMBOS • The Asymptotic Theory of Extreme Order Statistics
GIBBONS, OLKIN, and SOBEL • Selecting and Ordering Populations: A New Statistical Methodology
GNANADESIKAN • Methods for Statistical Data Analysis of Multivariate Observations
GOLDBERGER • Econometric Theory
GOLDSTEIN and DILLON • Discrete Discriminant Analysis
GROSS and CLARK • Survival Distributions: Reliability Applications in the Biomedical Sciences
GROSS and HARRIS • Fundamentals of Queueing Theory
GUPTA and PANCHAPAKESAN • Multiple Decision Procedures: Theory and Methodology of Selecting and Ranking Populations
GUTTMAN, WILKS, and HUNTER • Introductory Engineering Statistics, *Third Edition*
HAHN and SHAPIRO • Statistical Models in Engineering
HALD • Statistical Tables and Formulas
HALD • Statistical Theory with Engineering Applications
HAND • Discrimination and Classification
HARTIGAN • Clustering Algorithms
HILDEBRAND, LAING, and ROSENTHAL • Prediction Analysis of Cross Classifications
HOAGLIN, MOSTELLER, and TUKEY • Understanding Robust and Exploratory Data Analysis
HOEL • Elementary Statistics, *Fourth Edition*
HOEL and JESSEN • Basic Statistics for Business and Economics, *Third Edition*
HOLLANDER and WOLFE • Nonparametric Statistical Methods
IMAN and CONOVER • Modern Business Statistics
JAGERS • Branching Processes with Biological Applications
JESSEN • Statistical Survey Techniques
JOHNSON and KOTZ • Distributions in Statistics
Discrete Distributions
Continuous Univariate Distributions—1
Continuous Univariate Distributions—2
Continuous Multivariate Distributions
JOHNSON and KOTZ • Urn Models and Their Application: An Approach to Modern Discrete Probability Theory
JOHNSON and LEONE • Statistics and Experimental Design in Engineering and the Physical Sciences, Volumes I and II, *Second Edition*
JUDGE, HILL, GRIFFITHS, LÜTKEPOHL and LEE • Introduction to the Theory and Practice of Econometrics